"Eric Topol is the perfect author for this book. He has a unique understanding of both genomics and wireless medicine and has a remarkable track record as a charismatic pioneer, visionary, and change agent in medicine. I'm sure this book will reach a very large number of people with information that can both empower and help transform their lives for the better."
—Dean Ornish, M.D., Founder and President,
Preventive Medicine Research Institute, and author of *The Spectrum*

"Eric Topol outlines the creative destruction of medicine that must be led by informed consumers. Smart patients will push the many stakeholders in health to accelerate change as medicine adapts to a new world of information and technology."
—Mehmet Oz, M.D., Professor and Vice-Chair of Surgery,
NY Presbyterian/Columbia University, author of many You books,
Dr. Oz Show

"In an upbeat, comprehensive volume, Dr. Topol has woven the prevailing technological undercurrents of the post-PC world; its power of many; its Gucci of gadgets; its cloud ecosystem; its 'Arab Spring' of apps; and its ubiquitous, calm computing; with the disruptive innovations of biomedicine, to create a compelling account of how this *bio-digital transformation* will hasten personalization of the highest quality of medical care."
—Eric Silfen, M.D., Chief Medical Officer, Philips Healthcare

"Eric Topol has written an extraordinarily important book at just the right moment. Drawing upon a unique and impressive array of convergent expertise in medical research, clinical medicine, consumer and health technological advancements, and health policy, Dr. Topol opens the door for an essential discussion of old challenges viewed through an innovative lens. In the context of increasingly unaffordable health care costs, suboptimal quality of care delivery, a tsunami of preventable chronic illness, and new accountabilities for consumer health choices and behaviors, this book helps all of us to think about solutions in new and exciting ways!"
—Reed Tuckson, M.D. Executive Vice President and
Chief of Medical Affairs, UnitedHealth Group

"Dr. Topol is the top thought leader in medicine today, with exceptional vision for how its future can be rebooted. This book will create and catalyze a movement for the individualization and democratization of medicine—and undoubtedly promote better health care."

—Greg Lucier, CEO, Life Technologies

Eric Topol offers a new and intriguing perspective on how the intersections of medicine and technology could further transform the delivery of healthcare and the role of a patient. He advocates for a future world of medicine where informed consumers are in the driver seat and control their own healthcare based on genomic information and real-time data obtained through nanosensors and wireless technology."

—John Martin, Ph.D. Chairman and CEO, Gilead Sciences

"Eric Topol is that rare physician willing to challenge the orthodoxies of his guild. He recognizes that in the US health care business-as-usual is unsustainable. But he does not despair. He bears witness to the rise of *Homo digitus* and the promise it holds to upend the inefficiencies and dysfunction so entrenched in clinical medicine. *The Creative Destruction of Medicine* is a timely tour de force. It is a necessary heresy."

—Misha Angrist, Ph.D., Duke University Institute for Genome Sciences & Policy, author of *Here Is a Human Being: At the Dawn of Personal Genomics*

"*The Creative Destruction of Medicine* is an engaging look into how the discoveries in genetics and biology will change the landscape of medicine. Along the way, Dr. Topol provides a fascinating compendium of stories about the shortcomings of medicine as it is currently practiced and how the revolutionary discoveries coming since the first sequencing of the human genome a decade ago will shape the delivery of healthcare in the 21st Century."

—William R. Brody, M.D., Ph.D., President,
The Salk Institute for Biological Studies, La Jolla, California

"Much of the wealth created over the last decades arose out of a brutal transition from A,B,C's to digital code. While creating some of the world's most valuable companies, this process also upended whole industries and even countries. Now medicine, health care, and life sciences are undergoing the same transition. And, again, enormous wealth will be created and destroyed. This book is a road map of what is about to happen."

—Juan Enriquez, Managing Director Excel Venture Management
and author *As the Future Catches You and Homo Evolutus*

"Dr. Topol believes that medicine, catalyzed by extraordinary innovation that exploits digital information, is about to go through its biggest shakeup in history. In *The Creative Destruction of Medicine*, he calls for a "jailbreak" from the ideas of the past. In the next phase of medicine, powerful digital tools including mobile sensors and advanced processors will transform our understanding of the individual, enabling creative "mash-ups" of data that will spark entirely new discoveries and spawn ultra-personalized health and fitness solutions. And with over 6 billion mobile connections worldwide, the mobile technology platform will have a major impact on that vision—leading to what Dr. Topol describes as nothing less than a "reboot" of the health care system. And we share Dr. Topol's view that individual consumers have the opportunity, and the power, to increase the pace of the titanic change that's coming."

—Paul Jacobs Ph.D., Chairman and CEO, Qualcomm

"Our sequencing of the human genome eleven years ago was the beginning of the individualized medicine revolution, a revolution that cannot happen without digitized personal phenotype information. Eric Topol provides a path forward using your digitized genome, remote sensing devices and social networking to place the educated at the center of medicine."

—J. Craig Venter, Chairman and President, J. Craig Venter Institute

"Eric Topol has been a longtime innovator in healthcare. In *The Creative Destruction of Medicine*, he cites the big waves of innovation that will save healthcare for the future. Real healthcare reform has not yet begun, but it will. *The Creative Destruction of Medicine* lays out the path."

—Jeffrey Immelt, Chairman and CEO of General Electric

100001000001TTTCAAAAGATCTCTCTCTCTCAG01001111000TCGAGATCCTGATCGG1011
1110000001010101001CTCTCAGGGGGAATTCCGCGCTCTC0100001111000111CGCGCAT
ATAT000001110000TGAAAAATGCGCGCGC000100001CGCGCG000011100TGATACCACAC
TGTGTG0001010101011100TCTCAAATTTTCCCCTCGTGTGTGTG10110000111CCAAGGG
TT11000TAAAATTTGGCCTCTCTC1010101010100000111T

THE CREATIVE DESTRUCTION OF MEDICINE

How the Digital Revolution Will Create Better Health Care

ATTTCGGGTCCATTCCCGTGTAT00110011TTCCGG00111000100TT0011CCGGTTAAAGGGG
TGA10011100TGATCGGGGTTTATATGCGCATTT00111000110010TCCTAAGGCTAAAAG
GGCACATG01011100000TCGA1001GTCAAA1010100001TTTCAAAAGATCTCTCTCATTTCGGG

ERIC TOPOL, M.D.

BASIC BOOKS

A Member of the Perseus Books Group

New York

Published by
Basic Books, A Member of the Perseus Books Group
387 Park Avenue South
New York, NY 10016

Books published by Basic Books are available at special discounts for bulk purchases in the United States by corporations, institutions, and other organizations. For more information, please contact the Special Markets Department at the Perseus Books Group, 2300 Chestnut Street, Suite 200, Philadelphia, PA 19103, or call (800) 810-4145, extension 5000, or e-mail special.markets@perseusbooks.com.

Designed by Pauline Brown
Typeset in 10.5 point Berling LT Std by the Perseus Books Group

Library of Congress Cataloging-in-Publication Data

Topol, Eric J., 1954–
  The creative destruction of medicine : how the digital revolution will create better health care / Eric Topol.
       p.   cm.
  Includes bibliographical references and index.
  ISBN 978-0-465-02550-3 (hardback : alk. paper)—ISBN 978-0-465-02934-1 (e-book)
  I. Title.
  [DNLM: 1. Medical Informatics Applications. 2. Biomedical Technology. 3. Delivery of Health Care—trends. 4. Diffusion of Innovation. 5. Health Communication. 6. Internet. W 26.5]

  610.285—dc23

                                          2011041162

10 9 8

CONTENTS

*To my wife, Susan, for her tireless support,
love, and unmatched patience with me*

*To our children, Sarah and Evan, for their honesty and
love and for having molded me possibly more than
I have them*

*And to Gary and Mary West for their
unquestioning support of my ideas and their
generous charitable commitment to humankind*

INTRODUCTION

In the mid-twentieth century Joseph Schumpeter, the noted Austrian economist, popularized the term "creative destruction" to denote transformation that accompanies radical innovation. In recent years, our world has been "Schumpetered." By virtue of the intensive infiltration of digital devices into our daily lives, we have radically altered how we communicate with one another and with our entire social network at once. We can rapidly turn to our prosthetic brain, the search engine, at any moment to find information or compensate for a senior moment. Everywhere we go we take pictures and videos with our cell phone, the one precious object that never leaves our side. Can we even remember the old days of getting film developed? No longer is there such a thing as a record album that we buy as a whole—instead we just pick the song or songs we want and download them anytime and anywhere. Forget about going to a video store to rent a movie and finding out it is not in stock. Just download it at home and watch it on television, a computer monitor, a tablet, or even your phone. If we're not interested in getting a newspaper delivered and accumulating enormous loads of paper to recycle, or having our hands smudged by newsprint, we can simply click to pick the stories that interest us. Even clicking is starting to get old, since we can just tap a tablet or cell phone in virtual silence. The Web lets us sample nearly all books in print without even making a purchase and efficiently download the whole book in a flash. We have both a digital, virtual identity and a real one. This profile just scratches the surface of the way our lives have been radically transformed through digital innovation. Radically transformed. Creatively destroyed.

Some will argue the predigital era was a better and simpler one. We were not connected and distracted all the time—even when driving a car. We wrote handwritten notes to one another and communicated much more deeply and effectively, albeit less frequently. We spoke on the phone to each other and did not rely on texting and instant responses. We had much more privacy, and there was no digital, immutable archive of our lives for everyone to peer at via a few clicks. We used maps to find our way from place to place instead of global positioning systems. But those days are truly past tense, and our world has irrevocably changed. The cumulative effect of extraordinary innovation that exploits digital information has turned our world upside down. Essentially, there is no turning back.

But the most precious part of our existence—our health—has thus far been largely unaffected, insulated, and almost compartmentalized from this digital revolution. How could this be? Medicine is remarkably conservative to the point of being properly characterized as sclerotic, even ossified. Beyond the reluctance and resistance of physicians to change, the life science industry (companies that develop and commercialize drugs, devices, or diagnostic tests) and government regulatory agencies are in a near paralyzed state, unable to break out of a broken model of how their products are developed or commercially approved. We need a jailbreak. We live in a time of economic crisis because of the relentless and exponentially escalating costs of health care, but we've done virtually nothing to embrace or leverage the phenomenal progress of the digital era. That is about to change. Medicine is about to go through its biggest shakeup in history.

This book is about the creative destruction of medicine, of how medicine will inevitably be Schumpetered in the coming years, and why it is vital for consumers to be fully engaged. Without the active participation of consumers in this revolution, the process will be inexorably slowed. All the other forces that could come to bear—doctors, the life science industry, government, and health insurers—are incapable of catalyzing this transformation. At the same time, the democratization of medicine is taking off. You, the consumer, are going to be needed to make it happen.

There is one theme, one reason, why this creative destruction is ready to go. It is because for the first time in history we can digitize humans. You know about digitizing pictures and information like books, newspapers, and magazines. It seems that everything now is digitized and widely transferable. You can download a two-hour movie in seconds. But that is a world apart from digitizing a human being.

Digitizing a human being is determining all of the letters ("life codes") of his or her genome—there are six billion letters in a whole genome sequence. It is about being able to remotely and continuously monitor each heart beat, moment-to-moment blood pressure readings, the rate and depth of breathing, body temperature, oxygen concentration in the blood, glucose, brain waves, activity, mood—all the things that make us tick. It is about being able to image any part of the body and do a three-dimensional reconstruction, eventually leading to the capability of printing an organ. Or using a miniature, handheld, high-resolution imaging device that rapidly captures critical information anywhere, such as the scene of a motor vehicle accident or a person's home in response to a call of distress. And assembling

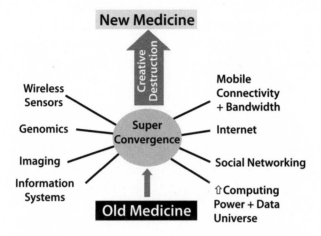

FIGURE INTRO.1: The transformation from medicine today (old, dumbed down) to new, individualized medicine that is enabled by digitizing humans.

all of this information about an individual from wireless biosensors, genome sequencing, or imaging for it to be readily available, integrated with all the traditional medical data and constantly updated. We now have the technology to digitize a human being in highest definition, in granular detail, and in ways that most people thought would not be possible, if even conceivable, for many decades to come.

This is a story about an unprecedented super-convergence. It would not be possible were it not for the maturation of the digital world technologies—ubiquity of smart phones, bandwidth, pervasive connectivity, and social networking. Beyond this, the perfect digital storm includes immense, seemingly unlimited computing power via cloud server farms, remarkable biosensors, genome sequencing, imaging capabilities, and formidable health information systems.

Think of the cell phone, which is a hub of telecommunications convergence but also a remarkable number of devices all rolled into one gadget: camera, video recorder, GPS, calculator, watch, alarm clock, music player, voice recorder, photo album, and library of books—like a pluripotent stem cell. Armed with apps it carries out diverse functions from flashlight to magnifying glass. Then connect it to a wireless network, and this tiny device is a web surfer, word processor, video player, translator, dictionary, encyclopedia, and gateway to the world's knowledge base. And by the way, it even texts, emails, and provides phone service. But now picture this device loaded for medicine, capable of displaying all of one's vital signs in real time, conducting laboratory analyses, sequencing parts of one's genome, or even acquiring ultrasound images of one's heart, abdomen, or unborn baby. This embodies a technological convergence, a coalescence of distinct and far-ranging functionalities, from elemental forms of communication to the complexities of medicine.

These are the collective tools that lay the groundwork for digitizing humans. This is a new era of medicine, in which each person can be near fully defined at the individual level, instead of how we practice medicine at a population level, with mass screening policies for such conditions as breast or prostate cancer and use of the same medication and dosage for a diagnosis rather than for a patient. We are each unique human beings, but up until now there was no way to establish one's biologic or physiologic individuality. There was no way to determine a relevant metric like blood pressure around the clock while a person is sleeping, or at work, or in the midst of an emotional upheaval. This represents the next frontier of the digital revolution, finally getting to the most important but heretofore insulated domain—preserving our health.

We have early indicators that this train has left the station. The first individual—a five-year-old boy—who had his life saved by genome sequencing was only recently documented. And this led to the first health insurance coverage of genome sequencing. But it's not just about finding the root molecular cause of why an individual is sick. We can now perform whole genome sequencing of a fetus to determine what conditions should be watched for postnatally. At the other end of the continuum of life, we can do DNA sequencing to supplant a traditional physical autopsy, to determine the cause of death. We can dissect, decode, and define individual granularity at the molecular level, from womb to tomb.

That's just the start of illuminating the human black box. Recognizing that we are walking event recorders and that we just need biosensors to capture the data, and algorithms to process it, sets up the ability to track virtually any metric. Today these sensors are wearable, like Band-Aids or wristwatches. But soon enough they will also be embedded into our circulation in the form of nanosensors, the size of a grain of sand, providing continuous surveillance of our blood for the earliest possible detection of cancer, an impending heart attack, or the likelihood of a forthcoming autoimmune attack. Yes, this does ring in the sci-fi concept of cyborgs, the fusion of artificial and biological parts in humans. We've already been there with cochlear implants for hearing loss, a trachea transplant, and we're going there in the creation of embedded sensors that talk to our cell phones via wireless body area networks in the future. With it comes the familiar "check engine" capability that we are accustomed to in our cars but never had before for our bodies. Think true, real prevention for the first time in medical history.

While this may seem a bit too futuristic, in the context of the information era it may appear to be eminently more realistic. We live in an extraordinary data-rich universe, a world that had only accumulated one billion gigabytes (10^9 or 1,000,000,000 bytes of data) from the dawn of civilization until 2003. But now we are generating multiple zettabytes—each representing one trillion gigabytes—each year and will exceed thirty-five zettabytes by 2020, roughly equivalent to the amount of data on two hundred fifty billion DVDs.[1] Sensors are now the dominant source of worldwide-generated data, with 1,250 billion gigabytes in 2010, representing more bits than all of the stars in the universe.[2] The term "massively parallel" is an important one that in part accounts for this explosion of data and brings together the computer, digital, and life science domains. Note the convergence: from single chips that contain massively parallel processor arrays, to supercomputers with hundreds of thousands of central processing units, to whole-genome sequencing that is performed by breaking the genomes into tiny pieces and determining the life codes in a massively parallel fashion.

In 2011, the Watson IBM computer system beat champion humans in the game of Jeopardy. Watson is equipped with a 15-terabyte (10^{12}) or 15,000,000,000,000-byte databank and massively parallel 2,880-processor cores.[3] So beyond its television premiere and victory, where is Watson first going to be deployed? At Columbia University and the University of Maryland medical centers to provide a cybernetic assistant service to doctors.[4] David Gelernter's op-ed in the *Wall Street Journal*, "Coming Next: A Supercomputer Saves Your Life," introduced the concept of a WikiWatson, which could bring together the whole world's medical literature and clinical expertise.[5] Putting a massive databank to use to improve health care is emblematic of the overlay of the digital and medical worlds.

By now I hope I have made my preliminary case for super-convergence abundantly clear. But just having these technological capabilities will not catapult medicine forward. The gridlock of the medical community, government, and the life science industry will not facilitate change or have the willingness to embrace and adopt innovation. The U.S. government has been preoccupied with health care "reform," but this refers to improving access and insurance coverage and has little or nothing to do with innovation. Medicine is currently set up to be maximally imprecise. Private practice physicians render medicine "by the yard" and are rewarded for doing more procedures. Medical care is largely shaped by guidelines, which are indexed to a population rather than an individual. And the evidence from clinical research is derived from populations that do not translate to the real world of persons. The life science industry has no motivation to design drugs or devices that are only effective, however striking, for a small, well-defined segment of the population. At the same time, the regulatory agencies are entirely risk-averse and, as a result, are suppressing remarkable innovative and even frugal opportunities to change medicine. The end result is that most of our screening tests and treatments are overused and applied in the wrong individuals, promoting vast waste. And virtually nothing is being done to accelerate true prevention of disease.

But the practice of imprecise medicine has not yet emboldened consumers to demand more, despite increasing awareness of the problem. Many patients now trust their peers on social networks—online medical communities such as PatientsLikeMe—more than their physicians. In some health care systems, patients can directly download their laboratory reports and medical records, which they were never allowed to do in the past. Any consumer with adequate funds can have his or her genome scanned or even wholly sequenced. In parallel and intersecting with super-convergence, we are now finally moving toward the democratization of medicine.

When the revolutions were occurring in 2011 in Tunisia and Egypt, predominantly propelled by the young oppressed citizens who could express and organize themselves via social networks and exploit the digital world, sharing pictures and videos, I tweeted: "Tunisia . . . Egypt . . . American medicine?" In fewer than forty characters, this conveyed my sense of urgency for consumers to provide the impetus for new medicine—a new medicine that is no longer paternalistic, since the doctor does not necessarily know best anymore. The American Medical Association has lobbied the government hard for consumers *not* to have direct access to their genomic data, that this must be mediated through physicians. We know that 90 percent of physicians are uncomfortable and largely unwilling to make decisions based on their patients' genomic information. But it is your DNA, your cell phone, and your right to have all of your medical data and information. With a medical profession that is particularly incapable of making a transition to practicing individualized medicine, despite a new array of powerful tools, isn't it time for consumers to drive this capability? The median of human beings is *not* the message.[6] The revolution in technology that is based on the primacy of individuals mandates a revolution by consumers in order for new medicine to take hold.

Now you've probably thought "creative destruction" is a pretty harsh term to apply to medicine. But we desperately need medicine to be Schumpetered, to be radically transformed. We need the digital world to invade the medical cocoon and to exploit the newfound and exciting technological capabilities of digitizing human beings. Some will consider this to be a unique, opportune moment in medicine, a veritable once-in-a-lifetime Kairos.

This book is intended to arm consumers to move us forward. In the first section, I review the overall digital landscape—how the digital world has evolved and changed our lives outside of medicine; how our information in medicine is grossly deficient and population-based; and how consumers, despite progress toward convergence of health information, are too often poorly informed.

In the second section, I drill down into each of the four areas of digital medicine—wireless sensors, genomics, imaging, and health information—and lay out a vision of how these technologies will converge. In the last section, I preview the impact that digitizing humans will have on doctors and hospitals, on the life science industry and regulatory agencies, and, ultimately, on the individual.

As with any revolution, there are important downsides to consider. Here the concerns include the reduction of direct human contact and healing touch that may accompany increasing reliance on remote monitoring and avoidance of hospitalizations or even in-person office visits. It will be increasingly tempting for physicians to treat the virtual human being—the scan, the DNA results, the biosensor data—instead of the real patient. There is legitimate worry about adoption of new technologies before they have been adequately vetted and validated, or proven to be cost-effective and ideally cost saving. And certainly data deluge and the inability to efficiently transform the massive data sets into information and knowledge loom large. An extension of data flow issues brings us to the worry about security and privacy of digitized medical information. Ironically, the technological triumph of being able to digitize human beings creates a convergence of the real and virtual individual, and there will be legitimate worries about depersonalization, about treating the digital information instead of the individual. Ultimately, you will have to decide about the trade-offs of medicine Schumpetered. This book is intended to put you in position to be ready and knowledgeable to make that decision.

Doctors prescribe medicine of which they know little,
to cure diseases of which they know less,
in human beings of which they know nothing.
—François-Marie Arouet Voltaire, about 250 years ago

PART ONE · SETTING THE FOUNDATION

THE DIGITAL LANDSCAPE
Cultivating a Data-Driven, Participatory Culture

In this electric age we see ourselves being translated
more and more into the form of information,
moving toward the technological extension of consciousness.
—Marshall McLuhan, *Understanding Media*, 1964[1]

WHEN MARTY COOPER invented the cell phone in 1973, he could not have dreamed or estimated that there would be over six billion cell phones by 2012 and that this platform would ultimately have a major impact on the future of health and medicine. The invention of the personal computer by Michael Wise in 1975, followed soon thereafter by the innovations of Steve Jobs and Steve Wozniak the following year, led to over one billion personal computers in use by 2008 and an anticipated two billion in 2014.[2] The Internet began to hit its stride by the mid-1990s, and now well over two billion individuals are connected with such expanded bandwidth that video files have become the dominant medium of exchange as measured by file size.[3]

But the biggest leap came in the first decade of the twenty-first century. The six billion bases of the human genome were sequenced, and this led to the discovery of the underpinnings of over one hundred common diseases, including most cancers, heart disease, diabetes, autoimmune disorders, and neurologic conditions. While scientists were busy sorting out the genome's zip codes, engineers were building on the wireless phone platform to add emails, texting, cameras, multimedia, global positioning, and access to the Internet. Concurrently, the bandwidth of the Internet was quickly expanding, and the ability to rapidly search it was increasing exponentially. The unprecedented transformative impact and uptake of mobile digital devices in the same decade, from the introduction of iPods in late 2001 and Blackberries in 2002 to the iPhone and Kindle e-reader in 2007, cumulatively changed the way we listen to music, communicate by text or phone, surf the Web, move from place to place, take pictures, make videos, play, read, and think.

In that same decade the number of discrete mobile phone users increased from five hundred million to over three billion, representing almost half of all people and the vast majority of adults on the planet.[4] And they're now sending over two trillion text messages a year.[5] Our ever-increasing computing power is exemplified by unfathomable data storage capacity. Last year we stored enough data to fill 60,000 Libraries of Congresses, and we can now purchase a device for $600 that will store the entire collection of the world's recorded music.[6]

The global growth of cameras as a result of being embedded in cell phones has been logarithmic: from a few million in 2000 to well over a billion in a decade.[7] Digital cameras can be considered the most widely available sensor since they are incorporated in most mobile phones; as O'Reilly and Battelle pointed out in their "Web Squared" white paper, "Our cameras, our microphones, are becoming the eyes and ears of the web."[8]

Even our games have remarkable digitizing capability. In late 2006, the Nintendo Wii came out with wireless accelerometer sensors and infrared to detect an individual's motion in three dimensions. By 2010, gaming had made major advances, such as Microsoft's Kinect for recognizing faces and gestures, responding to voice commands, and being able to play the games that display on-screen avatars with body motions instead of needing to use controllers or any button pushers. Five million of these were sold in the first two months of availability.[9]

When Mark Zuckerberg started Facebook in 2004, how could anyone have predicted there would be over eight hundred million registered users by the end of 2011? Or a projected one billion by 2012? Over 25 percent of Internet users are connected through this particular social network—only third in population size compared with the countries of China and

India. Over 1.5 trillion messages are now sent per year via Facebook. In 2011 Facebook had substantially overridden Google as the dominant website, with Facebook users looking at 103 billion pages and spending an average of 375 minutes per month, compared to Google users viewing 46 billion pages over 231 minutes. More than 40 percent of us are "hyper-connected" as defined by "using 7 different devices and 9 different applications in order to stay as screen connected as possible, in restaurants, from bed, and even in places of worship."[10]

These extraordinary accomplishments, from dissecting and defining DNA to creating such pervasive electronic technologies that immediately and intimately connect most individuals around the world, have unwittingly set up a profound digital disruption of medicine. Until now we did not have the digital infrastructure to even contemplate such a sea change in medicine. And until now the digital revolution has barely intersected the medical world. But the emergence of powerful tools to digitize human beings with full support of such infrastructure creates an unparalleled opportunity to inevitably and forever change the face of how health care is delivered.

This really boils down to a story of big convergence: a convergence of all six of the major technologic advances, likely representing the greatest convergence in the history of humankind (see Figure 1.1). When we just had a cell phone, we could only talk to one another, but it could occur on the go. As personal computer hardware developed from a work station to a laptop, we gained mobility, but we were still not connected to one another. The Internet strikingly changed both of these platforms. Nicholas Negroponte wrote in his 1995 book *Being Digital*, "The information superhighway may be mostly hype today, but it is an understatement about tomorrow. It will exist beyond people's wildest predictions."[11] Clearly, that was a prescient call. Although the first BlackBerry devices were inadequate cell phones, they were extraordinary at receiving and sending emails. This new capability engendered such addictive behavior that the devices were quickly known as "CrackBerries." But it took almost five years before the morphed cell phone, powered with emails and texting, faithfully performed its original purpose of making voice calls and also became a wholly functional Web surfing tool.

This transition from a "mail and text" phone to a "smart phone" relied on a much greater Internet bandwidth, broad connectivity via networks such as AT&T, Verizon, and others in the United States, along with appropriate, tailored mobile operating system development. Late in 2007, Apple's introduction of the 3G iPhone was a veritable game changer, and most would even qualify it as a life changer. It was dubbed the "Jesus" phone, and Steve Jobs was later pictured in 2010 on the cover of the *Economist* as "The Book of Jobs."[12]

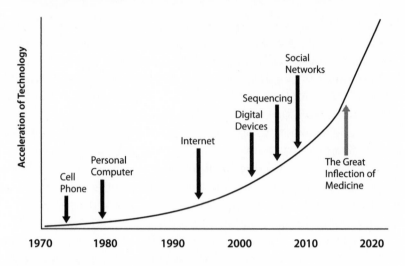

FIGURE 1.1: Timing of the big 6 major digital advances over the past 40 years that have set up the Great Inflection of Medicine.

For the first time, surfing the Web via a cell phone was performed rapidly with ease, enabling many other newfound pocket mobile functions such as global positioning and most importantly a sea of applications—downloadable software that runs on mobile devices. By 2011, just for the iPhone there were over 300,000 apps downloaded over 6.5 billion times.[13] This was made possible because writing code for the mobile phone had become an open platform. In a way the app era represents a reverse Web surf. Instead of any individual searching the Web, the Web was surfing the population for "worldwide developers," a new term denoting the search for people to create apps. Within a matter of months, hundreds of thousands of apps were created that accelerated the capabilities of smart phones in ways most people could not have imagined: from using the phones to determine a bird species by its picture or call, to instantly translating a page in English to Spanish or vice versa, to discerning colors correctly for people who are color-blind, to playing thousands of new games, to accessing or playing music. One can even convert the iPhone to a stethoscope to listen to heart and breath sounds.

The hybridization of the maturing Internet and the mobile phone were the two most vital components of the convergence. As Nicholas Carr aptly wrote, "With the exception of alphabets and number systems, the Net may well be the single most powerful mind-altering technology that has ever come into general use."[14] Just the ability to instantly search virtually anything on the Web, in itself—a peripheral brain—is still awe-inspiring. Nevertheless, it is hard to be intimate with the Internet per se. By contrast, almost 70 percent of individuals sleep with their cell phone. That figure goes up to 90 percent for digital natives, as defined by people under age thirty.[15] There are more mobile phones in the world than toothbrushes, and far more than toilets.[16]

We are preternaturally on the move, a peripatetic culture, and our phones are always with us. Many rank the mobile phone above food, shelter, and water as their most essential possession. The *Economist* put it simply: "mobile phones have made a bigger difference to the lives of more people, more quickly, than any previous technology."[17] *Nature*, the leading biomedical peer review journal, pointed out that we will have six billion mobile phones by 2013, with over 85 percent of the world's population having access to a mobile signal, and that "we've really never had a technology other than human observation that is as pervasively deployed in the world."[18] The extraordinarily rapid uptake of phones helps to tell the story. In 2001, it took ninety-one weeks for one million iPods to be sold, a record for digital devices at that time. By 2009, it took only three days for one million iPhone 3GS to be purchased.[19] By 2013, it is projected that the number of smart phones will surpass personal computers, not counting tablets.[20]

In 2009 judges at the Wharton School of the University of Pennsylvania were asked what were the biggest innovations, the "life changers," of the past thirty years. Their response in rank order was: (1) Internet, broadband; (2) PC and laptop; (3) mobile phones; (4) email; and (5) DNA testing and sequencing.[21] The smart phone had already captured four of five life changers and, as it continues to evolve, is working its way to incorporating the fifth.

Like a syzygy with alignment of the sun, the moon, and the Earth, we have a propitious convergence of a maturing Internet, ever-increasing bandwidth, near-ubiquitous connectivity, and remarkable miniature pocket computers in the form of mobile phones. And with data storage and processing fortified with cloud computing, under the stars, most of the people on this planet have been quickly and deeply affected in ways few really recognize.

THE WAYS WE HAVE CHANGED: THE C'S

Constant Connectivity

Attention deficit disorder (ADD), with or without a hyperactivity (ADHD) component, is a disorder that is diagnosed in 3–5 percent of school-aged children globally.[22] It is the most common behavioral abnormality in children and adolescents and is characterized by inattentiveness, resulting in difficulties in listening, sitting still, finishing tasks, and being easily distracted. A similar condition is well recognized in adults without hyperactivity (adult ADD). The root cause is uncertain, but many studies have reinforced that there is abnormal function, imbalance, or dysregulation of neurotransmitters, especially dopamine. Neurotransmitters carry signals across nerve cells in the brain at their connection points, known as synapses. Paradoxically, the treatment for the inability to stay focused on a particular task is the use of a stimulant drug such as Adderall and Ritalin, which are amphetamines. The condition is not believed to be related to the impact of the era of the Internet and digital devices.

In addition to the inborn, biologic form of ADD, the digital era has ushered in an environmentally induced form. In the current digital era, there is little need for a stimulant drug. We are in a state of near-constant connectivity (equivalent to near-constant dopamine squirts), and digitally induced ADD (DADD) is a widespread problem.[23] This is not neurotransmitter-based, but our neurotransmitters are surely revved up; DADD represents an outgrowth of immediate access to all of our communication, our social networks, and the world's happenings via the Internet. Bilton coined the term "consumnivores," referring to those who are "collectively rummaging, consuming, distributing, and regurgitating content in byte-size, snack-size, and full-meal packages."[24] In *Hamlet's BlackBerry*, William Powers described the view from some pessimistic quarters that "digital natives (age <30) who have grown up with screens, are effectively a new species of human being, innately incapable of holding a sustained conversation or thought."[25] Further evidence of the digital dalliance comes from data at work: employees with computers check their email or change windows thirty times per hour, which has been correlated with diminished work productivity.[26]

During a typical workplace meeting, people continually check their email, surf the Web, or tweet their followers with their latest insight or experience. When alone, when we are not looking at a screen, there is a sense of disconnectedness reflecting addictive behavior. Much has been written about the impact of the Internet and constant connectivity on the brain. In *The Shallows*, Nicholas Carr reviewed the data on neuroplasticity of the brain—the potential deleterious effect of continual digital stimulation to actually change our brain tissue and reroute neural activation and pathways. Bill Keller, former executive editor of the *New York Times*, wrote, "Basically, we are outsourcing our brains to the cloud."[27] "The Google Effect," a recent study testing the cognitive consequences of using Internet search engines, documented poor memory for information that was obtained through an electronic search. Although it remains controversial whether the Internet and connectivity directly lead to adverse anatomical and functional brain effects, few would debate whether it has affected our behavior. One of the most important aspects is the blurring of the sea of content without discrete elements of information sources such as a newspaper, TV, or radio or for that matter pictures, words, video, or music—it's all a continuum. The term *Homo distractus*, used by Powers, captures our relative inability to be focused or stay on any particular task in the midst of a blitzkrieg of data flow.[28]

Beyond multitasking and a shortened attention span, the constant connectivity affects the way we think. In order to process the considerable body of data that comes from so many sources, we are less apt to be linear and much more likely to be networked in our thinking. For example, we click on links while we are reading text that may take us far away from our line of thinking before we get back to the book or article. Whether it is rapidly digesting text, graphics, links, photos, or videos, we are constantly scanning an extraordinary body of data. With Google, Bing, and other search engines, our peripheral brain is just one click away. Two portmanteaus help provide context: "Netizen," combining Internet and citizen, and "digerati," derived from the fusion of digital and literati. Whether we are Netizens, the term applied to any person actively involved in online communities, or digerati, referring to the influencers in the digital community, we are perpetually adapting to and contending with a highly dense, data-rich environment.

Collaboration and Crowdsourcing

Collaboration requires all parties to actively participate. The world of the Internet, mobile phones, personal computers, and social networks has created a participatory culture that has greatly exceeded expectations. Media of yesteryear was unidirectional, centralized, and completely controlled by the entities that created and delivered it. The media of today is multidirectional. Every Netizen can be viewed as an informant, an e-activist. We no longer live in the third person; we live on the Web in the first person.[29] Data and information generation have been democratized.

In 2011 the *Huffington Post* had only 150 paid staff but more than 12,000 volunteer "citizen journalists."[30] Because of YouTube any individual can be considered a videographer.[31] At least twenty-four hours of video content are uploaded every minute, which leads to more than three billion videos viewed per day. The average Internet user watches about two hundred videos per month. Twitter, with over 160 million registrants, has amassed over 15 billion tweets and over 2,000 are added every second— over 100 million tweets per day in 2010 and over 200 million per day in 2011. There are over twenty billion pieces of content and more than two hundred million photos added to Facebook every day—over ninety billion photos on the site. The retail industry has been rapidly transformed by such social networking sites as Groupon and Living Social. Groupon, known as the "world's fastest growing company ever," had over 4,000 staff and over fifty million subscribers in six hundred cites.[32] Each Netizen has equal rights to using the Internet as a platform for content, and this truly represents the democratization of communication.

This point was brought home to me when I had an e-counter with a *New York Times* science journalist who covers genomics. In June 2010, at the ten-year anniversary of the first human genome sequence announcement, Nicholas Wade wrote a front page article in the *Times* entitled "A Decade Later, Gene Map Yields Few New Cures: Despite Early Promise, Diseases' Roots Prove Hard to Find."[33] I had previously emailed Nicholas Wade many times, arguing that he was much too pessimistic about the progress that was being made in medical genomics. But instead of a private email to him this time, I tweeted, "NY Times Nicholas Wade discounting progress in genomics for about the 10th time." Within a couple of days, I got an email from Nicholas Wade. He wrote, "You can perhaps imagine my surprise on finding myself rebuked . . . by yourself or someone tweeting under the same name." This led to a brief moment of satisfaction that I had been able to post something on the Web that reverberated back to him. Much more effective than private email contact!

Social networking is a collaborative experience in sharing information, photos, links, and videos—the social graph—to one's circle of friends. The percentage of social networking users age fifty-five to sixty-four shot up from 9 percent at the end of 2008 to 43 percent by mid-2010. With Facebook, the sharing operates in over seventy-five languages, or reaching 98 percent of the world's population. The time Netizens spend on social networking now greatly exceeds email time and searching—in the United States over six hours per month and still rising. In *The Facebook Effect*, David Kilpatrick reflected on how communication was changed via Facebook's News Feed, which became active in late 2006, in which any individual could broadcast information to his or her friends: "It turned 'normal' ways of communicating upside down. Up until now, when you desired to get information about yourself to someone, you had to initiate a process or 'send' them something, as you do when you make a phone call, send a letter or an email, or even conduct a dialogue by instant message." Beyond the ability to broadcast using the social network platform, the transparency and openness changed. For developing what has become the largest social network and the dominant force on the Web, Zuckerberg was recognized by *Time* magazine as the Person of the Year in 2010. To ante up in 2011, Google introduced Google+, but it remains to be seen whether it can catch up after initially missing "the friend thing."[34]

One of the greatest accomplishments of the digital era that was largely unforeseen was to bring people together for common laudable goals. The wiki world exemplified by Wikipedia has laid the groundwork for an exceptional number of open platforms. Linux is considered the exemplar of free and open source software collaboration, installed on a large proportion of mobile phones, tablets, computer mainframes, and supercomputers—it even runs the top ten fastest supercomputers in the world. The authors of the widely acclaimed *Wikinomics* and more recently *Macrowikinomics*, Don Tapscott and Anthony Williams, described Linux, the prototype of mass participation, as "the quintessential example of how self-organizing, egalitarian communities of individuals and organizations come together— sometimes for fun and sometimes for profit—to produce a shared outcome."[35] Other familiar examples of companies that adopted an open platform are Google, Amazon, and Facebook.

But the number of wiki communities that have been spawned in recent years is mind-blowing. As reviewed in depth in *Macrowikinomics*, these range from simple carpooling entities like iCarpool, PickupPal, Carticipate, GoLoco, and Zimride, to environmentally conscious communities such as Carbonrally, Earth lab, Better Places, and GreenXchange, to a mass astronomy wiki known as Galaxy Zoo, financial wikis such as Open Models Company, Ven Corps, Zopa, Prosper, and Lending Club, innovation wikis such as Innocentive and Nine Sigma, education communities like Academic Earth, Open Course Ware, and Wikiversity, and hundreds more. Relevant to the medical space are PatientsLikeMe, WeAre.Us, MedHelp, Sermo, and others, which will be discussed in depth.

Innovation has been propelled to new heights as an outgrowth of open, collaborative, nonproprietary networking. As Steven Johnson categorizes the "Fourth Quadrant" of innovation in *Where Good Ideas Come From*, the global scientific collaboration on understanding the human genome exemplifies this concept[36]—the original Human Genome Project, the International HapMap, ENCODE, and 1000 Genomes. While the incentives for this form of innovation are considerably different from what drives entrepreneurs or private corporations, the barriers are almost nonexistent.

Collaboration also sets up the world of crowdsourcing, the principle of tapping into a brain trust in real time, which was previously inaccessible. The pooling of minds and "wisdom" of the crowd can be readily accessed through social networking sites like Facebook, Twitter, Windows Live, MySpace, and Baidu. While this can be utilized for simple matters like picking a restaurant or finding a recipe, there are no limits. Twitter has been referred to as the "nervous system" of the Internet.[37] One of my favorite people to follow on Twitter is David Pogue, the tech guru for the *New York Times*. In preparing his column for the *Times*, he asked his 1.5 million Twitteratti followers for some new iPhone apps that hadn't been invented. Here are a few responses: Read2me app: it reads your email, texts, tweets, etc., aloud so you can do other things, like drive; Rejuvenator: aim iPhone camera at your face, snap picture, digitally subtracts five, ten, twenty years from your image; Switcher: switches your iPhone from AT&T to the Verizon network so you can make calls with your phone. Most people probably don't adequately appreciate the depth of the knowledge reservoir and creative solutions available from their social networks. Things have so drastically changed from the "Who Wants to Be a Millionaire?" TV show that originated in 1998, which allowed game show participants to call a single trusted friend for information. Now we can instantly access the knowledge base of thousands or tens of thousands of people.

The era of crowdsourcing social networks has changed whom we trust. In 2009, Nielsen surveyed 25,000 consumers in over fifty countries and determined that individuals trusted their friends, family, and peers for recommendations 90 percent of the time.[38] As Nick Bilton put it in *I Live in the Future*, "the shifting nature of trust is one reason I think we're moving toward investing more of our attention and confidence in individuals online and away from traditional companies and their brands."[39] As will be seen, our go-to source for health and medical information is moving away from our doctor—it is increasingly by crowdsourcing and friendsourcing our entrusted social network.

Customized Consumption

For us digital immigrants who are music aficionados, we had to buy the whole record, or in later years the CD, which inevitably had a bunch of "lemon" songs or fillers. When iTunes and the iPod became available, the music industry was profoundly disrupted as the consumer became empowered. One could easily listen to and download select digital recordings and no longer purchase the whole album. And one could access via the Internet virtually any music that had ever been recorded, regardless of its obscurity. This movement is captured by Chris Anderson in his book *The Long Tail: Why the Future of Business Is Selling Less of More*, in which he describes how the Internet radically transformed business.[40] Companies like Netflix, Amazon, and Rhapsody could quite profitably offer relative obscure film, book, and music selections, respectively, that you couldn't find anywhere else but online. The ability to cater to individual niche interests was catapulted by seemingly unlimited "deep" inventory and accessibility. This has in many ways set up an era of unprecedented choice, the notion that one size does not fit all, and a compelling alternative to the blockbuster model that has previously dominated our culture. Based on the data available through our online search engine use or our online social networks, personalized digital advertising has become embedded in our Internet experience.

As Eric Schmidt, the former CEO of Google, said, "The power of individual targeting—the technology will be so good it will be very hard for people to watch or consume something that has not in some sense been tailored for them."[41] The Internet and digital era have driven such customization and targeting to emerge as the norm in almost every industry today—not just advertising but media, retail, finance, travel. Note the exception of medicine and health care.

Hyperpersonalized is the theme of the day. We log on to the websites we are interested in, connect with the people and networks that matter to us, watch the video clips and shows that attract us, listen to the music we like, download the apps we want to use, follow the links and blogs we find interesting, and share the information and pictures we care to. While in the pre-digital era we could pick the radio station we wanted to listen to or the book we wanted to read, the choices and immediate access to content have increased exponentially. Overall a grand sense of individual empowerment is created. On the other end, the companies that are using the Web to target us have the capability of leveraging the rich content we have created. They use demographic and user preference data to create highly personalized advertising. Beyond the individual, the Facebook social network of people who click a button to "like" Patagonia or Pepsi is prime for promotion by Patagonia or Pepsi for its products. This ability to hyperpersonalize goes further with knowledge of the precise location of the individual—retail stores or restaurants can connect with individuals in close proximity by tracking GPS data. Symmetrically, then, the activity and operation of both individuals and companies are positioned to be more precise and efficient. While there are serious matters to consider regarding privacy, which I will fully discuss in a later chapter, the information flow of today facilitates exceptional customization in multiple directions—from the individual and network, to the individual and network, and between individuals and networks.

Cloud Computing

Massive server farms, many with several hundred thousand servers, called "clouds," can now be accessed from anywhere in the world and provide expansive computing infrastructure. One of the largest services is offered by Amazon (Elastic Compute Cloud), which stores two hundred billion digital objects (ranging from files to movies) and handles over 200,000 requests per second and generated about $700 million of revenue in 2010. A cloud's three main components are Web-based applications (like Gmail, Windows Azure, Apple iTunes); platforms, which allow developers to write applications; and core-computing services ranging from number crunching to data storage.[42] The availability of seemingly unlimited computing power, at very reasonable costs, has provided an extraordinary resource to catalyze all of the changes in the digital world. Eric Schadt, a highly accomplished genetics researcher and "master of information," was recently featured in *Esquire* for doing genomics computing via the cloud:

> Fortunately, he has the same access to supercomputers that every other American with an Internet connection and a credit card has. He waits till the plane climbs to a cruising altitude, waits for the pilot to allow electronic devices, and then uses the plane's WiFi to get on Amazon. Amazon sells a lot of stuff—books, washing machines, whatever the hell you want. What it sells Schadt is super-computing on the cheap. You see, companies like Amazon have a lot of computing power available, and now it's gotten in the business of selling some of that to guys like Schadt and whoever else might want it. A guy like Schadt doesn't have to work for a company like Merck anymore, because he has as much computing power available to him on an airplane as a scientist at Merck does on the company's multimillion-dollar supercomputer. More even. On cross-country flights he tells Amazon what data to crunch after takeoff, and for a few hundred bucks the job's done by the time he lands.[43]

These C's have, in aggregate, paved the way for the D's.

DISRUPTION AND DESTRUCTION

The cumulative effect of the six C's has more than fulfilled the concept of "creative destruction" that was originated by Werner Sombart and popularized by Joseph Schumpeter. There is a long list of examples of radical innovation that led to transformation. Specific to the digital world are the shutdown of most music stores like Tower Records, the replacement of video rental stores like Blockbuster by Netflix, the gradual demise of major chain bookstores like Borders and their replacement by online browsing and purchasing via Amazon and other sources, the attrition of print newspapers such as the announcement by the publisher in 2010 that in the future the *New York Times* would not be printed, even though since 1851 its motto has been "All the News That's Fit to Print."[44] In place of paid print newspapers there are very successful free online news sources like the *Huffing-*

ton Post. Television, as it exists today, is widely believed to be the next victim of creative destruction. Already sites like Hulu stream many popular shows online, so with the ongoing convergence of the Internet and television, individuals will no longer be subject to watching a program at a specific time. While digital video recording was a start, the disruption that has hit most other forms of media, including music, newspapers, and video, will likely continue to chip away at television. The number of homes with Internet-connected televisions is expected to reach forty-three million by 2015, up from two million at the beginning of 2010.[45]

DEALING WITH A DATA DELUGE

Another critical impact of the C's relates to the generation of a tsunami of data. The 1965 paper by Gordon Moore framed what is known as Moore's law. The doubling of capability of digital devices every eighteen months—such as transistors per mm^2 on an integrated circuit, memory capacity, or processing speed, or the size and number of pixels in digital cameras, has been remarkably in step with Moore's law for the past forty-five years. To appreciate the change in data available for one digital immigrant, some numbers can help provide context. Almost thirty years ago, I purchased my first personal computer. It was the IBM 5150, with a central processing unit of 4.77 MHz and 16 KB of RAM, which cost over $2,000. In 2010, I bought a MacBook Air with 1.8GHz (1800 MHz or 377 times the 5150) with 2 GB memory (2,097,152 KB =131,072 times the 5150) for a fraction of that price of adjusted 1981 dollars. This experience certainly follows Moore's law. In 1982 the transistor count in a central processing unit was about 50,000; today it is in excess of 2 million. Or as the futurist Ray Kurzweil recently put it, "a computer that fit inside a building when I was a student now fits in my pocket, and is a thousand times more powerful despite being a million times less expensive."[46]

Notably, and quite important for the topic at hand, there has been one digital technology that represents the exception to Moore's law—sequencing DNA. As shown in Figure 1.2, the throughput and cost of sequencing DNA has ratcheted down in a manner far exceeding Moore's law.[47] With the use of the so-called next generation of current sequencing platforms, such as Illumina's HiSeq or Life Technologies SOLiD 4, more than twenty-five gigabases (a giga is a billion) of sequence are generated per day.

Eric Schmidt of Google pointed out that from the dawn of civilization to 2003 there were a total of 5 exabytes (10^{18}) or 1 billion gigabytes of data. Today there are at least 5 exabytes every two days![48] And the tsunami of data is far from having reached a plateau. In 2010 the digital universe crossed

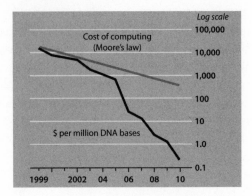

FIGURE 1.2: The marked reduction of the cost of computing (transistor/mm2) is exceeded by the reduction in cost of sequencing DNA.

the zettabyte threshold (10^{21}), or 35 trillion gigabytes (1 sextillion pieces) of data.[49] Supercomputers are now capable of performing 2,500 trillion operations per second.[50] Much of the next wave and amplification will come from human biologic and physiologic data sets. For each person who has his or her whole genome sequenced, with fortyfold coverage (sequencing an average of forty times to improve accuracy) there will be about 240 billion bytes of data generated. By the end of 2011, there is expected to be between 3,000 to 10,000 people fully sequenced.[51] That number will subsequently increase logarithmically as a result of the remarkable plummeting of the cost and efficiency of future sequencing technology platforms. Similarly, continuous monitoring of physiologic data via biosensors connected to a large cohort of individuals will contribute to data flooding.

This deluge of data to reckon with doesn't even take into account the output from sensors for smart energy grids and "smart cities," in which traffic patterns, water, refuse, power, and other systems are monitored and connected via sensors placed throughout many cities around the world. The number of wireless sensors was estimated to be 10 million in 2009 and projected to be 645 million by 2015. In late 2010, a special report by the *Economist* on smart systems portrayed "a sea of sensors," in which "anything and anyone—machines, devices, everyday things and particularly humans—can become a sensor, gathering and transmitting information about the real world."[52]

THE DATA-DRIVEN CULTURE

Even the uninitiated individual who was not particularly datacentric in his or her preceding analog life must now acknowledge that our current digital world has radically changed the landscape. The instantaneously accessible data coming in through texts, emails, photos, books, videos, searches, Web surfing click-click-clicks has changed us forever.[53] This combined infostructure and infrastructure yields a ubiquity of information.

This ubiquity is well exemplified by the "culturomics" project of Google. Over five million books published from 1500 to 2008, with more than five hundred billion words from six languages, were digitized and made available for searches of any word or group of words. The project is still expanding, and Google has already scanned two trillion words. When culturomics was first launched at the end of 2010 in the journal *Science*, the *New York Times* described it as offering "a tantalizing taste of the rich buffet of research opportunities now open to literature, history, and other liberal arts professors who may have previously avoided quantitative analysis."[54] It is astounding that such a simple online tool can provide access to trillions of words from the published books of over six centuries. It represents the "think big" approach to the sea of information.

We have markedly changed the way we communicate and interact with one another and, to an even greater extent, how we think and behave. All of these developments have been penetrating across our culture, affecting how we shop, travel, bank, invest, and consume information—and making us more apt to be quantitative. But interestingly and perhaps not surprisingly, given the sclerotic nature of the medical community, little has thus far touched the practice of medicine. While medicine is remarkably resistant to change, the ability to digitize any individual's biology, physiology, and anatomy, along with other elements—all things digital medicine—will undoubtedly reshape the future of medicine. The sense that a great inflection, from a time in medicine that shunned the digital era to one that will become dependent on it, is imminent derives from the unprecedented changes that have already taken place in virtually every other walk of life. Our health care system is broken and desperately needs to change. The recent flourishing of health and fitness digital devices and apps further lays the groundwork for the big changes that are destined to occur in medicine.

THE BIG PICTURE SUPER-CONVERGENCE

Marshall McLuhan, the visionary communication theorist, forecast the impact of the digitizing human in many ways. In his 1962 book, *The Gutenberg Galaxy: The Making of Typographic Man*, he coined the word "surfing" for "rapid, irregular and multidirectional movement through a heterogeneous body of documents or knowledge" nearly thirty years before the World Wide Web existed.[55] Add to that his concept of "the global village" to depict the interconnected culture via an electronic nervous system, presaging the social networks of today. Moreover, he wrote at length about the outward extension of the self through connective devices such as the typewriter, telephone, and television. In the prologue of that book, he characterized media as "extensions" of our human senses, bodies, and minds. David Gelernter, in his 1992 book *Mirror Worlds*, laid out the vision for digitally capturing a chunk of reality, a "brand new equilibrium," and even applied this concept to medicine:

> A Mirror World is an ocean of information. Fed by many streams . . . fed by automatic data-gathering and monitoring equipment, like the machinery in a hospital's intensive care unit, or weather-monitoring equipment, or traffic-volume sensors installed in roadways. These streams may be so fast-rushing that they threaten to overwhelm the main program with information tidal waves. The solution is to connect Mirror Worlds to fast-rushing data streams via a sort of software hydroelectric plant. Such programs are designed to sift through complex floods of data looking for trends and patterns as they emerge. They are constructed as layered networks. Data values are drawn in at the bottom and passed upwards through a series of data-refineries, which attempt to convert them into increasingly general and comprehensive chunks of information. As low-level data flows in at the bottom, the big picture comes into focus on top.[56]

The ability to digitally define the essential characteristics of each individual—the high-definition human—sets up a unique era of medicine. To lay the foundation, we begin to explore the status of medicine today and how it drastically needs to be upgraded by recognizing the primacy of the individual. Perhaps the most oft-quoted line by Marshall McLuhan is "the medium is the message." Medicine today relies on the median, whereas in the imminent future it can and will be anchored to the individual. The median will ultimately not be the message in medicine. Beyond the advances in technology, for that to happen, individuals—consumers—will need to step up and lead the way. The upcoming information will define individuals in unprecedented ways, and individuals need to exploit their information to transform medicine.

THE ORIENTATION OF MEDICINE TODAY

Population Versus Individual

*Care more particularly for the individual
patient than for the special features of the disease.*
—Sir William Osler, 1899

A BUZZWORD IN MEDICINE is "evidence-based." If something is evidence-based, then it has some kind of sanctified quality and must be a good thing for patients. A large proportion of tests and prescriptions used frequently in medicine have little or no supportive evidence of utility. A recent poll of Californians found that 65 percent believe that nearly all of the health care that they receive is based on solid scientific evidence. The Institute of Medicine, a prestigious group of physician experts and researchers, weighed in on this question and determined that any valid evidence supports "well below half" of the practice of medicine.

So let's consider the most widely used prescription drug in the world—Lipitor. Lipitor is in the family of medicines known as statins, which inhibit the liver enzyme HMG CoA reductase and, in most patients, achieve substantial lowering of cholesterol levels in the blood. Lipitor wasn't the first statin in common use. It was preceded by Mevacor back in the 1980s, and also by Pravachol and Zocor. But Lipitor became the number one statin, with over $13 billion in worldwide sales per year—the highest revenue in the history of prescription drugs—because it lowered cholesterol more than the statin drugs it preceded, it was tolerated well with only infrequent side effects, and it was marketed very effectively.[1]

The marketing of Lipitor attracted considerable attention in 2008, when the primary pervasive pitchman on television commercials and in newspaper, magazine, and radio ads was exposed. Although he was advising all listeners and readers to take Lipitor, Dr. Robert Jarvik, a pioneer of the artificial heart device, had never practiced medicine. The TV commercials portrayed Jarvik engaging in significant physical activity: rowing a racing shell across a mountain lake. But the rowing was actually performed by a stuntman who resembled Jarvik. Beyond that we learned that Jarvik only started taking Lipitor after he signed a contract with the drug manufacturer for at least $1.35 million over two years.[2] But there was something far more disturbing than an unlicensed physician giving medical advice to millions of people.

The advertisements stated that "Lipitor reduces the risk of heart attack by 36 percent*." A 36 percent reduction seems quite impressive. There are over a million heart attacks in the United States per year, and they represent the most frequent cause of death. Wouldn't reducing the number of heart attacks by more than a third prevent hundreds of thousands of such catastrophic events?

But the asterisk linked to a definition on the full-page ads that appeared frequently in the *New York Times, USA Today,* and the *Wall Street Journal* and read, "That means in a large clinical study, 3 percent of patients taking a sugar pill or placebo had a heart attack compared to 2 percent of patients taking Lipitor." Now we are talking about evidence-based medicine: of every one hundred patients taking Lipitor to prevent a heart attack, one patient was helped, and ninety-nine were not. So why would tens of millions of individuals take Lipitor or other statins every day for the rest of their lives? The drugs cost at least $4 per day or more than $1,500 per year for the unfortunate folks who do not have a prescription plan. This even prompted John Carey to write a feature article in *Business Week* in 2008, entitled "Do Cholesterol Drugs Do Any Good?"[3]

One major reason for the widespread use of statins is what is known as the "surrogate end point." When people see the term "surrogate," the first association is with surrogate mother—not the real mother. It's similar here. Even though the rationale for prescribing a statin is to reduce the likelihood of a heart attack, stroke, or death (the real end point), there is an intermediate measurement that is thought to correlate well with the primary goal—lowering blood cholesterol. It is considered the proxy, or surrogate end point, for improving patient outcomes. The thesis is that for each percentage point that bad cholesterol (low-density lipoprotein, or LDL) is lowered, there would be about 1 percent reduction of heart attacks. So these two end points, the blood cholesterol test and heart attacks, should track very closely.

Unfortunately, that is not the case. Almost everyone who takes Lipitor has a reduction in LDL cholesterol, and often the lowering is pronounced. The patient and the doctor are quite gratified to see an LDL cholesterol reading drop from 150 mg/dl to 90 mg/dl. Hospitals now even assess the quality of care of their doctors by examining the records to be sure that every patient with an LDL above 130 mg/dl has been prescribed a statin. If a physician does not prescribe a statin or record in the chart that this was not possible because of side effects, such as muscle inflammation, he or she is essentially given demerits for not following "evidence-based medicine." Typically a monthly or quarterly report is issued to the medical staff indicating the compliance of the physicians with the norms set for prescribing.

So almost all patients will have a great blood test result with Lipitor. But only 1 out of 100 without prior heart disease but at risk for developing such a condition will actually benefit. It therefore seems that the predominant benefit is cosmetic, normalizing an out-of-range blood test, at the risk of engendering side effects and adding to our current burden of $300 billion per year for prescription drug costs in the United States.[4] The statin benefit is certainly greater in those individuals who have already manifested heart disease and can be readily justified. But the wholesale use for primary prevention, beyond overwhelming regard for a surrogate end point, is the outgrowth of how we interpret clinical trials.

The holy grail of evidence-based medicine is the large-scale randomized, double-blind, placebo-controlled clinical trial performed under the most rigorous conditions. This means that typically 10,000 or more patients are randomly assigned to take a drug or placebo without the patients or their doctors knowing what they actually received, with extended follow-up to see if major adverse events were diminished with the drug. Figure 2.1 represents the event curves for heart attack, stroke, or death from a recent major trial of Crestor, another statin that is even more potent for lowering cholesterol than Lipitor. Over 17,800 patients were enrolled. The reduction from 4 percent of events in the placebo group to 2 percent in the Crestor

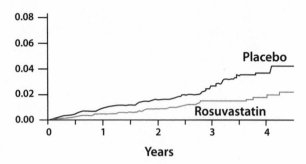

FIGURE 2.1: The risk of death, heart attack, and stroke among 17,800 patients randomly assigned to receive either Rosuvastatin or placebo for over four years. Source: P. M. Ridker, "Rosuvastatin to Prevent Vascular Events in Men and Women with Elevated C-Reactive Protein," *New England Journal of Medicine* 359 (2008): 2195–207.

(Rosuvastatin) group is statistically significant. There is a probability of less than 1 in 10,000 that this result could have occurred by chance. But is helping only 2 out of every 100 patients who take lifelong Crestor worth it? How about the 98 out of every 100 patients who don't derive benefit? And what about the unexpected trade-off of developing diabetes in 1 of every 200 patients treated? A recent global consortium known as the Cochrane Collaboration reviewed all the data from fourteen randomized trials and over 34,00 patients and concluded there was no net overall benefit of statins for patients without preexisting heart disease.[5]

This is as good as it gets. The trials of Lipitor and Crestor are state-of-the-art and considered exemplary proof for the broad use of these medicines for preventing heart disease. There is nothing particularly unusual about Lipitor or Crestor; they are commonly used drugs and heavily promoted. But what this represents is really population medicine, the antithesis of medicine directed at and for an individual. The push for "statins for all" has been dubbed "mass medicalization" and accounts for $26 billion a year of statin prescription costs.[6] Instead of identifying the 1 person or 2 people out of every 100 who would benefit, the whole population with the criteria that were tested is deemed treatable with sufficient, incontrovertible statistical proof. Even the term "NNT"—numbers needed to treat—has been coined to denote how many people have to be given a therapy to identify the few who will derive the expected benefit. You can even look up the data for the numbers needed to treat for many medicines on the website www.theNNT.com. What constitutes evidence-based medicine today is what is good for a large population, not for any particular individual.

Another major deficiency of medicine is the use of experts to make recommendations or "guidelines" for a large proportion of decisions for which no or minimal data exists. These guidelines, typically published in major specialty journals, have a pronounced impact, as they are believed to represent the standard of care, even though they are based on opinion with a paucity of facts. In fact, this should be considered "eminence-based medicine." As we are able to accrue more meaningful data and information on individuals, the hope is that we can override our dependence on such recommendations.

In the meantime, flawed evidence-based medicine of today is being advocated to provide immunity from medical malpractice. In a *New York Times* op-ed, Peter Orszag, formerly the director of the White House Office of Management and Budget, suggested that we could "provide safe harbors for doctors who follow evidence-based guidelines."[7] The 2009 U.S. economic stimulus act promoted comparative effectiveness in medical research and the new Patient-Centered Outcomes Research Institute. Unfortunately, the funding is disproportionate with the relative void of information, and the efforts that are being mounted are tied to population medicine. It's ironic that the new initiative is called "Patient-Centered"—if only that were the case![8]

It was August 1995, and at the Hotel de Crillon, one of the most extravagant hotels in Paris, the steering committee of the largest trial ever to be conducted had gathered to review the data for the first time. The trial, known as CAPRIE and involving more than 19,000 patients, was intended to test the efficacy of a new drug for treating vascular disease. It had to beat aspirin. Both drugs can prevent blood clots, albeit by different mechanisms. But the most important difference was price: Aspirin costs pennies per tablet. This new drug, if effective and ultimately commercially approved, would cost $4 a day. There were more than twenty experts in vascular disease in the room, as well as the senior management of the company that sponsored the trial, Sanofi. Everyone knew the stakes: the drug had cost hundreds of millions of dollars to test, but the rewards, if it was successful, were much bigger.

I was one of those experts in the room. Eager to review the data, I quickly leafed through the pages of the results book to get to the bottom line: did Plavix work better than aspirin? The answer was, in my mind, quite disappointing. There was only an 8.7 percent improvement in the end point, with only 2 patients per 100 benefiting.[9] Still, it was considered really first-rate evidence-based medicine and potentially good for the population of patients who have vascular disease, and within a year the Food and Drug Administration (FDA) along with regulatory authorities all over the world approved the use of the drug—known as Plavix—for patients with vascular disease.

By 2010, Plavix had become the second-largest prescription drug by dollar sales in the world, just after Lipitor, with $9 billion of sales a year. But the FDA placed a black-box warning on Plavix to inform doctors that the drug may not work in patients if they are carrying particular gene variants.[10] It turns out that at least 30 percent of people cannot normally metabolize Plavix to convert it to an active drug. Without the intact functioning gene in the liver cells responsible for metabolism, known as cytochrome 2C19 (or CYP2C19), Plavix does not adequately suppress the platelets or prevent blood clots. So if a patient has a stent put in his coronary artery because of a blockage, and he carries an allele (alternative form) of the gene that does not allow the metabolism of Plavix, the risk of developing a blood clot of the stent is at least 300 percent greater. Although clotting a stent is not a frequent phenomenon, if it occurs, it is catastrophic and usually results in a heart attack or death. For individuals who carry one so-called loss-of-function allele, the problem can sometimes be overridden by doubling the dose of Plavix. With two copies of the CYP2C19 loss-of-function gene variants, however, the chance of not metabolizing the drug approaches 100 percent. Interestingly, there are also alleles that lead to faster metabolism of Plavix; if an individual carries two copies of the alleles, she would actually require a lower dose of Plavix.[11] But a lower dose of Plavix does not exist, and higher doses were never tested in a large trial.

Why did it take more than two decades for the realization that a significant proportion of individuals do not or cannot respond to the drug? It was actually known by the late 1990s that the response to this medication was quite variable. There was even an early study of healthy volunteers that tested the CYP2C19 gene alleles and showed that the 25 percent of people who did not respond to Plavix were more likely to have a loss-of-function DNA variant.[12]

There are two explanations for why the heterogeneous response to Plavix was not discerned twenty or more years ago. The first was population, evidence-based medicine. From the CAPRIE trial, with very modest impact, there was enough benefit to get the drug commercially approved and on track to be widely prescribed and promoted. Imagine if the patients in that trial had been genotyped, and the individuals who could not metabolize Plavix were excluded from participation. But the threshold for demonstrating benefit and proof that a drug works is relatively meager for a population, and there were enough responders in the big cohort for Plavix to pass the test and show overall improvement.

A second explanation is the use of the same dose for all patients, which is a common problem in the development of pharmaceuticals. How could it be possible that all individuals who take a medicine would respond in the same way to the same dose? With the extraordinary differences in age, gender, and weight, not to mention metabolism and genes, is there any drug that would be uniform in its effect with respect to dose? Yet the assumption that the same dose works for all patients is quite typical for a drug company. The priority is to keep it simple. If there are multiple doses, it means much larger trials and much more complex marketing. Pharmaceutical companies are aware of the lack of modification or tweaking of a dose, since when a doctor starts a patient on a medication it is relatively unusual for the dose to be changed. The starting dose typically represents the default dose. The convenience of a "one dose fits all"[13] mentality could not be more advantageous for both physicians and the drug manufacturer. Consequently, it is one of the essential elements of population medicine today.

On the morning of Saturday, February 11, 1984, a fifty-seven-year-old, gray-haired woman of Polish descent, wife of a steel worker, who smoked a pack a day and had a family history of heart attack, arrived at the emergency room having suffered an hour of severe, crushing chest pain. She was sweating profusely, appeared very pale, and was terrified. Her electrocardiogram showed she was in the midst of evolving a large heart attack that could damage more than 40 percent of her heart muscle. At the time, the usual care for such a patient was oxygen, morphine to alleviate the pain, lidocaine to prevent abnormal, life-threatening heart rhythm, and hope for the best. But this lady's timing was exquisite—she was about to become the first person to receive the genetically engineered clot buster tissue plasminogen activator (t-PA). This is a naturally occurring clot dissolving enzyme in the body, but when a blood clot underlying a heart attack or stroke forms, there is not nearly enough of it to dissolve the clot. The hope of the doctors treating her—including me—was that by rapidly supplying the protein to the vein while the patient was in the throes of a heart attack, it might be possible to restore the blood supply to the heart and prevent some or all of the damage to the heart muscle that was otherwise destined to occur.

On a Saturday morning at Baltimore City Hospital there were usually only a few doctors-in-training making rounds, and it's most unusual to see the senior, attending physicians. But because this was a history-making event, and a team had been called in to do the emergency cardiac catheterization, more than twenty doctors came out of the woodwork. We proceeded to introduce a tiny tube or catheter through the artery in the leg and threaded it up to the heart via X-ray guidance. At the spot where the arteries to the heart take off, a bit of contrast dye was shot through the catheter to see if the left anterior descending artery was cut off. Indeed it was. Now we were to give t-PA in her intravenous line, which had been placed in her right forearm. Since no patient had ever received human recombinant t-PA, we had no idea of the effective dose; the protocol called for giving 20 mg and then taking additional pictures of the artery to see if the blood supply had been restored. After we gave t-PA, we took a considerable number of pictures with dye shot down the vessel, and about fifteen minutes later the blood flow came back. The patient's chest pain was abating, her color was restored, her EKG was improving, and the large group of doctors and nurses, in the lab and the adjacent control room, were cheering. I started to cry. Little did I know that the dose that we used of t-PA was totally insufficient to open the artery, and that it was the repeated injections of the dye that likely was responsible for restoring the patient's blood supply to her heart. Later we would learn that it took an average of forty-five minutes and 100 mg of t-PA to open up the heart attack arteries in most patients. One more chapter of medical serendipity had been written.[14]

A few years later, when there were enough data showing that the blood clots were rapidly dissolved with t-PA in a high proportion of patients, the FDA advisory panel, consisting of many physician experts in the field of cardiology, weighed the evidence for approval for treating heart attack patients. At the hearing in 1987, the morning was dedicated to considering streptokinase, an inexpensive clot-dissolving agent derived from streptococcal bacteria, which had been around for decades. A large trial of almost 12,000 patients, involving most of the hospitals in Italy, had shown that streptokinase saved lives—about a 20 percent reduction in deaths—compared with not giving any clot-dissolving therapy.[15] The panel voted unanimously to approve intravenous streptokinase for evolving heart attacks. But in the afternoon, when t-PA was considered, there were only data for surrogate end points. The studies showed that t-PA dissolved clots and restored blood supply, far better than streptokinase, and a small trial performed at Johns Hopkins took this a step further by demonstrating that patients receiving t-PA, compared with placebo, wound up with better heart function. But there were no data to say that lives were saved, and so the panel declined to recommend t-PA for approval.

The aftermath was a memorable one. The meeting was held at a large auditorium in the days prior to cell phones. There were pay phones in the back of the room, and one could hear stock brokers on the phones calling in—and yelling—to sell shares of the manufacturer of t-PA, Genentech. The *Wall Street Journal* published an editorial entitled "The Flat Earth Committee."[16] Here one of the original biotechnology companies, with its darling genetically engineered product t-PA, was going down because of the unacceptability of surrogate end point data.

Eventually, through heavy lobbying and submission of more data related to the avoidance of congestive heart failure, t-PA did garner FDA approval the following year. But it cost $2,200 a dose, as compared to $300 for streptokinase, and such a cost difference didn't meet the standard of heightened efficacy, a cornerstone of evidence-based medicine for heart attack therapy. The vulnerability of t-PA was further highlighted when subsequent large trials from Europe directly compared t-PA with streptokinase but did not show any meaningful difference for survival. Doctors began to question whether dissolving clots more rapidly and efficiently really made a difference. Maybe the clots reaccumulated after t-PA. Maybe streptokinase did other things that were beneficial besides dissolving clots. Maybe the regimen of t-PA was inadequate. Confusion set in for doctors treating heart attacks and the medical experts who were left without a clear explanation for why t-PA didn't prevail. But one thing was clear: t-PA was not going to be used unless it could be definitely shown to incrementally save lives.

By 1990, the big t-PA trial known as GUSTO was planned, and after a pilot study assured feasibility, a trial of over 41,000 heart attack patients in twenty countries was launched in 1991 and completed in 1993.[17] At the time, this was the largest clinical trial to have ever been organized and initiated in the United States. The end point was dead or alive status of the heart attack patients at thirty days after treatment. T-PA turned out to be better than streptokinase by 15 percent, reducing the death rate from 7.3 percent to 6.3 percent. That means that for every 100 patients, 1 would benefit more from t-PA than from streptokinase. The only problem was identifying who those patients were. But we didn't know that at the time of the trial, and we still don't.

––––––

Several years ago a colleague of mine at the Cleveland Clinic had an abnormal prostate-specific antigen (PSA) blood test result during his annual physical examination. After it was confirmed, he underwent a prostate biopsy to determine whether he had developed prostate cancer to correlate with the elevated PSA. This is not an easy procedure to go through: a catheter is inserted in the genital tract through the penis, and a bioptome, a small pincer-shaped tool, takes multiple tiny pieces of the prostate tissue (which involves considerable pain) to be studied for any cancerous cells. But when it was found that there was no evidence of prostate cancer (a "false-positive" PSA), he had to have serial prostate biopsies every six months for the next year to be sure that any prostate cancer was not missed. And my friend is only 1 of the 250,000 men every year in the United States who has a false-positive PSA test and then undergoes multiple prostate biopsies. Thirty million men in the United States continue to have their PSA test every year.[18]

Dr. Richard Ablin, a pathologist now at the University of Arizona, discovered the PSA in 1970. Forty years later he wrote an op-ed in the *New York Times* entitled "The Great Prostate Mistake," in which he stated, "The test's popularity has led to a hugely expensive public health disaster."[19] I have never known the inventor of a clinical test to declare a public health disaster based on that test. Inventors notoriously support their invention without limits. So why in this case did Ablin issue a public health warning?

Prostate cancer is extremely common in men, with over 15 percent eventually carrying the diagnosis. But only 3 percent of men actually succumb to prostate cancer.[20] So there is considerable prevalence of non-aggressive prostate cancer, and when this is diagnosed, typically surgery is performed to remove the prostate gland, followed by radiation and other cancer-ablating procedures. Just the PSA test alone costs the United States $3 billion per year—and billions more when the cumulative costs of all the biopsies, surgeries, treatments, and the complications of the surgery such as urinary incontinence or impotence are factored in.[21] Ablin's editorial plea ended with: "The medical community must confront reality and stop the inappropriate use of PSA screening. Doing so would save billions of dollars and rescue millions of men from unnecessary, debilitating treatments."

Nonetheless, mass screening for early detection of cancer is one of the most accepted rituals of health care in the United States. The current recommendation is that all men over age fifty should have their PSA checked every year. Even though there has been some recent public debate, the current recommendations are that all women should undergo a yearly mammogram after age forty. Everyone should undergo a colonoscopy at age fifty and every five years thereafter. Similar to the problem of "one dose fits all," how can we possibly be viewed as all having the same risk profile, given the differences in our biology and environmental exposures?

The PSA in men has a parallel in breast cancer screening using mammography in women. For mammography, the trade-offs of screening and false-positive results are now broken down by age. For every 1,000 women who undergo mammography between ages forty and forty-nine, 98 (about

10 percent) will have a false-positive mammogram, 60–200 women (the range indicates variability in multiple studies) will undergo an unnecessary biopsy, and 84 per 1,000 women screened will have to undergo additional imaging, typically involving magnetic resonance or ultrasound. Only 1 per 2,000 women screened might avoid death from breast cancer with screening. In later decades of life, the numbers don't change very much. Even in women ages seventy to seventy-nine, false-positive mammograms occur in 69 out of every 1,000, and additional imaging is necessitated in 64 per 1000. Proven overdiagnosis, representing unnecessary surgery, chemotherapy, or radiation, rises from 1–5 per 1,000 at ages forty to forty-nine, to 1–7 per 1,000 women ages fifty to fifty-nine.[22]

Here is more evidence that population medicine, in this case mass screening, disregards individual variability and promotes considerably more unnecessary medical testing and procedures. Beyond the tabulation of data on false-positive results, the emotional toll to the woman and her family after notification of an abnormal mammogram is considerable and impossible to quantify.

————

Back in 2003, two British physicians, Nicholas Wald and Malcolm Law, published a paper on the "polypill." One pill with six drugs in it: a low-dose statin for lowering LDL cholesterol; three blood pressure medications, each at half the regular dose; the vitamin folic acid; and low-dose aspirin. Wald and Law asserted that if everyone over fifty-five took this polypill, there would be 88 percent reduction of coronary heart disease and 80 percent reduction of stroke. A notably strong statement was contained in the report: "It would be acceptably safe and with widespread use would have a greater impact on the prevention of disease in the Western world than any other single intervention."[23] This assertion of profound benefit, completely theoretical and untested, ignited controversy in medical circles. Further, the idea of giving a pill containing six medications to all people over a certain age seemed quite foreign.

One could liken this concept to the use of fluoride in the water supply, which was initiated in the United States in 1945 and has been documented to benefit the population, with an overall 15 percent reduction of dental cavities. Nevertheless, in many places around the world, there continues to be opposition to water fluoridation for ethical concerns: it is seen as forced use of medication on the masses.[24] The only known side effect of fluoride in the water is the condition of dental fluorosis, which is usually characterized by tiny white stains of the teeth. Still, with an inadvertent overdose the teeth can be grossly discolored. Here, if properly regulated and administered, the balance of benefit and harm weighs in favor of the practice of water fluoridation. But how does that compare with a polypill?

It took several years after the initial Wald and Law proclamation before a polypill could be produced, ultimately made in India, as such a product would not likely be a manufacturing target for large, traditional, pharmaceutical companies. Clinical trials testing got under way in 2010, so it will be years before we know whether this population medicine strategy has any merit.[25] Perhaps by promoting adherence and decreasing the cost of drugs it will have merit in parts of the world where taking and affording multiple medications is especially problematic. One can even make a good case for this being the profile in the United States, since only 50 percent of patients actually adhere to their prescriptions, and the costs continue to skyrocket.

If successful, we may someday have the modern version of fluoridation of the water in the form of a multidrug pill taken every day. (We already have ample evidence of current prescription drugs in our water supply.)[26] However, the notion of encapsulating multiple drugs into one, at unconventional and fixed doses, appears to be the virtual opposite of medicine for individuals. Instead of identifying the condition to prevent or treat, every person is given multiple drugs that carry known side effects, compounded by the potential for drug interactions. Rather than attempting to focus on directed, tailored prevention or therapy for a particular person, the polypill approach capitalizes on benefits shown at the population level, but with the same problems that confront all other population-based medicine.

———

As problematic as these population-based investigations have been, they at least have conformed to the best-in-class model for medical studies. Consumers, unfortunately, are typically getting data from small, observational studies, published in obscure journals or not at all, in which there is no real control group or no randomization, and shaky end points. For example, one study compared two hundred individuals who took vitamin E for two years and had less heart disease than two hundred people who said they didn't take vitamin E. Even very large-scale observational studies have led us astray. A study of 87,245 nurses suggested that using vitamin E supplements would reduce the incidence of heart disease by 30 to 40 percent. And a Finnish study of 5,133 men with fourteen-year follow-up suggested the same benefit. But when the vitamin E story was subjected to several randomized placebo-controlled, double-blind trials, there was no indication of benefit whatsoever. In fact, in one such trial of 10,000 patients, the participants actually had a surprising 21 percent higher rate of developing heart failure.[27]

Observational studies have misled more than once. One of the most impressive mistakes relates to hormone replacement in women, which for decades was widely recommended to reduce heart disease. The leading manufacturer of hormone replacement, through the use of ghostwritten articles in medical journals, propagated some of this practice. When randomized trials were finally performed, the recommendation turned out to be completely off base. The Women's Health Initiative trial of over 16,000 healthy postmenopausal women compared the combination of estrogen and progestin to placebo and found significant increases in breast cancer, heart disease and heart attacks, strokes, and dangerous blood clots—far overriding the benefit of less colon cancer and fewer hip fractures. The results of the trial were so negative that it was stopped prematurely, at 5.6 years (instead of the planned 15 years) of follow-up. New results released in 2011 continue to engender confusion, suggesting disparate outcomes with hormone replacement as a function of what age the treatment was initiated.[28]

In 2005 John Ioannidis, now at Stanford University, published the article "Why Most Published Research Findings Are False" in the journal *PLoS Medicine*, sending chills through the academic medical community.[29] His conclusions are in keeping with what has been reviewed here: (1) the smaller the studies, the less likely the research findings are to be true; (2) the smaller the effect, the less likely the research findings are to be true; (3) the greater the financial and other interests, the less likely the research findings are to be true; (4) the hotter a scientific field (with more scientific teams involved), the less likely the research findings are to be true. A similar note was sounded by Jonah Lehrer in the *New Yorker*, in a piece entitled "The Truth Wears Off." The significant issues in science of replicability, the subsequent "regression to the mean" after initial results are impressive (often referred to as "the winner's curse"), the bias of the researchers, and the bias of what is actually published (mainly positive results) led him to the following conclusion: just because an idea can be proved doesn't mean that it's true.[30]

I don't want to be excessively negative, but the right assumption in reviewing any new data presented to consumers is to question it. Unlike our legal system, in which a defendant is considered innocent until proven guilty, new scientific or medical evidence has to refute and transcend the "null hypothesis." In other words, consider the new findings null and void unless you are thoroughly convinced that the evidence is compelling. I coined the term "litter-ature" to denote that too much of the medical literature is littered with misleading and false-positive findings. That there is simply too much literature (and litter-ature) is evidenced by the statistic that only 0.5 percent of the 38 million published papers are cited more than two hundred times by others, and half were never cited. Moreover, when pooled analyses of prior studies are published, many relevant papers are excluded.[31]

All of these problems can be seen in the case of the medicines Zetia and Vytorin, which are the trade names, respectively, for ezetimibe and ezetimibe plus simvastatin. These prescription drugs help to lower LDL cholesterol in the blood. Ezetimibe was approved in 2002 on the basis of small (but randomized) studies demonstrating only a surrogate end point—that showed LDL cholesterol was reduced 19 percent—rather than a proven decrease in incidence of disease or death.[32] The FDA, in approving the drug, simply assumed that lowering LDL cholesterol by any means would be a good thing for patients. The major outcome trial (notably dubbed "IMPROVE-IT"), which tests whether the drug actually benefits people, was not started for several years and will not be complete before 2012 at the earliest. Meanwhile, in 2008, small, randomized studies began to show that ezetimibe had no effect on artery plaque development; moreover, there were signs that this drug was linked to a higher risk of fatal cancers.[33] Media reports have been panicky and remarkably inconsistent, but professional organizations such as the American College of Cardiology and the American Heart Association, which receive large financial support from the manufacturers of the medicine, simply proclaimed that ezetimibe was safe. Annual sales in the United States reached $5 billion.

How do we get out of this mess? Better studies are part of the solution—although not all of it. We need real evidence based on individuals, not populations. Fortunately, our ability to get just that information is rapidly emerging, beginning an era characterized by the right drug, the right dose, and the right screen for the right patients, with the right doctor, at the right cost. Medicine for the common good is not good enough. Now let us see how to get something better.

TO WHAT EXTENT ARE CONSUMERS EMPOWERED?

Clicks and Tricks

Medicine has built on a long history of innovation, from the stethoscope and roentgenogram to magnetic resonance imaging and robotics. Doctors have embraced each new technology to advance patient care. But nothing has changed clinical practice more fundamentally than one recent innovation: the Internet.

—Pamela Hartzband and Jerome Groopman,
New England Journal of Medicine[1]

MORE THAN SEVEN BILLION people on the planet
Over three million doctors
Tens of thousands of hospitals
6,000 prescription medicines, 4,000 procedures and operations[2]
Countless supplements, herbs, alternative treatments
Who gets what, when, where, why, and how?

When a fifty-eight-year-old, active, lean, intelligent financier from Florida came to see me for a second opinion, I should not have been surprised. For Valentine's Day the prior year, his wife's present was a computed tomography (CT) scan for his heart. She heard about it on the radio and also saw heart scan billboards on the highway. There was even a special deal of $100 off for Valentines.

But her husband didn't have any symptoms of heart disease, didn't take any medications, and played at least two rounds of golf a week. On the other days, he worked out on an elliptical machine for thirty to forty minutes—until he got the heart scan.

My patient was told that he had a high calcium score of 710, and his physician had told him that he would need to undergo a coronary angiogram, a roadmap movie of the coronary anatomy, as soon as possible. He did that and was found to have several blockages in two of the three arteries serving his heart. His cardiologists in Florida immediately implanted five stents (even though no stress test or other symptoms had suggested they were necessary) and put him on a regimen of Lipitor, a beta-blocker, aspirin, and Plavix.

Now, in my office four months later, this patient was not doing well at all. He was worried that he might have a heart attack if one of the stents became clotted. He felt profoundly tired and had muscle aches so disturbing he could neither play golf nor do his usual exercise. He complained of marked depression and an inability to have or sustain an erection. A fit individual who had taken good care of himself and was enjoying his life, he was now debilitated and depressed. The cardiology trainee who saw this patient with me asked, "How could this have happened?"

Unfortunately, this individual's story is not so uncommon. Think predator and prey: the physicians and hospital advertise, leading to a high volume of heart scans, billed directly to the patients at some $500 each. Then, should an abnormal score come up, the patient may be quickly referred for first a diagnostic procedure and then the implantation of metal stents in the arteries on the surface of the heart. Naturally the cardiologist who put in multiple stents feels gratified to have saved the patient's life with unsuspected, advanced coronary disease. Overall, however, these cases are like riding a train to the last stop, regardless of the most logical destination. All procedures are performed even though, as likely as not, the outcome is not a saved life but a "cardiac cripple."

I didn't enjoy telling the patient that he should probably never have had the stents. I could see the cholesterol buildup in the two arteries on an angiogram he brought with him, but the case was not severe. Of course, it was too late to do anything about the stents, which can't be removed, except to reassure him that he was not in any imminent or real danger, but I could get him off some of his medications, which would help his current symptoms and get him back to golf and exercise.

Mark Twain is reported to have said, "To a man with a hammer, everything looks like a nail." Surgeons are notorious for a similar bias: "When in doubt, cut it out." My patient was the victim of the same tendency. As badly as he got pounded, it could have been worse: in 2010 the "Olympic record" of stenting was published. One patient had sixty-seven stents placed throughout his coronary arteries and bypass grafts, in the course of twenty-eight coronary angiograms over a ten-year period.[3]

This problem of inappropriate use or overuse of medical procedures is a difficult nut to crack. For one, physicians, hospitals, and the life science industry are all aligned and incentivized to do more procedures. Even at the subconscious level, as graphically portrayed in Atul Gawande's 2009 *New Yorker* article, "The Cost Conundrum," patients can be seen as representing an ATM.[4] Certainly not all those procedures are carried out in the patients' best interests; the profound regional variability seen in the use of all sorts of procedures and operations across the United States reinforces the fact that appropriateness and need are not the sole determinants of whether patients are subjected to them. And it isn't just across the United States. For every 1,000 people in France, 192 will have an angioplasty or stent procedure. In the United States, the number is more than double, at 437.[5] Too few in France or too many in the United States? The difference can't simply be attributed to Americans drinking less French red wine.

In the case of my patient, of course, it didn't just start with the unnecessary procedure but with the initial response to an advertisement, followed by his trust in his original physicians to make objective recommendations about what the proper course of care would be. Similar problems confront anyone trying to navigate all the medical procedures, operations, prescription medications, vitamins, supplements, herbs, alternative treatments, over-the-counter products, and home devices that confront them. The key to the problem is an empowered, knowledgeable patient, but as we shall see, extra information need not lead to empowerment. Whether information is pushed to consumers (by the news media or by direct-to-consumer advertising) or pulled out of the system by consumers themselves (by, for example, visiting Google Scholar or a social networking site developed for people with a particular disease), if a consumer can't make the best, most intelligent use of it, all sorts of trouble can unfold.

———

Before there was Viagra, Levitra, or Cialis, practically no one had heard of erectile dysfunction. Even if they had, of course, there would have been nothing to be done about it. Now, however, middle-aged men, especially those who watch televised sporting events, are likely to think they suffer from it. Over the years, countless patients of mine have requested prescriptions for these medications. Interestingly, the patient usually does so over the phone or at the end of the office visit, when his spouse or significant other has already exited and is walking down the hall. When I hear "Hey, doc," I know what the next few words are going to be and get the prescription pad out.

Direct-to-consumer, or DTC, advertising of prescription drugs has been legal in the United States since 1997; besides New Zealand, it is the only country that allows this form of promotion, not just of medicines but also devices, screening tests, and biologic agents.[6] And it's a remarkably large industry, now well over $5 billion per year.[7] It also has an excellent return on investment, generating at least $2–3 in revenue for every dollar expended. In a recent *Consumer Reports* survey of over 1,150 adults, 20 percent reported requesting a prescription that they heard about through DTC advertising.[8] Ads can even create new medical conditions—a practice some have labeled "disease mongering"—sending thousands of people, self-diagnosed after seeing DTC ads, to their doctors with "restless leg syndrome" and "social phobia."

There is a considerable literature and debate about DTC drug advertising, often separating lifestyle drugs such as Viagra from potentially life-saving drugs like Lipitor or Plavix. Some have argued the education for the latter category is good for consumers, but it is hard to think that a thirty- to sixty-second TV ad that glamorizes a medicine and then quickly lists off the potential side effects could possibly provide meaningful education.

Although many side effects might be discovered while treating closely observed participants in a carefully controlled study, when a new medicine is released to the public, the situation is vastly different. Persons of any age, taking a variety of different medications with multiple other chronic medical conditions, are now getting exposure. Where a trial might involve a few thousand patients, an approved mainstream drug could be used in millions of individuals; the result is that rarer but serious, and potentially fatal, side effects might only then come to light. But when a new drug is DTC promoted, the mass exposure can happen so quickly that the unexpected, untoward adverse events constitute a new drug-induced epidemic.

Say a drug has a 0.5 percent risk of severe liver inflammation, so 1 in 200 patients would develop the condition of fulminant hepatitis. In the trial that tested the drug, only 400 patients were enrolled, and by chance there was no case of hepatitis (which actually would happen about 13 percent of the time). Then the big DTC campaign begins, and in the first month 20,000 patients are treated. Over 100 patients are hospitalized or dying with liver shutdown. But with the weak surveillance mechanisms that exist, it may take months before the cause-and-effect relationship of the new drug and life-threatening hepatitis is established, and by then hundreds of people could be affected. For years there has been debate about setting a two-year moratorium on any newly approved drug before a DTC campaign could be initiated. The plan makes a lot of sense, but nothing yet has been solidified. Several years ago I worked with then Congressman (now Senator) Sherrod Brown of Ohio to introduce legislation in the House of Representatives to ban DTC advertising.[9] That got as much response on Capitol Hill as an offer for free root canals.

DTC advertising is the epitome of population medicine. Millions of TV viewers are pounded with infomercials, and some ask their doctors for a drug they don't really need. The major prescription excess, as with many aspects of population medicine, adds to the profound waste of resources.

———

A patient recently sent me this email about his blood pressure (BP) medications:

> I went off the BP meds and statins a month ago to see how I felt and to also take a bunch of supplements my wife found, like Oregano, Purge, Royal Jelly etc.
>
> I took my BP this am for the first time and it was 188 over 108. I guess I will go back on the BP meds.

The patient is a highly educated, intelligent, and affluent individual with a serious blood pressure problem; he was taking two different medications at maximal doses for this condition. He had already had a small stroke a few years ago. How could he jeopardize himself to have another stroke or a heart attack by stopping his medications for oregano or Coptis Purge Fire, a Chinese herb promoted by holistic naturopathic doctors? How about Royal Jelly tablets, the "highly complex substance secreted from the glands of nursing bees, hermetically sealed in soft gels to enhance stability"?[10] Is this preposterous?

I guess not, since the use of supplements and herbs continues to skyrocket. In the United States, the out-of-pocket expenditures for vitamins, supplements, and herbs are now exceeding $30 billion per year and $60 billion worldwide.[11] Even with the problems of evidence-based medicine outlined in Chapter 2, the problem with most of these products is they haven't been tested or, if they have, they simply don't work. In 2011, the *Economist* reviewed this topic and concluded "Virtually all alternative medicine is bunk." That review reiterated that "95% of the industry was hokum," offering nothing more than a placebo effect and that "the alternative-medicine industry plainly excels as a placebo delivery service."[12] To date, the only randomized, rigorously performed trial of a supplement was of glucosamine's effect on knee arthritis.[13] It was quite effective for reducing pain and increasing mobility, but there are hundreds of different preparations of glucosamine at a wide variety of doses. And those problems are just the tip of a very large iceberg: There are more than 54,000 dietary supplements in the Natural Medicines Comprehensive Database.[14] Do any of them work? Which one? At what dose?

Then we can consider all the vitamins and herbs, such as the case of vitamin E we saw in the last chapter, that were supposed to provide considerable benefit but when put to the acid test, a randomized, placebo-controlled trial, they flopped. Selenium to avoid prostate cancer, Saint-John's-wort to avoid depression, gingko to improve memory, echinacea for prevention of colds—none of these has worked.[15] The same was true for B vitamins such as folate and vitamin B_{12}. A trial of over 12,000 individuals found that every examined end point—including adverse heart and blood vessel events and the occurrence of cancer—was worse under treatment with vitamin B supplements, just as was shown with vitamin E and antioxidants.[16] The vitamin D story, which has been hot in recent years, took a big hit when the results of an Australian trial of 5,504 women over age seventy were reported. Ironically, those randomly assigned to vitamin D had a *higher* rate of falls and fractures than the women who received placebo.[17] Despite a 2011 report by the U.S. Institute of Medicine that took the wind out of the purported "silent epidemic" of vitamin D deficiency and found no data to substantiate protection from cancer, diabetes, immune disorders like multiple sclerosis, or heart disease, the debate remains active among consumers on the Web.[18] The Institute of Medicine recommended a blood level for vitamin D of 50 nmol/L, which has engendered intense controversy because the data to support this threshold were so limited (another fine example of population medicine, as reviewed in Chapter 2). Nevertheless, the sales of vitamin D in 2009 exceeded $425 million.

Other examples abound. Omega-3 (fish oil) supplements are all the rage and can help reduce triglycerides in the blood. But the only real proof so far in randomized trials with replication is for prevention of a heart attack in patients who have already suffered one, and even more recent trial data have questioned this assertion.[19] Most consumers are unaware of the downsides of fish oils, including gastrointestinal side effects, increase in bad LDL cholesterol, and thinning of the blood, which can add on to prescribed blood thinners and pose a heightened risk of bleeding.

More men than ever before are taking testosterone supplements, either by injection or in the form of gels. Perhaps this is an outgrowth of the DTC advertising that promotes self-diagnosis of erectile dysfunction. But whatever the cause, the problem is serious. A recent randomized study of testosterone gel in men sixty-five years and older demonstrated a fourfold risk of heart attacks and adverse cardiovascular events.[20] It is rare that physicians inform their patients about this risk. A significant fraction of my male patients with prior heart disease are taking testosterone preparations of one form or another. But unless I specifically ask them, they don't reveal it. When I inform them of the risks of testosterone supplements for their heart, it is almost invariably a surprise.

Sunscreens, as my wife—the prototypical informed consumer—recently pointed out to me, are also not terribly effective "supplements." Not only do they not protect us very well from skin cancer, but also some of the products may actually be contributing to it! The Environmental Working Group, a nonprofit consumer advocacy group, publishes a report about sunscreens each year. In their fourth annual report of five hundred consumer beach and sport sunscreen products, only thirty-nine (that's just 8 percent) were deemed safe and effective.[21] One of the main reasons is that sunscreens were historically measured by sun protection factor—SPF—on the basis of only Ultraviolet B rays. There have not been any standards for rating protection from Ultraviolet A (UVA) rays, for which most products in the United States offer little to no protection, and the rules at the Food and Drug Administration, which regulates sunscreen products, have not changed since 1978!

Of note, one of the reasons for the lack of updating the rules and acknowledging UVA rays has been heavy pressure from sunscreen manufacturers, which include Johnson and Johnson (Neutrogena), Merck-Schering Plough (Coppertone), Proctor and Gamble (Olay), and L'Oreal. Interestingly, in Europe products that provide solid UVA protection have been available for years. The concerns run even deeper because many of the products (41 percent in the United States) contain a form of vitamin A known as retinyl palmitate, which has been associated with increased likelihood of skin cancer. There are, however, no randomized studies, but biological plausibility and the observational findings of a rising incidence of basal cell carcinoma and melanoma, despite the widespread use of sunscreens. In mid-2011, the FDA finally unveiled some new rules about sunscreen claims.[22]

This issue really hit home when my wife brought out a tube of Neutrogena Ultra Sheer Dry-Tough SPF 30 Sunblock. It claims "Broad Spectrum UVA/UVB Protection" despite repeatedly failing UVA tests. But the real eye-opener is to find the American Cancer Society logo on the front of the tube with the message "Help Block Out Skin Cancer." Now what is the American Cancer Society logo doing on the tube of Neutrogena? The fine print on the bottom reads: "The American Cancer Society (ACS) and Neutrogena, working together to help prevent skin cancer, support the use of sunscreen. The ACS does not endorse any specific product. Neutrogena pays a royalty to the ACS for the use of its logo."

Beyond the claims made on the labels of many such products, a new concern about herbal supplements has been developing, since the FDA does not regulate them. Recently, it was found that sixteen of forty supplements tested contained pesticides that exceeded the legal limits, as well as heavy metals such as cadmium, mercury, and arsenic at subthreshold, but still worrying, levels.[23] Another study showed that the desired ingredients in supplements were actually found at much lower levels than promised in more than a fourth of 2,000 dietary supplements from three hundred manufacturers. There are also some important interactions with conventional medications. For example, if Saint-John's-wort and the commonly prescribed blood thinner warfarin are taken together, the former interferes with the effect of the latter. But patients are largely unaware of these interactions and largely remain unwilling to share their intake of supplements and herbs with their doctors, owing to the concerns of bias among conventional medicine practitioners. (What might be worse, only 2 percent of physicians even take time to ask if patients are taking them.[24]) Despite all these problems, more than half of Americans take vitamin supplements, and more than 25 percent take herbal supplements.[25]

When I used to practice medicine in Ohio and Michigan, it was striking how patients felt that if they did not get a prescription at the end of the office visit, there was something missing. Then, in late 2006, I came to California and was introduced to a new culture of prescription avoidance. The contrast was striking, but it wasn't just about geography. It is a sign of the times. Consumers have developed progressively stronger distrust of conventional medicine—doctors and the pharmaceutical industry. Consumers are exercising their own right and authority to independently seek out natural remedies, which are also promoted, albeit not through DTC TV ads. Instead, consumers highly regard TV doctors such as Dr. Andrew

Weil, who promote the use of such herbs and supplements and are happy to suggest purchase via their websites. Even Dr. Mehmet Oz of Oprah fame, who has his own TV show and has authored multiple "how to" books, and whom I know well and highly regard, pushes the use of vitamins and supplements far beyond what the data support. The pervasive use of these agents reflects, at least to some degree, a consumer rebellion against conventional medicine. Whether right or wrong, it is an important sign of their emerging empowerment.

The mistrust and rebel spirit are also manifest by consumers turning to all sorts of alternative treatments—about 40 percent of Americans have tried some form of alternative medicine.[26] Most physicians are skeptical of acupuncture, biofeedback, reiki, homeopathy, the manipulation therapy by chiropractors, aromatherapy, hypnosis, Ayurveda, and various other forms of complementary medicine. One physician labeled it "quackademic medicine."[27] In a recent interesting case that reinforces skepticism about alternative medicine, Simon Singh, a British scientist and author, wrote a book on alternative medicine treatment called *Trick or Treatment?* It challenged the claim of chiropractors that they could manage childhood asthma and resulted in a libel lawsuit by the British Chiropractic Association. Singh spent two years and incurred substantial legal expenses to defend the lawsuit; he ultimately prevailed.[28] Clearly, the stakes—both monetary and medical—are high, and it's important to sort out the outrageous claims versus the true, positive impact of such therapies.

My own direct experience is quite illuminating. Right after my wife, Susan, and I had moved to San Diego, she had to undergo an unplanned, fairly urgent hysterectomy. She was petrified that she might have advanced disseminated uterine cancer. She couldn't sleep, and even with no prior elevation in her life, her blood pressure soared to systolic levels of 180 mm Hg.

Where I work as a cardiologist at Scripps, there is a superb Integrative Medicine Center. At my wit's end, I walked over there to see if there was a therapy such as biofeedback to help the situation. I felt a bit like an atheist soldier praying in a foxhole. The therapists at Scripps really came alive. On the next day Susan visited one and underwent about thirty minutes of hypnosis to relax her. When I went to meet her just after the session, it was like being introduced to a new person, at peace and altogether secure about moving onto the surgery in just a couple of days. Furthermore, she was now equipped with guided imagery tapes to listen to and help her relax, especially at night. I will never forget her smile and the confidence she expressed to her surgical and anesthesia team as she was wheeled off to the operating room.

I got to watch the hypnosis in action just after her surgery, when she was in pain and again became very tense. The same therapist came by to deal with her anxiety and blood pressure of 170 mm Hg systolic. Just into the session of hypnosis that I watched, her BP dropped to 120, and she was restored to her usual calm, relaxed state. It was truly eye-opening. Rigorous randomized trials have shown that biofeedback, acupuncture, hypnosis, and guided imagery can lead to better outcomes.[29] The first two techniques, for example, can lead to control or better management of high blood pressure, particularly for mild cases, and reduced frequency or severity of migraine headaches for many individuals. Guided imagery, or tapes that play relaxing music and help with control of breathing and body relaxation, have improved postoperative outcomes for patients undergoing open heart surgery. Biofeedback is a fully reimbursed therapy for patients with urinary or fecal incontinence, as it has been shown to be a highly effective intervention for regaining sphincter control for some individuals.[30] So as opposed to limited data to support many herbs and supplements, some good studies help anchor the use of particular "healing touch" interventions in certain circumstances.

My patient's email at the beginning of this section raises one last matter of patient unrest: poor compliance. Countless studies now have documented that only 50 percent of patients actually follow their prescriptive plan.[31] That doesn't mean that half of all patients don't take the medicines prescribed, and the other half are fully compliant. The problem is diffuse. Surprisingly, it has relatively little to do with level of education or intelligence, and it does not clearly track with socioeconomic status. The reason for the noncompliance of this immense consumer base is not clearly related to the cost of the prescriptions, either, even as proprietary drugs can typically cost $200 per month, and generics themselves are often expensive as well.[32] Recent data—from the *Consumer Reports* 2010 survey referred to earlier on DTC response—indicated that about 70 percent of consumers think that pharmaceutical companies have too much control of doctors' prescriptions, with more than 80 percent believing that physicians get rewarded for writing prescriptions for a particular drug and over 70 percent concerned that physicians are getting paid for providing testimonials or serving as spokespersons.[33]

––––––

Medical procedures, just like products, can be heavily promoted by their creators, even when validation is wholly lacking. In recent months, I have had two acquaintances with family members who developed symptoms suggestive of multiple sclerosis (MS). I referred both of them to a colleague at the University of California, San Francisco, Dr. Stephen Hauser, whom I consider to be the leading authority in MS. Both of them eventually had their care mapped out by Dr. Hauser, even though it was provided, on a day-to-day basis, by their local neurologists.

Looking into the state of MS treatment myself, I was somewhat horrified to read of the new "vein opening procedure," a real sign of the times.[34] Dr. Paolo Zamboni of the University of Ferrara in Italy has developed a "liberation procedure" in which he uses a balloon to dilate the veins in the neck and the chest, which are involved in the drainage of blood back from the brain. Supposedly, the veins in the neck are more likely to be narrowed in MS patients; Dr. Zamboni claimed to have found vein blockages in 100 percent of 109 patients with multiple sclerosis but zero in 177 patients without MS. Even if this is widely confirmed, it certainly does not make the case that dilating the veins will help treat the disease.[35] Indeed, the well-established science of MS indicates that a self-attack by the immune system on neuronal tissue, not the lack of blood flow in the veins from the brain, is the basis of the disease. Despite any real scientific evidence, the vein-opening procedure became the rage of the Internet, whereby affected individuals were pulled in by YouTube videos of testimonials or the procedure itself; sites like Facebook, in which five hundred groups were organized, created a buzz about Chronic Cerebro-Spinal Venous Insufficiency (CCSVI); and "liberation package" advertisements flourished throughout India, Poland, Jordan, and Bulgaria. Even a professor at Stanford University performed many of the procedures, taking the "liberation" a step further with the use of stents in the veins, and some of his patients have had serious complications, such as brain hemorrhage.[36]

Patient advocacy groups have put pressure on the Multiple Sclerosis Society to fund research on this procedure, and small, randomized studies are getting under way. In 2009, one of Canada's leading newspapers, the *Globe and Mail*, published a feature story on a dramatic improvement with the liberation procedure, and the Canadian Television Network program *W5* described CCSVI as "a revolutionary treatment for a most debilitating disease that could free MS patients from a lifetime of suffering."[37] By 2011 an article appeared in *Scientific American* entitled "The YouTube Cure," highlighting this as one of the first Internet-mediated popularized medical procedures—illustrating "the growing power of social media to shape medical practice—for good and ill." More recent studies have suggested that vein blockages in MS are a result rather than a cause of the disease and are found in only about 25 percent of individuals with MS and especially those who have had the condition for an extended period of time.[38] It will be several years before we know whether this far-fetched remedy has any value whatsoever, but in the meantime it represents an excellent case study of the dangers of the Internet, which can be used to hype an unproven procedure in an unprecedented way.

There are many other examples of procedures that were never validated in medicine but were heavily promoted. The pattern is quite remarkable, since the predator-prey relationship figures prominently. Step number one is a diagnosis that is debilitating with a considerable number of patients who are desperate, despite current treatments. Multiple sclerosis is a perfect target. Amyotrophic lateral sclerosis (ALS, Lou Gehrig's disease) is another. In 2010, *60 Minutes* used hidden cameras to expose a scam stem cell program that was heavily marketed in the United States for patients with ALS to be treated in Mexico.[39]

Back in the late 1990s, a surgical operation for another condition of desperation—heart failure—got legs. It was known as the "Batista procedure" and was glamorized on the television newsmagazine show *20/20*. Dr. Randas Batista, a Brazilian heart surgeon, invented an operation to remove a significant amount of heart muscle from patients who suffered from intractable heart failure. The *20/20* segment showed him riding around Brazil on his horse and presented multiple testimonials from patients who had outstanding restoration of their quality of life and exercise capacity by having the extensive open-heart surgery. Yet there was no good rationale, no control group, no validation for the procedure. Ultimately, when it was studied more thoroughly, it proved to be a big bust, with acceleration of deaths. Even the idea of "tailoring" the main pumping chamber (left ventricle) got large funding support from the National Institutes of Health with a randomized trial known as STITCH, which showed no benefit at all for the extensive heart muscle resection or what was called "remodeling."[40] Fortunately, the Batista fad occurred before widespread use of the Internet, so far fewer centers and patients were adversely affected by a major procedure without any evidence of benefit.

The last major aspect of patient empowerment I want to discuss is what you might call do-it-yourself medicine. About twenty years ago, the management team from a start-up company came to my office to inform me that they were going to manufacture automated external defibrillators (AEDs) for consumers. I thought they were absolutely irrational. How could the public handle how and when they would deliver a high-energy (up to 400 joules) electric shock to someone? Having been part of countless "Code

Blue" cardiac arrest resuscitation efforts throughout the years in a hospital setting, I couldn't imagine it was wise for consumers to have defibrillators in their homes. How would they know when it was appropriate to shock someone? Would they also wind up shocking themselves? What a rude potential awakening for an individual in a deep sleep!

I wasn't the only skeptic. The first demonstration for the AED traces back to 1979, but it took almost twenty years before the devices became widely available in public places, and in 2000 the large drugstore chain CVS started selling AEDs with a prescription.[41] It turns out that they work exceptionally well and are as close to "idiot-proof" as one could ask for, with the device providing auditory guidance each step of the way, and there is even a version made for people who are deaf or hard of hearing. The chance of a patient who suffers cardiac arrests (because of ventricular fibrillation) to be successfully resuscitated is almost 90 percent if the AED is used in the first minute.[42] Unfortunately, most arrests, when occurring in the home, happen while the person is asleep, and thus the spouse or family members are not alerted, so it isn't clear who should buy one. Furthermore, patients with heart disease with known risk or prior history of ventricular fibrillation usually get an implanted defibrillator.

Still, the topic of AEDs brings two patients to mind. The first was an exceptionally charismatic investment banker who had suffered a heart attack when he was thirty-nine years old. He had a family history of heart attack—his father had one at age forty-five—didn't exercise, and was stocky but not markedly overweight. I had my "big-time" motivational talk with him, and he started a new lifestyle, lost considerable weight, and got on a durable, rigorous exercise program. He traveled frequently but was faithful about cardiac workouts wherever he went. In 2005 one fall afternoon he suddenly keeled over on a treadmill in a gym. There was no AED, and by the time the paramedics came, he was dead. While he was en route to the hospital undergoing continuous CPR to no avail, his wife called me frantically for advice. I felt completely helpless.

The second patient wasn't mine but an icon in political media coverage—Tim Russert. I had met him once at the little airport at Nantucket and realized he was quite overweight. When I approached him to say hello and that I was a big fan, he radiated warmth and friendliness, even though I was a perfect stranger.

In June 2008, Mr. Russert was at the NBC studio in Washington and suffered a cardiac arrest. This occurred a few weeks after he had moved his father to a nursing home and in the midst of intense presidential campaign election coverage. Several weeks before the incident, he had undergone a stress test that lacked abnormalities. He was fifty-eight and faithfully taking a statin and medicines for high blood pressure. That morning in the studio when he collapsed, the people around him could not find the AED. It took seventeen minutes before the paramedics arrived, at about the time the studio defibrillator was found. The paramedics delivered the first shock, but it was much too late.

Both these events remind me of how I once had the impression that consumer, nonmedical personnel application of a defibrillator was a bad idea. Now, in marked contrast, I see it as a tragedy that these deaths were preventable. The Russert case could have at least become a major teachable moment for employers and the public. But the potential of a layperson to save another person's life with this do-it-yourself defibrillator device is still a remarkable symbol of public enablement and progress.

At-home medicine isn't just about emergencies. In fact, it can play a major role in making more everyday medical scenarios far less cumbersome. Taking the medicine warfarin (Coumadin), for example, is extremely difficult because it requires a blood test every week or two to check on how much the blood is thinned. Yet there are over twenty million prescriptions filled per year,[43] mostly for avoiding stroke in patients with the heart rhythm disturbance of atrial fibrillation, for those with mechanical heart valves, or those who have developed significant blood clots. If the blood becomes too thin, there is a high risk of bleeding. If it doesn't get thin enough, clotting can occur. It's a fine line, and it's no wonder that warfarin is the drug that everyone loves to hate. Moreover, it's the same agent used for rat poison. Most of the time patients go to have their blood drawn at a clinic or lab facility. But now many patients have a home device, and when they travel, the device goes with them. Not only is this more convenient, but studies have shown that self-testing leads to better regulation of blood-thinness compared with having to go to a lab or clinic.[44]

This is just one example of home medical testing that has been rapidly evolving since the 1970s, when home early pregnancy tests (EPTs) became widely used. From EPTs, to blood glucose or hemoglobin A1C, to blood cholesterol and triglycerides, to self-testing for HIV, do-it-yourself (DIY) medicine continues to evolve.[45] My father had insulin-dependent Type 1 diabetes; he was diagnosed in his teens and went blind from retinopathy by age forty-nine. As I grew up, I watched the progression of home adjunctive testing for how he handled sugar and regulated his insulin dose. When I was a teenager, he used urine dipsticks; when I would come home from college and medical school, he was using finger sticks with very accurate blood glucose readings. Surely this represents remarkable progress for DIY diabetes management, with much more yet to come.

Based on the discussion so far, empowering patients might seem like something of a mixed bag. Advertising of drugs, supplements, and procedures can lead to seriously negative outcomes, but in other ways the ability of patients to choose how and with what they are treated can have major benefits. The second, and more important, part of the equation is the quality of the information used to drive those empowered choices, making them well-informed choices. Now we will look at how information is pushed to consumers and pulled out by them, and see how to make the best of both.

Having access to the wealth of information on the Web has indeed been transformative. The data for how many patients look at health information on websites, like WebMD, health.nih.gov, healthfinder.gov, intelihealth.com, and mayoclinic.org are staggering. WebMD alone gets more than eighty million unique visitors every month. More than 8 out of 10 Americans have looked up health-related questions on the Web. The latest data from the Centers for Disease Control and Prevention showed over 50 percent of adults researched health information on the Internet in the past year, and in every age category women exceeded men. More than 20 percent of adults have posted on an online forum related to health care.[46]

Nowadays patients arrive to see me with a list of questions from surfing the Web. Many new patients have researched information about me online and know more about me than I could ever imagine. They have looked up their medications to see what kind of side effects they could experience. (That's a bit dangerous, since the average drug has a list of over seventy potential side effects.) They are far more savvy about their conditions than they would have been, even just a few years ago. This is a good thing—a step in the right direction.

The step, though, is really just on average in the right direction. I've shown repeatedly in this chapter that information can often be bad, misleading, or deeply dangerous to patients who find it. One of the problems is the quality of the websites from which information is derived. Before delving too far into a site's content one should consider a number of things: what are its purpose, sponsors, governance, sources of information, and frequency of updating? The predator-prey dynamic should always be kept in mind. Physicians as well as reporters have expressed increasing concern about the interaction between health information websites, like WebMD, and the pharmaceutical industry; commonly, ads for related drugs appear on the page for the condition the consumer is reading about.[47] And as we've seen, many supplement companies lie about their products. Even with limited, well-established, trusted, and user-friendly sites, it's a good idea to be suspicious of whatever you are reading.

As an example, let's look at the two top categories of disease—heart disease and cancer. At the American Heart Association website, the information for consumers is fairly good, but it is difficult to navigate and not up-to-date. There has been much written about Plavix and getting genotyped, but you don't find a word on it about that subject. How about the American Cancer Society? This is the organization that licensed its logo for Neutrogena sunscreens, the ones that didn't pass any tests for adequate sun protection. The society's website, with the tagline "the official sponsor of birthdays," is more user-friendly and has some comprehensive information on all the different cancers and treatments, but there is relatively little updated information on the site relative to the remarkable progress being made on new cancer therapies. Another major cancer website is cancer.net, created by the American Society of Clinical Oncology. It has some useful, easy-to-find information, with colorful icons to click on, and appears to serve as a complementary source to the American Cancer Society site. By comparison, the U.S. government sites for these two disease categories, the National Heart, Lung, Blood Institute (NHLBI) and National Cancer Institute (NCI), are difficult to use; while the information, once found, has some value, it is not updated adequately. Often it's worthwhile to check Wikipedia, as its information is presented in a straightforward manner, but updating is not consistent across medical conditions.

Organized consumer health advocacy organizations, such as Public Citizen and Consumers Union, and patient advocacy groups such as AdvoConnection, My Nurse First, and Patients Not Patents, have become increasingly prominent and important in recent years. Many of these patient advocacy organizations are foundations specifically dedicated to conditions such as breast cancer, colon cancer, and Parkinson's disease. Their websites, such as the National Breast Cancer Coalition, Colon Cancer Alliance, and the Michael J. Fox Foundation for Parkinson's Research, also provide quite useful, independent information. Patient organizations dedicated to funding research of rare diseases, such as lymphangioleiomyomatosis, Duchenne's muscular dystrophy, cystic fibrosis, and Huntington's disease, support both basic and clinical research to better understand and treat the specific illness.[48]

Fortunately, another layer of information on the Internet can be far more helpful and is gaining traction: social media health sites. Undifferentiated social media sites like Twitter and Facebook can be helpful for crowdsourcing health-related information. In fact a recent case was published, "How Facebook Saved My Son's Life," in which a mother documented that she obtained her son's diagnosis of the life-threatening condition Kawasaki's disease through one of her Facebook friends after posting his pictures on the website. Beyond this capability, the Internet highway was paved for specialized social health networks. In these highly interactive online sites, consumers are teaching each other.[49] This feeds the lack of trust in the medical profession (a poll in 2011 showed nearly 1 in 3 users do not share health information with their doctors,)[50] but it has positive attributes as well: electronic connections between individuals affected with the same condition, of the same age and sex, can provide extraordinary emotional support.

One of the sites that have attracted the most attention is PatientsLikeMe, which hosts patients with any chronic condition. It is striking to go to the site and see how many patients are taking a particular drug (sometimes 3,000 or more) and the list of indications, side effects, dosages, and individual tips from the large community. CureTogether.com targets more than five hundred different conditions and has initiated some interesting clinical research, such as response to medication by patients with migraine headaches. It was launched in 2008 to help people "anonymously track and compare health data, to better understand their bodies, make more informed treatment decisions, and contribute data to research." Diabetic Connect has close to 300,000 registered members, with monthly traffic of over one million unique visitors.[51] One should note the sources of funding for these sites: many dedicated to diabetes seem to be either sponsored or originated by pharmaceutical companies, and PatientsLikeMe has been called out for selling their de-identified membership data to pharmaceutical companies and third parties.[52] Many other health-networking sites are being created and rapidly gaining membership, like Inspire.com, with different disease-specific targets and financial models. Despite the bumps in the road, the rapid and enormous success of social health networking appears to conceptually validate many features of the Web 2.0 model—collaborative, interactive, virtual communities with peer-to-peer generated content. Discussions with my patients and many individuals who use these resources have caused me to conclude a large proportion of users find the sites indispensable.

———

Not a week goes by without a friend, patient, or acquaintance asking me for a referral to the right doctor for a specific condition. This turns out to be one of the most privileged pieces of information I have. For the five years I served on the Board of Governors at the Cleveland Clinic Foundation, I had frequent contact with the sixty-plus-member Board of Trustees of the hospital, who volunteered their time and efforts to meet frequently and help the health system. The Executive Committee of Trustees, a subgroup of nearly twenty of the trustees, met on a monthly basis and were called on quite regularly for specific support, be it a financial donation, networking, or guidance. I would ask them why they were willing to devote so much of their time, since most were still active as CEOs or senior management of their companies. I always got the same answer: "In case anyone in my family or I get sick." They saw their volunteer position as an insurance policy to get VIP access, particularly to the right doctor.

It might seem ironic that in the information era knowing which doctor is the best would be so difficult. After all, hasn't *U.S. News & World Report* been ranking hospitals by medical specialty since the early 1990s? Nevertheless, the basis for its rankings has been contentious, as the dominant factor is a "reputational score" derived from a relatively limited sample of physicians in that specialty. The other metrics include death rate, statistically adjusted to complexity of the patients; patient safety; patient volume; level of nurse staffing and whether the center has been designated a Nurse Magnet hospital; technology score; and patient services score.[53]

A recent study on how well the top U.S. heart hospitals (based on the magazine's ratings) held up to scrutiny of their data concluded: "A number of the *U.S. News & World Report* top hospitals fell short in regularly applying evidenced-based care for their heart patients. At the same time, many lesser known hospitals routinely provided cardiovascular care that was consistent with nationally established guidelines."[54] So what does all this hospital ranking really mean, other than bragging rights for a particular hospital about a medical specialty?

To be candid, not much. In 2011 Malcolm Gladwell wrote a *New Yorker* article, "The Order of Things," in which he called the *U.S. News* rankings a self-fulfilling prophecy. He pointed out that *U.S. News* chiefly relies on reputational scores: "But reputational ratings are simply inferences from broad, readily observable features of an institution's identity, such as its history, its prominence in the media, of the elegance of its architecture. They are prejudices."[55] For a few years when I was at Cleveland Clinic, it was on the top ten *U.S. News* list for geriatric medicine even though we didn't even have a geriatrics department! This demonstrates the magazine's overriding reliance on perception and reputation score, which outstrips reality. Studies of outcome measures for such diagnoses as heart attack have generally failed to distinguish a meaningful difference between top-ranked institutions and hospitals that are not ranked in the top forty.[56]

Regardless, the results, which are quite similar year after year, are the substrate for massive marketing campaigns that include local and regional TV commercials, billboards on highways and at heavily traveled intersections and airports, direct mailings, and radio, magazine, and Internet ads.

For marketing purposes, there is no real hindrance to being ranked forty-eight out of fifty—it is simply positioned as "one of the best hospitals in the United States."

Other agencies provide rankings, such as Thomson Reuters, which in 2010 ranked Scripps Health as one of the top ten health systems in the United States, along with Mayo Clinic.[57] Several of the others on their top list were not to be found anywhere in the *U.S. News* rankings. Unlike *U.S. News*, Thomson Reuters does not use a reputational score. Its metrics are risk-adjusted mortality, complications, readmission to the hospital, length of stay, and patient rating of performance. All of these data are available through public data sources. Of the top 10 "America's Best Hospitals" from *U.S. News*, 8 are not on Thomson Reuters's top ten list.

Overall, it is fair to say that at a top-ranked hospital for a given specialty there is a higher density of outstanding doctors than one would find at any community hospital. But there are "lemons" or weak-hitter doctors at every hospital, even those widely considered elite. So if you were to think you would like to get a second opinion at a top-ranked hospital and not specify whom that would be with, watch out. The doctors with the least referrals at these tertiary centers would be the ones you'd most likely see. By contrast, a trustee of the hospital would likely be able to find the right doctor.

This type of information is not on the Internet. For many routine matters, there are websites like ratemd.com, HealthGrades.com, docboard.org, and Angie's List that provide some information about a prospective physician. But it is essential, if at all possible, to have a go-to physician expert and authority when one has a newly diagnosed, serious condition, such as a brain tumor, neurologic conditions like multiple sclerosis and Parkinson's disease, or a heart valve abnormality. How do you find that individual doctor?[58]

In order to leverage the Internet and gain access to state-of-the-art expertise, you need to identify the physician who conducts the leading research in the field. Let's pick pancreatic cancer as an example of a serious condition that often proves to be rapidly fatal. The first step is to go to Google Scholar and find the top-cited articles for that condition by typing in "pancreatic cancer." They are generally listed in order by descending number of citations. Look for the senior, last author of the articles. The last author of the top-listed paper in the *Journal of Clinical Oncology* from 1997 is Daniel D. Von Hoff, with over 2,000 citations ("cited by . . ." appears at the end of each hit). Now you may have identified an expert. Enter "Daniel Von Hoff" into PubMed (www.ncbi.nlm.nih.gov/sites/pubmed) to see how many papers he has published: 567. Most are related to pancreatic cancer or cancer research.

Now go back to Google Scholar and enter his name, and you'll see over 24,000 hits—this number includes papers that cite his work. There are some problems with these websites, since getting citations by other peer-reviewed publications takes time; if a breakthrough paper is published, it will be years to accumulate hundreds, if not thousands, of citations. Thus, the lag time or incubation phase of citations may result in missing a rising star. If it is a common name, there may be admixture of citations of different researchers with the same name, albeit different topics, so it is useful to enter in all elements including the middle initial and to scan the topic list to alleviate that problem. For perspective, a paper that has been cited 1,000 times by others is rare and would be considered a classic. In this example, the top paper by Von Hoff in 1997 is a long time ago, and he is no longer at the University of Texas, San Antonio—he moved to Phoenix, Arizona. How would you find that out? Look for Daniel D. Von Hoff using a search engine such as Google or Bing, and look up his profile on Wikipedia. Without any help from any doctor, you will have found the country's leading authority on pancreatic cancer. And you will have also identified some backups at Johns Hopkins using the same methodology.

This is one DIY method for finding a leading authority on a specialized medical condition. The other main strategy, which can be viewed as complementary, is to ask your doctor. The problem with that is the answer will likely be a colleague in the same hospital, which is rarely the right answer. Physicians stick to their own specialty, so it would be quite unusual for any doctor to know the national authority in a different discipline. Doctors have to hunt to find the best people to provide care or weigh in with guidance even for themselves or their own family members. This is particularly true for a serious medical condition that is either life threatening or could prove to be if the operation is performed by someone who is not skilled in that particular surgery or procedure. The unique physician expertise comes only from the combination of seeing and treating a large number of patients with the specific condition and publishing insights, discoveries, advances, and critical observations about it. So asking your doctor needs to be taken to a more assertive, proactive level: Whom would you see for this condition? Who is the best person in the world for diagnosing and treating patients with this condition? If you're asking a primary care doctor these questions, he or she may not have answers but should seek them from the appropriate specialists in their network.

The heterogeneity of the quality of care is not adequately appreciated, and all too often consumers accept the convenient, easy alternative. I am not suggesting that your care be completely transferred to a physician and hospital geographically remote from where you live, where your family and support network reside, at considerable expense. But I want to emphasize the importance and invaluable contribution of getting a second opinion. It is best to do this at the outset, rather than after failing a standard therapy that is initiated locally. Getting access to promising, cutting-edge research programs, like whole genome sequencing of a tumor to biologically target the medicines, or a new procedure that preempts open heart-surgery by using a stent to repair a valve, is ideally initiated straight away. If this involves a physician or surgeon who does procedures or operations, it is essential to ask for the exact number of procedures performed per year and cumulatively over his or her career, and the precise data on complications for this individual operator (not the hospital or the group). If the procedure or operation involves a particular manufacturer's device or equipment, it is essential to ask if the expert has any financial interest in the company and to discern whether this conflict of interest, if present, might provide an incentive that is not aligned with the best treatment for your condition. It is vital to do a detailed online search in advance to learn when and where the doctor was educated and trained, what he or she has published, and what has been written about him or her in the lay press.

What constitutes the level of a "serious" or important medical condition varies with the individual. For me, it would include any joint replacement; any neurologic degenerative condition such as mild cognitive impairment (the precursor to Alzheimer's), movement disorder like Parkinson's disease, and MS, most cancers, and significant immunologic diseases like lupus, inflammatory bowel disease (Crohn's disease or Ulcerative colitis), and rheumatoid arthritis.

Back in 2002, I witnessed a unique "individualized medicine" experience involving a patient with a serious condition. A billionaire got a new diagnosis of a brain tumor, glioblastoma multiforme. Characteristically this is a cancer with one of the worst prognoses, and few live even a year from the time the diagnosis is made. This patient used his resources to bring in all of the international experts—not just the physicians who were known for looking after such patients but those doing clinical investigation with experimental agents, and basic scientists who were studying the molecular and cellular biology of the tumor. All of these authorities were gathered from around the world, including researchers from North America, Europe, Australia, and Asia, for a summit, a unique situation involving the world's cognoscenti. For two full days, his case was presented and discussed, with the patient present to ask questions and moderate, to see if there might be an experimental approach suitable that would extend his lifespan. Many novel treatments were eventually tried, and his death was perhaps slightly decelerated.

At roughly the same time, Amy Dockser Marcus published "Hiring Your Own Scientist to Find a Cure" in the *Wall Street Journal*.[59] One of the patients she featured was Stephen Heywood, a successful carpenter who developed Amyotrophic Lateral Sclerosis (ALS) at age twenty-nine. Stephen Heywood's case catalyzed the ALS field in many ways. He was the subject of an important book by Jonathan Weiner, *His Brother's Keeper: A Story from the Edge of Medicine*,[60] and the documentary film *So Much So Fast*. His two brothers, James and Ben, actively worked to improve Stephen's chances of surviving and the outlook for patients with ALS. Ben started the health social networking site PatientsLikeMe, and James set up the ALS Therapy Development Foundation.

One affected patient, cited in Marcus's article, donated $500,000 to the ALS Therapy Foundation and proclaimed, "Now I have scientists working for me." Another patient with Huntington's disease provided more than $1 million to Aurora, a biotech company, to test drug compounds "essentially hiring Aurora scientists to focus on finding a cure."[61]

Marcus subsequently won a Pulitzer Prize for her series of articles that characterized the extremes of patient activism.[62] In another piece, a third-year medical student at Tulane University with a very rare cancer, sinonasal undifferentiated carcinoma, went to work in the laboratory of a top cancer scientist to understand the biology of his tumor.[63] Another article featured a nurse who had an exceptional number of surgeries for treatment of her lung cancer after being told that her condition was "inoperable."[64]

The 2010 movie *Extraordinary Measures* was based on the book *The Cure* by Geeta Anand, a *Wall Street Journal* reporter.[65] In this true story about a child with Pompe disease, a rare genetic disorder of the heart and muscles, the father starts a biotech company to find a cure and pushes exceptionally hard to get a drug developed and tested to ultimately save his daughter's life.

While representing outliers in many respects, these cases demonstrate that the current level of being informed is often not adequate. They capture the sense of inspiration and independence, the hunt for innovation, and the primacy of the individual. Although today they may seem like extremes of patient activism, in the years ahead they will be viewed more as the norm. They represent the precursors for the next phase of medicine, in which powerful digital tools will provide data that was heretofore unavailable. Whether it is sequencing the genome of cancer tissue to determine a specific driver mutation and the effective drug to counter it, or being able to antic-ipate the likelihood of a fatal disease before it has ever manifested, our ca-pabilities will greatly expand.

————

Throughout this chapter I have tried to provide many examples of how consumers are increasingly accessing health and medical information and progressively getting empowered. Nevertheless, there are clear-cut limitations of both access and quality of information. Even if one is a billionaire, or a trustee of a major referral hospital, too much of the requisite information is lacking. As you will see in Part 2, this gaping deficiency is about to be filled in by various digital and highly informative tools. And consumers will have an upper hand, a driving force capability, as they have never had before.

On the other hand, the top-tier media coverage of new medical studies and discoveries demonstrates remarkable convergence on the way informa-tion is reported. In this book I often cite a *New York Times* or *Wall Street Journal* article alongside the actual paper published in *Nature, Science,* or the *New England Journal of Medicine*. Over time, journalists have begun to use more scientific language and cover topics in depth approaching that of scientific articles (at least in the Introduction and Discussion sections). Rather than the "dumbed-down" versions common in the past, we are seeing increasing respect for the consumer's ability to understand the principal re-sults and implications. That trend will surely continue, as the convergence of medical information proceeds, and the lines become blurred between the medical community, patient online communities, and the broad base of consumers.

PART TWO · CAPTURING THE DATA

PHYSIOLOGY
Wireless Sensors

*When the devices we use to capture and process data are
sparsely distributed and intermittently connected, we get an incomplete, and often outdated, snapshot of the real world. But
distribute billions and perhaps trillions of connected sensors
around the planet—just as we are doing today—and virtually
every animate and inanimate object on Earth could be generating data, including our homes, our cars, our natural and man-
made environments, and yes, even our bodies. Although our
bodies are not connected to the Internet today, they will be, as
biochips embedded in patients report their vitals back to a central database that is monitored remotely by physicians.*
—Don Tapscott and Anthony Williams, *Macrowikinomics*[1]

IF THERE IS ANYTHING to marvel about, it ought to be the human body.
During one's lifetime, the heart beats about three billion times with minimal

irregularity and consistent force of contraction. We breathe over six hundred million times; our lungs nurture our cells with oxygen and dispose of carbon dioxide. Our brain, with its hundred billion neurons connected in a quadrillion synapses, exhibits constant intricate electrical activity, even when we are in deep, restorative sleep. Via the pancreatic beta ß-islet cells that secrete insulin on demand, our blood glucose is normally under exquisite, tight regulation with food ingestion, exercise, and stress. Our kidneys, gastrointestinal track, and liver are all hard at work on a continuous and harmonious basis dealing with excretion, digestion, and metabolism. And when we need our muscles, joints, and skeletal system to come through, they are at our beck and call. Our highly integrated bodily functions are far more complex than this description can convey, and for the most part during our lifetime the systems are intact, reliably performing to preserve our health and well-being. We take all this for granted. We hardly ever measure our body's functions, and neither do our doctors—not because they wouldn't want to get more data, but until now that really wasn't feasible.

Once a year some individuals see a physician and get a checkup. At that time their blood pressure is measured, the heart is listened to, and laboratory tests are done to check things like liver and kidney function, fasting blood glucose, and electrolytes. But in essence medicine in the current era is really like the fable of "The Blind Men and the Elephant." These are spot checks: a snapshot of the blood pressure and heart rhythm, a one-off measurement of blood glucose, or any other laboratory test. But in between such measurements or occasional appointments, what is going on? What is the person's blood pressure doing during a stressful situation or in sleep during a nightmare? How is the glucose responding to the individual's diet and lifestyle, or how does it fluctuate during the night? Are certain foods or snacking putting undue burden on a weak pancreas to churn out insulin? What exactly is the heart rhythm or the glucose level when the person is feeling lightheaded? What is the oxygen concentration in the blood while the patient is sleeping? The list of such questions goes on and on. The point is that we have such limited insight into the physiology of each and every individual—a nonrepresentative, fleeting, pinhole view, through an artificial environment prism.

The other factor that keeps us from getting a more complete picture of a person's basic biology is patients themselves. Most healthy patients are reluctant to take such measurements on more than an occasional basis. For example, I have advised each patient of mine with high blood pressure to frequently check their blood pressure and chart their home and work readings so that we can see if acceptable control has been achieved using the medicines they are taking or the recommended lifestyle changes. Compliance is mixed, and I frankly feel fortunate when I get a patient to send me a few readings over a week. But the world is changing fast, and we are at the cusp of being able to get all these measurements done automatically. If there was no effort involved, and one would not even have to remember to take out a blood pressure cuff, that would make things easier. The new medicine promises capabilities far beyond making things easier. Instead, the future will bring with it the ability to obtain measurements continuously, even during sleep and times of substantial stress, which, as you might expect, are periods that represent essential gaps in our ability to track things today. As the *Economist* put it in a recent report, in the future "everything will become a sensor—and humans may be the best of all. . . . Anything and anyone—machines, devices, everyday things and particularly humans—can become a sensor, gathering and transmitting information about the real world."[2] Medicine will be revolutionized by the "Internet of Things," a world of interconnected, sensor-laden devices and objects.[3] As Thomas Goetz captured in "The Feedback Loop," published in *Wired*, sensors have the power to measure our every action, and as self-regulating organisms, we can profoundly change our behavior once we are provided with the relevant data.[4]

––––––––

In my twenty-five years as a cardiologist—and even before the Internet of Things was even a notion—I have had some patients who were diligent about data gathering. One current patient of mine, who is eighty-three years old, sends me an email every two weeks that includes the following data: his blood pressure (both systolic and diastolic), heart rate, and blood oxygen levels (O_2, measured as a percentage of total possible saturation), each measured three or four times a day, every day. He also tracks the number of steps he takes and the number of miles he has walked each day (see Table 4.1).

Note that he even does these measurements while traveling to the Middle East and aboard a cruise ship. This patient has a history of a common heart rhythm abnormality, atrial fibrillation, but is otherwise perfectly healthy. Few other patients are this exhaustive, and even his measurements don't involve anything invasive and don't even cover the full suite of vital signs (which also include respiratory rate and body temperature). How many other individuals would be willing to track their own data like this? And how do we go further?

Central to the Internet of Medical Things is the cell phone. Invented in 1973, it—like many other innovations—took a while to become firmly rooted in everyday life.[5] Steven Johnson calls this the "10/10" rule: a decade

TABLE 4.1: Representative sample of physiologic data from my patient.

Day	Time	SBP	DBP	HR	HR	Avg HR	Avg/d	O₂ percent			Place
5/10	3:00 AM	114	79	72	73	72.5		96			
Tue	7:00 AM	145	86	64	65	64.5		95	12,297	Steps	
	3:00 PM	109	63	70	71	70.5		96	2.5	ft/step	
	11:00 PM	115	70	67	67	67.0	68.6	97	5.82	Miles	Dubai
5/11	7:20 AM	114	82	66	66	66.0		96	8,857	Steps	
Wed	4:18 PM	116	69	74	75	74.5		95	2.5	ft/step	
	11:30 PM	111	70	67	68	67.5	69.3	97	4.19	Miles	Dubai
5/12	6:10 AM	111	70	72	73	72.5		96	10,058	Steps	
Thurs	3:17 PM	110	66	67	67	67.0		97	2.5	ft/step	Board
	12:00 PM	124	71	67	68	67.5	69.0	97	4.76	Miles	Crystal
5/13	1:05 PM	108	67	70	70	70.0		95	5,000	Steps	
Fri	6:10 PM	116	71	72	73	72.5		95	2.5	ft/step	
	9:05 PM	101	56	67	68	67.5	70.0	96	2.37	Miles	
5/14	9:30 AM	94	56	69	70	69.5		97			
Sat	12:00 PM	112	69	65	66	65.5		98	5,555	Steps	
	6:00 PM	112	66	65	66	65.5		96	2.5	ft/step	
	8:30 PM	106	59	65	66	65.5	66.5	96	2.63	Miles	On Ship
5/15	9:00 AM	121	68	60	61	60.5		97			
Sun	2:00 PM	110	62	73	74	73.5		96	9,287	Steps	
	4:00 PM	112	71	65	65	65.0		96	2.5	ft/step	
	9:00 PM	116	70	65	66	65.5	66.1	96	4.40	Miles	On Ship

to build the new platform and a decade for it to find a mass audience.[6] In 1990 there were a million cell phones in the world. By 2010, the number of cell phones exceeded five billion. Now, more than 85 percent of the world's population has access to a mobile signal. By 2012, it is expected that the number of cell phones will catapult beyond six billion.[7] The device-ification of cell phones—making them pluripotent stem cells, capable of acting as our calculator, alarm clock, photo album, watch, camera, video and voice recorders, flashlight, and more—set the stage for this medical revolution.[8] Now we can use them to monitor virtually any physiologic metric from any place, any time, or even all of the time.

Wireless healthcare really got its start late in the last decade in the fitness and health fields.[9] One early device was the Withings WiFi body scale, which records weight, fat mass, lean muscle mass, and body mass index. As is the case for most devices, the information can be stored on a private website secured by a password, sent to an iPhone, or even tweeted—or all of the above. The Nike + shoe and wireless accelerometers are other examples,

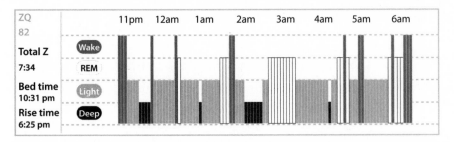

FIGURE 4.1: A night of my sleep, with the phases of sleep and duration of time in each shown, along with a ZQ score that integrates all this information, including time to get to sleep and time awake during the night.

recording data as one moves. Other athletic shoe manufacturers such as Adidas (miCoach Pacer) have launched similar products. Alternatively, the app RunKeeper on a smart phone with GPS and an accelerometer can graphically display data for duration of exercise, distance, velocity, and even the route taken. Wireless accelerometers such as the Fitbit and DirectLife record each step a person takes during the day to encourage users to take the recommended 10,000 steps per day. These devices are quite small, even smaller than an average thumb drive; somewhat larger devices, such as the Bodybugg and BodyMedia, connect to a smart phone to provide data on how many calories are burned throughout the day. GreenGoose enables cyclists to track their exertions. The Q-sensor from Affectiva, worn on the wrist, tracks emotional arousal and provides feedback through one's phone when it's time to take a break and calm down.[10] Sharing this data via social networking might be a bit over the top, but I frequently recommend the individual use of some of these tools for tracking aerobic exercise and to promote more activity.

One of the first consumer medical monitors was the Zeo, using sensors on a headband to monitor brain waves during sleep. Originally designed by three students at Brown University to help them avoid waking during deep, restorative sleep, the device morphed to track the phases of sleep every few minutes, sensing one of four phases: awake, light sleep, rapid eye motion (REM) dream sleep, and deep sleep (as shown in Figure 4.1).

Figure 4.1 shows the quantitative display of a night of my sleep. My aggregate score, called a ZQ, was 82, reflecting a composite of total duration of sleep and times and duration of being awake, and weighted for REM and deep sleep, which are most valuable. I slept for a total of seven hours and thirty-four minutes, waking seven times (I was unaware of most of these). Most deep sleep came in the first third of the night, which is typical, despite the popular notion that sleep gets deeper as the night wears on. This explains why going to sleep later than usual often results in a diminished time in deep sleep.

One night I experienced an interesting side effect of sleep monitoring. My wife is a night owl, so not infrequently she'll retire an hour or two after me. One night, in the first week after I started monitoring my sleep, she walked in the room, looked at the Zeo clock, and noted I was in a wake cycle. She said, "Eric, I know you're awake, and I'd like to talk." It's hard to play possum with a sensor displaying your real time brain waves.

Like the fitness wireless devices and apps, the Zeo has a mobile app, and the data can be shared with one's social network.[11] This may have practical value, say, for warning office staff that a real grump is going to arrive at work. The other side of social networking with sleep monitoring is the ability to compare your results with peers of the same age group, since our quality of sleep deteriorates with age. And of course one can have a competition with one's spouse, family members, neighbors, and friends. The real goal of monitoring is not simply to quantify one's sleep phases and quality but to improve it, whether by avoiding alcohol in the evening or caffeine at any point during the day, keeping pets or light out of the bedroom, or by "powering down"—avoiding electronic stimulation for at least an hour before going to sleep. Many of these recommendations seem obvious, but the ability to enact them and quantify the effect is what is different.

Wireless health and fitness monitoring has helped create the Quantified Self movement, started by two *Wired* magazine editors. Its purpose is to use data about ourselves to improve our lives. A 2011 feature article by the *Forbes* journalist Kashmir Hill, entitled "Adventures in Self-Surveillance, aka Extreme-Navel Gazing," provided a first-person, blow-by-blow account of trying most of the health and fitness wireless gadgets that are currently available.

The next logical step in this evolution is for sensors to help manage chronic disease, which affects more than 140 million individuals in the United States, which account for more than 75 percent of our health care expenditures—that's $2 trillion.[12] We have already had a preview of the success of remote monitoring via permanently implanted devices, such as pacemakers and implantable cardiac defibrillators, that enable remote monitoring.[13] Permanent implantable devices are not likely to be the solution for most individuals, however, given that they involve a significant procedure, high cost, and serious potential complications that include infection and device failure.

BLOOD GLUCOSE AND DIABETES

The first chronic disease for which continuous wireless monitoring became commercially available for patients was diabetes. We are in the midst of a diabetes epidemic, in part stemming from the global obesity problem; more than 1 of every 10 individuals in the United States has the diagnosis, with a global total of about 350 million people. The escalating toll on children is especially profound. Central to the management of diabetics is achieving good control of blood glucose; high levels contribute to many of the major complications of the disease, and low levels can lead to fainting or loss of consciousness. Until a few years ago, the only way to monitor blood glucose at home or on the go was via finger sticks, providing intermittent assays of the glucose level. The procedure is somewhat painful, hard to do in public, and expensive, with test strips costing $3 or more per day. Even with three or four measurements at different times during the day, the information on blood glucose levels is limited; having a very high or low blood glucose during the night would probably go unnoticed, and even during the day, glucose levels can vary considerably in response to exercise, food intake, fluids status, and medications, such as insulin, that are taken to manage the disease. A randomized study using a mobile app to alert patients to check their blood glucose and provide automated feedback has shown superior glucose control at one year compared with conventional, self-directed diabetes management.[14]

Transcending apps, sensor technologies can relieve some of the need for fingersticks and provide continuous monitoring. The principal contemporary method, known as continuous glucose monitoring (CGM), relies on a sensor inserted with a 27-gauge needle just below the skin of the abdomen. This device measures the glucose in the interstitial fluid just below the skin; a dedicated transmitter sends the information to a receiver every five minutes. CGM can provide nearly continuous information about blood glucose and is particularly helpful for individuals who struggle with glucose control, which is fairly common. Even CGM, however, requires calibration with a fingerstick every twelve hours, and the CGM sensors are only approved for three- to seven-day use before they must be replaced. The disposable sensors cost more than $100 each and the receiver more than $1,000. As a result, CGM is used infrequently and mostly for insulin-dependent diabetics, but the potential for this technology for managing or even preventing diabetes in the future is quite remarkable. I actually tried a CGM sensor for a week. I was surprised by how easy it was to have in place, with no inhibition of exercise or showering, and by how marked could be the effect on my glucose levels from eating something like crackers or pizza.

At a conference known as TEDMED, held in San Diego in late 2010, one session featured Walt Mossberg, who is the consumer technology critic for the *Wall Street Journal* and the organizer of the All Things Digital annual conference. I hadn't met Walt before, but I have been reading his columns for years and have learned quite a bit from his insightful reviews. I have also followed him on Twitter for his remarks on digital devices. Walt is an articulate fellow in his sixties, bald and white-haired, bearded, and fairly thin— and, as he told the audience that day, a diabetic frustrated with the archaic nature of glucose regulation. Where, he wanted to know, was the smartphone app that took care of all of this? All he wanted, he explained to me afterward, was an app whereby he could shine his phone on a blood vessel and it would give him a glucose reading.

Many companies are trying to develop such sensors—they're the holy grail of diabetes—but the delicacy of measuring glucose has made them impossible so far. Efforts by manufacturers to have the glucose data relayed to cell phones have been stymied by the FDA because of concerns about a patient getting inaccurate data, affected by the operations of the cell phone, and acting on it (such as taking insulin as a response to an errant high blood-sugar measurement). This obstacle should be overridden soon. Future prospects for alternative sensors are bright; for example, contact lenses can be embedded with particles that change color as the blood sugar rises or falls or the glucose level can be assessed through tears. Another imaginative solution has been dubbed a "digital tattoo" in which nanoparticles are injected to the blood that bind glucose, and emit a fluorescent signal that is quantified by a reader on a smart phone.[15]

One other aim of technology in this space is to combine CGM and insulin pumps in a "closed loop" such that the insulin delivery is fully integrated with the glucose measurements to simulate an artificial pancreas. While many people with insulin-dependent diabetes wear both devices, this integration is not yet operative. Considerable research efforts are dedicated to making this a reality.

THE ELECTROCARDIOGRAM AND HEART RHYTHM MONITORING

With the heart beating roughly three billion times in a person's lifetime, it's not surprising that many individuals have abnormal heart rhythms at some point along the way. Atrial fibrillation is the most common electrical disturbance; this arrhythmia occurs when the normal pacemaker gives way to a chaotic, irregular heart rhythm emanating from or near the atria (the collecting chambers of the heart). As one of my recent patients described it, the atria go into "quivering" mode. As a result of this loss of forceful contraction of the atria, blood stagnates in these chambers, and there is risk of a blood clot forming. If, from this stagnation, a tiny blood clot is ejected from the heart and reaches the brain, it can cause a stroke. The only way to definitively diagnose atrial fibrillation is to record the heart rhythm, and ideally over a prolonged period since this rhythm disturbance is typically intermittent. Some individuals who develop atrial fibrillation may feel dizzy or lightheaded, but those same symptoms can be due to a very slow heart rhythm. The slow arrhythmias are caused by a block of conduction of the electrical impulse of the heart; if they are associated with symptoms (and not attributable to medications), the person may require a permanent pacemaker.

The other major class of heart rhythm disturbances is known as malignant ventricular arrhythmias—ventricular tachycardia and ventricular fibrillation—and, as the term "malignant" implies, are even more worrisome. They usually occur in patients who have coronary heart disease with prior damage to the heart muscle or patients who have a significantly weakened heart muscle. But some genetic mutations predispose even young, otherwise healthy individuals to having these rhythm disorders. And several medications have the potential, in susceptible individuals, to induce serious arrhythmias.

Hence, there is a real need for heart rhythm monitoring. Norman Holter, a physicist, invented the first portable device for continuously monitoring heart rhythm back in 1949, and it took more than ten years before it began to have clinical use.[16] The Holter hasn't changed much since. The monitors consist of a bulky box and multiple wires that must be attached to the patient, and the information downloaded, in a clinic. The cumbersome box worn on a belt along with several leads on the chest make it impossible to shower or exercise, so it is not practical to get multiple days of recording. In 1999 I saw a device meant to improve this state of affairs. Brook Byers, a longtime friend at the venture-capital firm Kleiner Perkins, asked me to evaluate a new technology from a company called CardioNet. Their device was intended to monitor heart rhythm in real time over the Internet for extended periods of days or weeks. I was excited enough by what I saw to become CardioNet's first of several outside medical advisors.

Ultimately, a clinical trial of approximately four hundred patients, with half randomly assigned to Holter and half to CardioNet, validated the concept that more arrhythmias, especially atrial fibrillation, could be picked up using extended mobile cardiac rhythm monitoring.[17] The FDA approved the device in 2002, and the company went public in 2008. Still, the technology is not perfect. In particular, the real-time recording requires human oversight at a large center rather than automatic signal processing, and the insurance reimbursement is not proportionate to the labor involved. Moreover, in most patients with a suspect heart rhythm abnormality, the data typically doesn't need to be accessed in real time.

There have been some other major innovations to capture heart rhythm. One is a simple patch known as iRhythm. Designed to archive the heart rhythm for at least seven days, the patch is a skin adhesive worn on the chest. It is mailed to the patient, who mails it back for interpretation with no need for real-time human oversight, unlike CardioNet (or its competitor, Life Watch). We started calling this the "Netflix model" of heart rhythm monitoring. Other technologies, such as AliveCor, capture a patient's real-time heart rhythm directly via a smart phone. A case with two exposed sensors that attach to the back of a smart phone immediately records the heart rhythm when the person places a finger from each hand on one of the sensors to complete a circuit with the heart. Or alternatively the case itself, or the case-phone unit, can be placed on the chest to record the heart rhythm. The ECG is displayed on the phone, and the recording can be made for several minutes. The heart rhythm data are archived in the phone and can be transmitted via the Web. Using this device, an individual with palpitations, lightheadedness, or faintness can use his or her phone to acquire heart rhythm data and send it directly to a physician for interpretation. The next iteration of this device features signal processing to rapidly diagnose the arrhythmia, if one is present, without any human oversight. Once the transmission is made and the data is automatically interpreted, a text is sent to provide the type of heart rhythm and appropriate advice for the patient.

VITAL SIGNS

Capturing vital signs noninvasively with real-time wireless transmission might appear formidable. Indeed, they are typically all captured simultaneously and incessantly only in an intensive-care unit. In 2009, I was introduced to a technology that seemed remarkably futuristic. Cameron Powell, son of a Texan oil baron, and a fully trained obstetrician-gynecologist, had given up his practice of medicine to start a company known as Airstrip

FIGURE 4.2: Airstrip Technologies' display of patient vital signs on a smart phone and the Sotera wrist transceiver, which directly displays the vital signs from the patient.

Technologies. Cameron's first product was known as Airstrip Ob. Most obstetricians are itinerants, serving multiple hospital facilities. Cameron's device transmitted information about an expectant mother's uterine contraction and fetal heart rate to a smart phone in real time, enabling the doctors to check on any high-risk patients regardless of distance. This app proved to be enormously popular, and tens of thousands of obstetricians started to subscribe to the service.

The next field for Airstrip's technology was critical care. The vital sign data was already being captured; Airstrip's task was to reformat the data so that it could be transmitted over the Internet and displayed on a phone. By 2010, Cameron and his team obtained FDA approval and started marketing Airstrip CC (see Figure 4.2). He appeared to be well on his way to becoming a wireless medicine baron of the future.

The big prize was capturing the vital signs outside the intensive-care unit. The toughest nut to crack here is blood pressure; accurately measuring it on a continuous basis without placing a sensor in an artery has proven to be challenging. Matt Banet, an inventive engineer from San Diego, decided to tackle this problem around 2005. Ultimately he invented something known as the Sotera wrist transceiver sensor, which is capable of detecting arterial blood pressure, although it does require occasional calibration with an arm cuff; attachments enable measurement of blood oxygen concentration and heart rhythm.

While the initial use of the Sotera device is for in-hospital monitoring, the implications go beyond one setting, making it technologically feasible today to simulate intensive-care unit monitoring anywhere. You get upset and want to know how that affects your blood pressure; you are breathing a bit heavy and want to know if your oxygenation is intact; you feel your heart flutter and want to see if you are experiencing an arrhythmia. The Sotera now makes it possible to know any of these things.

It doesn't take much imagination to figure out that this is a game changer in health monitoring—or that it might create a legion of e-hypochondriacs or cyberchondriacs. While many individuals who surf the Web start believing they may have the diseases they are reading about, continuous monitoring of one's vital signs takes this concern to a new level. Nevertheless, the upside of having such information available is considerable. We know that people who weigh themselves each day rather than once per week are more successful at losing weight and maintaining their weight. Similarly, mobile phone apps that count calories have facilitated weight loss for some people.[18]

Telemonitoring of blood pressure, likewise, has been shown to have a positive impact on management.[19] We also know that controlling blood pressure and especially avoiding marked shifts or variability are central to reducing the risk of stroke and heart attacks.[20] Remote monitoring also avoids "white coat hypertension"—elevated blood pressure brought on by anxiety about a doctor's appointment. Such data can provide the best guidance for titrating medications, helping reveal the right doses of the right drugs for the right patient. It can also help a patient discern how changes in diet, such as a high salt load, and exercise can affect the blood pressure.

The implications of having vital sign metrics are much broader than just optimal management or prevention of hypertension. Any home can theoretically be transformed into a mobile intensive-care unit, a term now used for an ambulance that provides extensive monitoring. Of course, a nurse and physician wouldn't be near at hand to provide treatment, but much of what a hospital provides for patients not in the intensive-care unit is simple monitoring—at considerable cost. Given that there are 100,000 fatalities in hospitals in the United States each year primarily as a result of patients receiving the wrong medicines, and that hospitals harbor the most serious, highly resistant bacteria and pathogens, there are some favorable trade-offs for monitoring people in their home.[21]

Responsive medical care isn't an impossibility, either. Currently, many homes are set up with alarm systems hardwired to alert the fire department and the police. The select home of the future can also be set up for continuous, real-time wireless transmission of the vital signs of its residents to the medical center. Automated processing of these signals could conceivably lead to the dispatch of an ambulance with paramedics, just the way fire fighters or police respond, and ideally without the problems of false alarms.

ASTHMA ATTACKS

More than twenty-three million Americans suffer from asthma, leading to more than 500,000 hospital admissions per year; an asthma attack can be fatal especially in children and young adults. Inhalers are a mainstay of treatment. The company Asthmapolis designed an add-on sensor to convert inhalers to enable wireless transmission; when a patient uses the inhaler, it sends a signal to his or her smart phone and over the Internet. This can alert a physician to a patient's frequent inhaler use and a community to particular locations associated with asthma attacks. These are the first steps of a dedicated wireless sensor solution for preventing asthma. Within a skin adhesive patch, which can be conceived as a large Band-Aid, sensors could be integrated and used to detect air pollution and pollen count. Vital-sign monitoring and a sensor to detect reactive airways could, if signs indicated an oncoming asthma attack, potentially prevent it either by helping the patient avoid an area with high pollution or prompting the patient to preemptively take additional medications. Likewise, patients with chronic obstructive lung disease—ten million Americans who are at even greater risk of needing hospitalization—could potentially benefit from a similar wireless sensor system.[22]

SLEEP APNEA

More than forty million Americans have a sleep disorder characterized by transient cessation of breathing, or extremely shallow ventilation, during sleep.[23] The predominant cause of this common problem is upper airway obstruction or abnormal regulation of ventilation by the central nervous system. Sleep apnea can lead to high blood pressure in the pulmonary artery, heart arrhythmias, and heart failure, and it can exacerbate diabetes. Treatment for upper airway obstruction uses continuous positive airway pressure (CPAP) administered throughout the night to keep the airway widely patent.

Sleep apnea is usually diagnosed in a hospital sleep laboratory. The patient comes in for the night and is hooked up to various devices to measure the oxygen concentration in the blood, respiratory rate, heart rate and rhythm, and brain waves. It is hard to think that anyone would have a "normal" sleep pattern under these circumstances. The charge for this service is approximately $3,000. If a patient has obstructive sleep apnea and requires CPAP, he or she has to come back to the sleep lab for at least one more night to determine the right level of CPAP. This is quite an ordeal and an expensive one at that. A wireless solution enabling patients to sleep at home is obviously desirable. It could be done serially to assess whether the CPAP at a particular level was working. We would quickly find out how much sleep apnea is underdiagnosed, since so many patients who potentially have this condition wish to avoid trying to sleep in a hospital laboratory.

MOOD DISORDERS

More than twenty million people are diagnosed depressive, take medications for the condition, or both.[24] All too often, depression is inadequately treated and leads to suicide or a suicide attempt. Serious depressive episodes can actually be detected remotely by quantifying mood via cell phone sensors, relying on tone of speech as well as the extent of phone and text communication. The company Cogito has developed a sensor that quantifies mood via the user's voice; another company, Affectiva, has developed a device that detects emotional state via skin conductance. GPS and accelerometers allow a physician to track a patient's mobility; less movement and communication indicate worse episodes.

––––––

This has not been an exhaustive listing of medical conditions that might be amenable to remote sensing; other examples include epilepsy, glaucoma, and movement disorders such as Parkinson's disease, among a long list of others.

AGING IN PLACE

The concept of wireless monitoring to create a "smart" medical home may be particularly well suited for select seniors, preserving their ability to stay at home by providing a greater safety margin. Nearly all seniors, more than 95 percent in surveys, want to stay in their own home rather than move to an assisted living facility or a nursing home. Naturally, there are risks: 40 percent of those over eighty fall at least once each year, which translates to about 300,000 broken hips per year.[25] These fractures carry a high risk for mortality and represent the number one cause of accidental death in the United States.

A recent report in the *Economist* focused on shoes embedded with sensors that give the person enhanced proprioception, helping to reduce the likelihood of falling.[26] Such wireless sensors detect unsteadiness of gait and could potentially prevent a fall. A "smart cane," developed by researchers at UCLA, has embedded force sensors, motion sensors, and accelerometers, providing guidance to the person using it. The applications for a such a device go well beyond use in seniors and may apply to many individuals who use canes because of uneven gait or other disabilities.[27]

Personal emergency response systems help seniors call for help if they do get hurt. One told the *New York Times*, "Without the sensors, I would probably be dead."[28] Other devices with wireless transmission that can be used to support aging in place include video cameras, motion sensors, tagging pills to monitor compliance, sensors on the mattress and doors, and vital-sign monitoring. One of the most formidable problems is integrating various sensors that have been developed separately.[29] With each of the sensors emanating from different companies with discrete and proprietary functions and signal processing, the lack of uniform standards is a major rate-limiting factor. There are multiple ongoing efforts to aggregate data from sensors, and eventually it is hoped this will not be an enduring obstacle. It will inevitably require cooperation among competing manufacturers.

MEDICATION COMPLIANCE

Some 50 percent of prescription medications are not taken as directed, mostly because the patient prematurely stops taking the drug, not that he or she misses doses. The recent economic downturn has prompted some patients to not even fill their prescriptions.[30] This is especially a problem with chronic diseases. For example, congestive heart failure is the leading cause of hospital admission and readmission each year in the United States. Expenditures for this condition exceed $37 billion annually, most of which are related to in-hospital charges. Often patients with congestive heart failure do not take their medications properly, and this is thought to be one of the principal reasons the readmission rate for people over sixty-five is 27 percent within thirty days and over 50 percent within six months.[31]

Wireless tactics can facilitate medication compliance in many ways, ranging from simple reminders by phone call or text messages to a skin patch embedded with both the medication and a wireless activator to release the dose. Wireless medication bottles, such as "Glow Caps," send a signal whenever the cap is removed.[32] A more elegant approach involves placing a digestible sensor in each pill that is activated upon contact with gastric juice. This electronic signal is then relayed via the sensor in an adhesive skin patch, placed anywhere on the body, to a mobile phone that transmits the data over the Web. Accordingly, the precise time that the tagged drug was ingested is captured. The manufacturer, Proteus Biomedical, is conducting several small clinical trials focusing on a variety of conditions frequently associated with poor compliance: tuberculosis, hypertension, heart failure, diabetes, and depression.[33] Phillips Electronics, a company active in wireless sensor and system development, has promoted the iPill,[34] which can be wirelessly activated to release a drug at a particular site in the gastrointestinal tract. We'll soon see whether wireless technologies can improve compliance and outcomes for patients.

THE EMERGING WORLD AND
THE MOBILE PHONE LABORATORY

In many emerging countries, there is nearly universal access to mobile phones and connectivity. A large randomized trial tested texting via mobile phones in sub-Saharan Africa by comparing texting plus standard care versus standard care alone; it demonstrated a marked improvement in compliance for antiretroviral medications directed against HIV infection.[35] Not only was adherence improved, but the clinical outcome of HIV viral suppression was significantly better.[36] These positive outcomes would likely extend to other diseases.

Multiple projects have been pursuing a lab on a chip, which uses a cell phone's SIM card as a biosensor for detecting malaria, sexually transmitted diseases, and other infectious disease pathogens.[37] The SIM card chip could also be used to analyze blood constituents such as electrolytes or perform complete blood counts. There are also many efforts to remotely diagnose conditions; for example, a cell phone app could be used to determine whether a cough is likely caused by pneumonia or heart failure. Efforts to exploit this mobile phone lab platform are focused on rapidly analyzing saliva, sweat, breath, and urine. The Skin Scan app can help differentiate a mole from a melanoma by simply taking a picture of the skin lesion and getting the image processed via a sophisticated algorithm. The rapid and convenient turnaround can help determine whether to recommend a biopsy. Digitizing breath turns out to be a billion times more sensitive than the breath analyzers used today to detect alcohol levels, and

breath sensors are being tested for detecting lung cancer and tracking pe-
diatric asthma.[38]

CONFRONTING THE
CHALLENGES OF WIRELESS MEDICINE

Although the technology has exciting transformative potential, there are
many significant challenges: data flooding, security and privacy, clinical validation, cost, and adoption into medical practice. Remote monitoring of
multiple physiologic metrics for even a single person generates a large
amount of data that typically needs to be processed via software algorithms
in order to be useful. Even if a system is programmed to only inform a physician if recorded data falls outside some preset "safe" values, it is easy to generate a deluge of data when hundreds or thousands of patients are being
monitored at any given moment for maybe a dozen different metrics. A
further problem is that much of the data generated by different sensors are
in proprietary formats, making integration difficult. Shared standards would
help and also make it easier to protect privacy. While electronic data are
never impervious to hackers or inadvertent mistakes, the wireless medical
industry should agree on and enact critical safeguards.

THE OFFICE VISIT
OF THE FUTURE

With remote sensing and the ready accessibility of video chat, virtual office
visits could soon replace the routine physical ones. A pioneering group of
doctors started a practice featuring electronic and video connections in New
York City called Hello Health. Patients pay out of pocket for access by
email, text, and video chatting and only actually visit with a physician in
person on an infrequent basis (usually less than once per year). Medical
practices in other cities such as San Francisco have also adopted this system,
and it was the subject of a 2009 feature article in *Fast Company*, "Doctor
of the Future," which described it as "part electronic medical record, part
practice-management system, and part social-networking site, complete
with profiles and photos of doctors and patients, all in a secure environment

that complies with federal privacy standards."[39] I will delve into this topic in much greater depth in Chapter 9.

WIRELESS SENSORS IN CARS

Several car manufacturers, including General Motors (On-Star subsidiary), Toyota, and Ford, are now developing sensor systems that physiologically monitor the driver in order to prevent accidents. Sensors that monitor heart rhythm either via the steering wheel (using two sensors like the AliveCor smart-phone case) or in the seat are being evaluated for drivers with a history of conditions, particularly heart arrhythmias. Automatic detection and audio readout of the blood alcohol level by breath and of the glucose by integrating with a continuous glucose-monitoring sensor (worn by some people with diabetes) are also being assessed. Such data can prompt the car ignition to shut down or relay a message to the driver, in the case of an arrhythmia or low glucose, to pull over. This provides further perspective on the upcoming ubiquity of wireless health sensors and their potential impact on our lives.[40]

BIOLOGY

Sequencing the Genome

*Just as personal digital technologies have caused
economic, social and scientific revolutions unimagined
when we had our first few computers, we must expect and
prepare for similar changes as we move forward from our
first few genomes.*
—George Church[1]

IN DECEMBER 2010 in Milwaukee, Wisconsin, Nicholas Volker, a five-year-old boy with a gastrointestinal condition that had not previously been seen, who had undergone over a hundred surgical operations and was almost constantly hospitalized and intermittently septic, was virtually on death's door. But when his DNA sequence was determined, his doctors found the culprit mutation. That discovery led to the proper treatment, and now Nicholas is healthy and thriving. Even though this was only the first clearly documented case of the life-saving power of human genomics in medicine, few could now deny that the field was going to have a vital role in the future of medicine. Some would argue that the treatment led to an even bigger breakthrough: health insurance coverage of sequencing costs for select cases.[2]

It took the better part of a decade from the completion of the first draft of the Human Genome Project for genomics to reach the clinic in such a dramatic way. To make treatment like Volker's common will likely take more time still. Even if that's the ultimate prize, the creative destruction of medicine still has various other, less comprehensive, genomic tools for us to use, based on investigations of things like single-nucleotide polymorphisms, the exome, and more. The material can be a bit heady, but it's worth pushing through: these tools could effect not just dramatic corrections of faulty genes but a better, more scientific understanding of disease susceptibility and what drugs to take. Moreover, as they empower patients and democratize medicine, they make medical knowledge available to all and deep knowledge of ourselves available to each of us. Nevertheless, at this level, perhaps more than anywhere else in this ongoing medical revolution, the resistance from the priesthood of medicine is at its height. The fight might be tougher than the material, but in neither case can we afford to give up.

GENOMICS 101

More than ten years ago, on June 26, 2000, a major ceremony was held at the White House to announce the first draft of the human genome sequence. President Bill Clinton said, "Today we are learning the language in which God created life. . . . It will revolutionize the diagnosis, prevention, and treatment of most, if not all human diseases."[3] He called it the "most important, most wondrous map ever produced by humankind." The *New York Times* headline was "Genetic Code of Human Life Is Cracked by Scientists," and *Time* featured on its cover Dr. Francis Collins, who led the NIH public-funded consortium, and Dr. Craig Venter, who led the private Celera Genomics effort, with the title "Cracking the Code!" The excitement was high, but even today—both among physicians and the public alike—we have yet to gain real knowledge about what was discovered, then and in the ensuing decade.

The human genome comprises twenty-three paired chromosomes with more than three billion bases each, arrayed in a double helix. Because we are "diploid," with two copies of each chromosome, one paternally derived and the other maternally derived, we have about six billion bases in our DNA. So while there are lots and lots of them, the DNA sequence per se is actually fairly straightforward. Each base can be one of only four molecules—the four "life codes"—adenosine (A), guanine (G), cytosine (C), and thymidine (T). Despite the simplicity of the code, it is difficult to read it, and developing the first machines that could do it was expensive, complicated, and time consuming. The other fundamental problem facing genomics, besides the sheer volume of code, is understanding what it all actually means. It has been a "fact" of elementary biological education for some time that a DNA string of three bases codes for an amino acid and that a collection of amino acids (like histidine and tyrosine) is assembled to form a protein—or more simply that genes code for proteins. Many of us wish it were that simple!

It turns out that of the whole six million bases, only 1.5 percent actually represent coding elements, known as exons of genes, that are capable of producing proteins. Collectively this portion of DNA is known as the exome. Before the Human Genome Project, the genomics community had expected there would be at least 100,000 genes in our DNA, and perhaps many more than that. Nevertheless, the final gene count is fewer than 23,000 genes in the human genome. That means that approximately 98.5 percent of the genome does not code for proteins and does not fulfill the classic definition of a "gene." If not coding for proteins, what is the vast majority of the genome doing?

The answer is regulating. Within genes there are promoters and introns, which also do not code for proteins. Promoters turn on or off the process of transcription, by which the DNA of the exome is copied into messenger RNA (mRNA), which itself gets read and translated into amino acids and proteins. Introns don't code for proteins either. Although they are transcribed to precursors of mRNA, they are ultimately edited (spliced) out in the process of making mature mRNA and translating it.

The rest of the genome, outside of the confines of genes, is chock full of regulatory elements that modify the function of genes that may be located thousands or even millions of bases away. That regulation can be achieved in many different ways, such as coding for RNA transcripts and affecting how much of a protein is manufactured or how it is edited. In fact, so much of the genome codes for RNA transcripts with no direct protein-coding function that it has become clear that RNA—in such diverse forms as microRNAs, small interfering RNAs, long noncoding RNAs, and more—is *the* big story of the genome. There may be more than 100,000 of these "RNA-only" genes (which are conceptually distinct from the original principle of gene as protein producer), more than four times the number of "traditional" genes. In 2007, the *Economist* featured RNA on its cover with the story "Biology's Big Bang: Unraveling the Secrets of RNA."[4] This headline was an outgrowth of the large-scale, government-funded ENCODE project, a collaboration of thirty-five groups in eighty organizations around the world, which focused on developing an *Encyclopedia of DNA Elements* (a reach for an acronym).

Thus our sense of the operational aspects and function of the DNA sequence has been radically altered from the dogma of decades ago—far fewer genes than expected, outnumbered almost a hundred to one by the regulatory complex that controls it. The system seems to be just the opposite of our banking system and its regulation, in light of the collateral debt obligations and financial meltdown. With our genome, governance is hyperemphasized. This positions the genome to make a person's biologic "meltdown" most unlikely, although not impossible.

Our DNA sequence comes in two categories, or versions: our germ-line DNA, representing the master DNA of the egg and sperm that created each of us and led to the formation of the cells in our body, and the copied, derived DNA, known as somatic DNA, found in our cells. The somatic DNA can develop mutations as cells replicate, as anyone familiar with cancer would know. It turns out, however, that not all mutations associated with cancer or other diseases are the cause of the condition; those that do are known as "driver" mutations, while others, known as "passenger" mutations, simply get dragged along by virtue of being physically proximate to the cause of the disease. The complexity here is that if one is looking for something that has gone biologically awry at the DNA level, it is necessary to zoom in on the DNA from the relevant cells. Accordingly, blood-derived (from white blood cells) or saliva DNA is typically assumed to be representative of germ-line DNA. But in a patient with a blood-borne malignancy, such as leukemia, this would not be the case.

For example, if we are trying to determine the root cause of a heart problem that was genetically based, we might look at the DNA from both the germ-line and somatic heart cells. In cancer, one might directly compare the DNA sequence from the tumor somatic cells to the germ-line sequence, known as paired sequencing.[5] This gets even more complex because in any tumor, there is heterogeneity of the DNA sequence, such that cells in different parts of the cancer tissue may have a different set of mutations. In another example of a brain disorder that may be genetically based, one would have to look at both the germ-line and brain tissue to determine whether any somatic mutations might account for the condition. To compound the complexity, genomic regulation varies at the local level. Micro-

RNAs, RNA transcripts, small RNAs, gene expression, and epigenomics (discussed below) all exhibit tissue and cell-specific effects. Not really Genomics 101!

What's more, the human genome varies from person to person. For any particular base in the genome there might be a difference in one person compared to another. The simplest and most common form of variation is known as a single nucleotide polymorphism (SNP).[6] Each variant is known as an allele. Each time another person's DNA is fully sequenced, hundreds of thousands of new SNPs are discovered. Genomic scientists around the world are asked to deposit any validated SNP (replicated and cross-checked for accuracy) into a publicly accessible database called dbSNP.

The large international project known as the HapMap (short for "haplotype map") was set up to determine the common human SNPs. In this project, 269 individuals representative of three major ancestries (African, Asian, and European) had their DNA genotyped to determine common SNPs, defined as occurring in more than 5 percent of the population.[7]

A haplotype is a key unit of the genome, a "bin" or string of bases that has been inherited as a unit. The theory behind the HapMap was that if one could identify common SNPs, this would serve as a locator or zip code directory for the genome.[8] The bin or block has SNPs or alleles that are in "linkage disequilibrium" (LD), a somewhat misleading term that actually denotes that they are linked together. So if you have information about a zip code of the genome, tagged by a common SNP, you could explore which haplotype (bin) is associated with any particular condition of interest (like eye color, height, or a disease). In the average individual of European or Asian ancestry, there are over 500,000 LD bins or haplotypes. With African ancestry there is much more diversity, reflecting that Africa is the origin and extended period of evolution of human beings, so that there are approximately one million zip codes. Thus, in order to "tag" the human genome, you would need to genotype at least 500,000 common SNPs, assuming you had at least one tag per zip code.

SNPs are not the only form of architectural or so-called structural variation in the genome.[9] There are about one-tenth as many insertions or deletions, known collectively as "indels," as there are SNPs.[10] (See Figure 5.1.) This simply means that a base or series of bases have been added or deleted (for example, in the figure, out of the four bases, G, A, and T are deleted).[11] There can be block substitutions or inversions, as depicted. An important type of structural variant is known as copy number variation (CNV). Here quite a large block of the genome, which can even represent millions of bases, can be duplicated or missing in one individual versus another. The

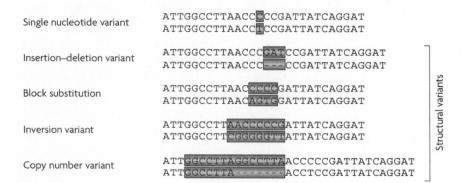

FIGURE 5.1: Types of variation that occur in the human genome. The bottom panel for each category provides an example of that variation compared with the top sequence of single nucleotide variant. Source: K. Frazer, "Human Genetic Variation and Its Contribution to Complex Traits," *Nature Reviews Genetics* 10 (2009): 241–51.

location of each of these variants could be determined relative to the original T substitution for a C.

A PEEK INTO THE GENOME (GWAS)

One important application of haplotyping is in genome-wide association studies (GWAS). These studies became possible because of jumps in the technology of genotyping. In 1997, we could only genotype one SNP at a time; by 2007 we could genotype one million SNPs in an individual using chips and automated robotic systems. GWAS is more than 99.99 percent accurate in correctly identifying the base (A, C, T or G) at a specific location. This is a remarkable feat of the technology, considering that the machines are handling a million SNPs, and it can be checked via other platforms that are designed to assess small numbers of genotypes with the highest level of accuracy. Coupling this technology with the zip code (HapMap) approach ushered in the GWAS era.

The first GWAS was published in April 2005.[12] That study was investigating age-related macular degeneration (AMD), the most common cause of blindness, which affects more than seven million Americans. The researchers examined 116,204 SNP tags in the haplotypes of ninety-six patients who had AMD and fifty suitable controls who were not affected. They found that variation in one specific zip code, a haplotype in a gene known as complement factor H (CFH) located on chromosome 1, was associated with more than sevenfold higher risk of being affected by AMD during one's lifetime.[13] Sequencing this bin of the genome pinpointed the variation in an exon—a simple change coding for the amino acid histidine instead of tyrosine—as a root cause of the disease. Of note, three independently conducted and reported studies replicated these findings.[14]

It was a stunning success for genomic science, with a limited number of patients and a minimal number of tag SNPs (at least 250,000 had been thought necessary for a cohort of European ancestry). The 700 percent increased risk of AMD disease susceptibility for the common SNP was striking, and a functional, culprit SNP was pinpointed by sequencing the incriminated genomic locus. Until this point, all we knew was that patients with the disease had inflammation of their retinal tissue, but there are thousands of genes that might have been responsible for such inflammation. A common, complex serious condition had been cracked by GWAS!

The term "complex" here is an important one. Before genomics, the only genetic diseases we had cracked followed simple Mendelian inheritance and expression patterns. They could be autosomal dominant (meaning only one copy on chromosomes 1 through 22 was necessary), autosomal recessive (requiring two copies for the disease to manifest), sex-linked (on either the X or the Y chromosomes), or mitochondrial (found in the power plants of our cells, which have their own DNA, reflecting their bacterial origins). These diseases are rare in the population and certainly not simple to understand at the genetic level. Examples are cystic fibrosis, Huntington's disease, and Tay-Sachs disease; these and more than 2,000 others have been catalogued by the Online Mendelian Inheritance in Man database.

Besides being rare or common, there are critical differences between Mendelian diseases and complex traits. For Mendelian diseases, typically a mutation or different mutations in a single gene accounts for the disease. If a person has the mutation, it is highly likely that the disease will develop—a so-called highly penetrant mutation and a more deterministic story. On the other hand, each complex disease is caused by many different genes. These diseases do not follow a classical Mendelian inheritance pattern from one generation to another, and the variants responsible for them each have relatively low penetrance; their pattern of expression is probabilistic. Thus, in the complicated case of macular degeneration, we can only talk about the identified haplotype being associated with a higher likelihood, instead of a certainty, of developing the disease.

Although the macular degeneration study became a cause célèbre for the genomics community, there are two clear caveats lest we get too carried

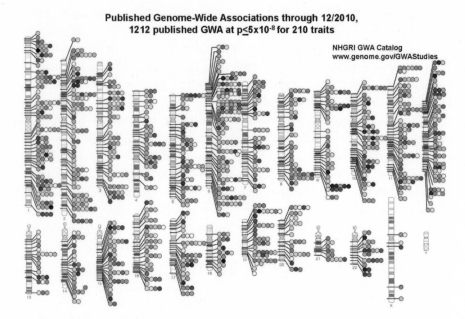

Published Genome-Wide Associations through 12/2010,
1212 published GWA at p≤5x10⁻⁸ for 210 traits

FIGURE 5.2: Output of the genome wide association studies (GWAS) from 2005 to 2010. The complex traits' "zip codes" are shown schematically by chromosome and location on each chromosome. Source: L. A. Hindorff et al., "A Catalog of Published Genome-Wide Association Studies," Office of Population Genomics, National Human Genome Research Institute, National Institutes of Health, n.d., www.genome.gov/gwastudies.

away. The first is simply that it was something of a lucky strike. Hundreds of subsequent GWAS of complex, polygenic diseases later, it has become clear that, with a few notable exceptions, lone haplotypes rarely signify such elevated risk for a specific disease. The second is that even one million SNPs represents only 0.03 percent of the genome, so this is simply a peek—even when we apply the haplotype approach. Nevertheless, GWAS has provided a veritable avalanche of data, the likes of which we have never seen in the history of genetics. In the early phase of GWAS reports, my colleagues and I published "The Genomics Gold Rush," about the unprecedented chain of discoveries and excitement in the field.[15] This era was defined by a major publication and discovery of disease susceptibility genes nearly every week. The papers started to appear in mid-2007 and have continued to appear in *Nature, Science, Nature Genetics*, and the *New England Journal of Medicine* (four of the highest impact journals in biomedical research) almost every week since. Figure 5.2 shows more than 1,200 GWAS studies with significant results published in leading peer-review journals; the genomic underpinnings for more than two hundred complex traits (predominantly diseases) have been mapped to the chromosome (and specific zip code) region.[16]

GWAS represents a unique type of science that is hypothesis-free. Rather than postulating that a particular gene or panel of gene "candidates" is related to a disease of interest, GWAS represents an unbiased approach that lets the genomes of people "talk." To start the hunt, there are no candidates, no nominations; if the results hold up, certain regions of the genome are elected and statistically incontrovertibly associated with the condition of interest.

Hypothesis-free research has been stunningly effective. Most of the genomic regions that have been discovered have never been previously theorized to have anything to do with the disease, such as the CFH gene in macular degeneration, FTO gene for obesity, TCF7L2 gene for diabetes, and a long list of examples.[17] GWAS has also revealed that multiple genes in a particular pathway may be involved. For example, Crohn's disease, a debilitating disease of the small intestine, can be the result of problems with "autophagy," or the self-digestion of our own cells.[18] Now through amassing serial, multiple large cohorts of individuals with and without Crohn's disease, with cumulatively over 15,000 cases and 14,000 controls without the condition, more than seventy susceptibility loci have been identified.[19] Only a few of these loci are related to the autophagy defect, reinforcing that there are clearly many molecular forms of what has been diagnosed as the same disease.

For so-called Type 2 diabetes mellitus, genes implicate several different pathways.[20] There could be problems with the production of insulin, the secretion of insulin, the transport of insulin, or the reception of insulin. GWAS may ultimately help us more precisely determine the molecular basis of the disease in a particular person. Hypothesis-free GWAS has also revealed that the same genes may be implicated in multiple diseases. In Type 1 diabetes, thought to be an autoimmune process, nineteen of the first twenty-six genes GWAS linked to this disease were immune regulation genes.[21] Surprisingly, gene variants indicating diabetes susceptibility were also implicated in prostate cancer,[22] for example, as well as in multiple other cancers and multiple other autoimmune diseases.

Nevertheless, GWAS has many deficiencies. For example, over 80 percent of the genome loci incriminated are not in exons. And simply knowing the zip code that is implicated is far from pinpointing the mechanism for the disease susceptibility. Furthermore, the macular degeneration case notwithstanding, the actual culprit or functional SNP variant(s) has been left undefined in almost all diseases. Most of the zip codes are only denoting a small risk, typically in the range of 10 to 20 percent, and thus are not particularly strong signals. Importantly, association with a disease is a different relationship from predicting a disease. The susceptibility figures are derived from large populations, so trying to use the data for any particular individual is a jump in extrapolation. Virtually all of the GWAS represent a snapshot in time, a binary yes or no of cases versus controls, such that any putative risk is for the lifetime of an individual rather than an assertion that something might happen at a particular age. Beyond these many uncertainties, there are tens of SNP variants that have been associated with each of a variety of complex traits: more than seventy for Crohn's disease, more than a hundred fifty for height, more than fifty for Type 2 diabetes, and so on.[23]

Most of these variants have a statistical association that holds up under multiple comparisons: the probability that the findings are random artifacts rather than true findings is on the order of 1 in 100 million (this is known as the P-value of a finding). The clinical value, however, is much more suspect. It is even more complicated since we do not know in most cases if the gene variants interact, such that there might be an additive or multiplicative effect (not to ignore the possibility of subtractive interactions) if someone carries two SNPs of different implicated genes. This phenomenon of gene-gene interaction is known as epistasis and represents a gaping hole in our understanding of the dynamics of the human genome. This notion—that it is rare for an isolated gene to be the sole major player in the output of a series of genes—is the underpinning of systems biology, which studies the interactions of networks of genes. These networks are a complicated affair; they may at times require perturbations in multiple elements of a pathway or a "node" in the pathway for the outcome to be meaningfully different.

Moreover, in multiple studies comparing GWAS with traditional prediction by analyzing family histories of diseases, GWAS has provided little information that we didn't already know.[24] Early studies in coronary artery disease and atrial fibrillation, in particular, highlighted this disappointment. Good old family history generally provided as much predictive capacity as genotyping in these conditions. Once again, this emphasizes the differences in advancing our understanding of the biology of diseases compared to prediction power of a "heredity horoscope."[25]

The difficulty of using GWAS data to try to predict susceptibility is in part related to the problem that the findings have only explained a small portion of heritability of the disease of interest. Heritability is usually defined by studying the differences between identical (monozygotic) and fraternal (dizygotic) twins. Most common diseases have an important heritable component, but GWAS has typically only explained about 10 percent of the story.

The unexplained 90 percent has become known as "missing heritability"[26] or, borrowing a term from cosmology, the "dark matter" of the genome. The prevailing thought before GWAS and the HapMap was that common variants would explain common disease. The macular degeneration study was an early fluke: because it accounted for most of the disease's heritability, it was the source of undue confidence. But if we define "common" to be occurrence at the 5 percent frequency level, SNPs do not largely explain the heritability of common diseases.

TEN YEARS LATER: THE ANNIVERSARY OF THE FIRST HUMAN GENOME SEQUENCE

Given the difficulties facing GWAS, it might not be surprising that in stark contrast to the exuberant coverage of the first draft human genome sequence in June 2000, the media handling of its ten-year anniversary in June 2010 was distinctly negative and sobering. The headline of the *New York Times* was "A Decade Later, Gene Map Yields Few New Cures; Medical Uses Limited; Despite Early Promise, Diseases' Roots Prove Hard to Find."[27] In its own editorial "The Genome, 10 Years Later," the *Times* wrote: "The task facing science and the industry in coming decades is as least as challenging as the original deciphering of the human genome."[28] In *Scientific American*, Stephen Hall's article "Revolution Postponed" stated that "the Human Genome Project has failed so far to produce the medical miracles that scientists promised."[29] *Fast Company*'s feature "The Gene Bubble" opened with the following: "When the human genome was first sequenced nearly a decade ago, the world lit up with talk about how new gene-specific drugs would help us cheat death. Well the verdict is in: keep eating those greens."[30] Victor McElheny's book *Drawing the Map of Life* emphasized the "struggle for medical relevance."[31] In the *Wall Street Journal* Matt Ridley opined in "The Failed Promise of Genomics" that "it's a curious fact that genomics has always been sold as a medical story, yet it keeps underdelivering useful medical knowledge."[32] *USA Today* was a bit more sanguine with "The Human Genome: Big Advances, Many Questions."[33] Interestingly, the *Economist* had a distinctly upbeat interpretation of the decade after the Human Genome Project. In a special report entitled "Biology 2.0," it proclaimed, "biological science is poised on the edge of something wonderful." In response to widespread disappointment elsewhere, the *Economist* declared, "Genomics has not yet delivered the drugs, but it will."[34]

The article "Major Heart Disease Genes Prove Elusive," published in *Science*, quoted me as saying, "At the end of the day, we have a bunch of loci and genes, but none of them do all that much to raise the risk of heart disease. Nor have they yet altered our understanding of how the heart falters— knowledge that will take time to develop." Another cardiologist researcher from the Gladstone Institute, Deepak Srivastava, was even more pessimistic: "People did studies with 300 or 500 people and didn't find anything, then did 1000 and didn't find anything . . . in retrospect, GWAS wasn't worth the expenditure."[35]

But virtually all of these articles missed out on a major wave of GWAS discoveries with the potential for immediate impact on the practice of medicine. These were not related to disease susceptibility findings but in fact GWAS probes of which genes were responsible for interacting with prescription medications—the field of pharmacogenomics.

PHARMACOGENOMICS

Hepatitis C is one of the most significant global health problems, affecting about 3 percent of the world's population, or approximately two hundred million people.[36] It is the leading cause of cirrhosis and cancer of the liver. The standard treatment for hepatitis C is to eradicate the virus by treatment with PEG-interferon-α, combined with ribavirin, and given for a year. Treatment in the United States costs $50,000. It makes almost everyone who takes it ill, with symptoms of a flu-like condition. What's worst about it, however, is that the drug only works in 50 percent of the people who take it, being generally more effective in those of European as compared with African ancestry.

In 2009, three different research groups did a GWAS to see if the drug response could be predicted—and it could.[37] A major signal was found at the gene IL28B with a single nucleotide variant that accounted for a twofold likelihood of a therapeutic response.[38] It also explained the disparity between ancestries, and the action of the drug fit precisely with a protein coded by the IL28B gene, also known as interferonλ3, which attacks pathogens. It was striking how small the number of patients was needed to generate this discovery—in one study, in Japan, GWAS assessed only sixty-four responders and seventy-eight nonresponders![39] Genotyping IL28 SNPs could, at least theoretically, be immediately used to help predict which patients were likely to respond to conventional therapy. Between some newly approved drugs and more than twenty other drugs being actively pursued for Hepatitis C therapy, there are many other choices for those individuals who are not likely to respond to PEG-interferon-α.

Actionable Information

The PEG-interferon-α example, unlike that of macular degeneration, turned out to be quite representative of the fates of many other pharmacogenomic genome-wide (genome peek) studies that followed. One involved Plavix, the trade name for clopidogrel, the second largest prescription drug in the world with over $9 billion in sales in 2010. It blocks a platelet receptor known as $P2Y_{12}$, which is an important mediator of the aggregation of platelets in clots. For many years, however, medical professionals were aware that the drug had quite variable effects in patients. Plavix worked well for some but in others had almost no effect.

As was well known at the time, the drug must be metabolized by the liver before it becomes biologically active. In 2006, a study in healthy volunteers showed that variants in the gene CYP2C19, which was involved in activating Plavix in the liver, played a role in the inconsistent effects of the drug.[40] A couple of years later multiple groups published definitive evidence that among patients who were having a stent placed in their coronary artery, the risk of the stent clotting was increased threefold if the patient carried a loss-of-function variant of CYP2C19.[41] And it was duly noted that these loss-of-function variants, particularly one known as the *2 allele, were extremely common—found in more than 30 percent of individuals of European ancestry, 40 percent of individuals of African ancestry, and about 50 percent of those of Asian ancestry.[42]

Stents are the stress test for a platelet blood clot, since this foreign body of metal placed in the artery attracts platelets on its surface. Patients are routinely given aspirin and Plavix together for several months, or even years, to prevent a clot in the stent. Most stent clots result in sudden death or a heart attack, so they can be considered a real catastrophe. Fortunately, although two million coronary stenting procedures are performed worldwide each year, stent clotting occurs only in 1 to 2 percent of patients.[43] But they are the same people who are much more apt to have the CYP2C19 variants that cannot normally metabolize or activate Plavix. And 1 percent of two million people represent a lot of heart attacks or fatalities. So the stakes are quite high with Plavix pharmacogenomics, even if it does not involve hundreds of millions of people like hepatitis C.

Before a GWAS was performed, the investigators of Plavix response had been using a candidate gene approach, genotyping people for CYP2C19 because they considered it a likely explanation of some of the heterogeneity

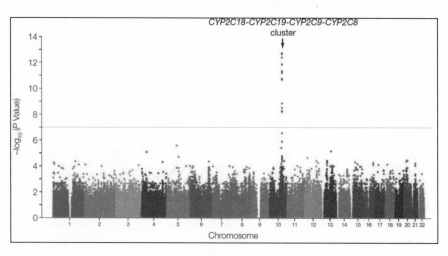

FIGURE 5.3: Genome-Wide Association Study of Plavix is shown schematically via a "Manhattan plot," which looks for skyscrapers, and in this case the cytochrome cluster, responsible for metabolizing Plavix, was the only common variant found. Source: A. Shuldiner, "Association of Cytochrome P450 2C19 Genotype with the Antiplatelet Effect and Clinical Efficacy of Clopidogrel Therapy," *Journal of the American Medical Association* 302 (2009): 849–58.

of Plavix response. But in 2009, Shuldiner and colleagues published a Plavix GWAS.[44] Their paper included a Manhattan plot (see Figure 5.3) on which the x-axis represents the position on all the chromosomes and the y-axis representing the P-values (actually the $-\log_{10}$ P, if one wants to get technical) of more than 300,000 SNPs throughout the genome. The phenotype—the clinical feature assessed—was the extent of platelet aggregation inhibited in more than four hundred individuals who had been treated with Plavix for a week. Such graphs are called Manhattan plots because they look something like the skyline of New York City. In this case, the GWAS revealed a lone "skyscraper" of many significant SNPs at a particular spot or locus in the genome, and it was in the CYP2C19 gene cluster.[45] This certainly didn't account for all of the Plavix variable response, but it anchored the multiple studies that had preceded it as the most important common gene variant influencing Plavix response.

This finding has had clinical applications in at least two places, Scripps and Vanderbilt University. At both hospitals, patients who are to undergo stenting are screened for loss-of-function variants founding the CYP2C19 cluster. If such variants are identified, patients can be switched to other medications (which do not rely on functional copies of these genes for activation) or given higher doses of Plavix, to which some patients respond.

Several other examples of GWAS have demonstrated the key genes involved in a drug's therapeutic action. Warfarin is a commonly used blood thinner, with over twenty million prescriptions in the United States per year.[46] The drug helps to prevent stroke in patients with artificial heart valves or atrial fibrillation and the formation of blood clots in patients who have experienced deep vein phlebitis and many other blood-clot disorders. A GWAS for warfarin effect has shown three major genes that are critical to the drug's response. One is VKORC1, which encodes an enzyme that enables vitamin K to play its role in blood clotting (warfarin inhibits the action of the enzyme). The other two are cytochromes, CYP2C9 and CYP4F2, which are active in the metabolism of the drug in the liver.[47]

Patient response to warfarin is remarkably variable. Some patients require only 1 mg per day, whereas others require 20-mg doses. Genotyping up front could help avoid inadvertent underdosing and the possibility that a blood clot would form, and also reduce the likelihood of bleeding from overdosing. Other GWAS studies indicating gene variants that modulate drug efficacy are metformin, the most commonly used drug to treat Type 2 diabetes, and methotrexate, which is used to treat cancer and autoimmune diseases. Studies investigating the efficacy of routine genotyping have so far produced mixed results and have left the prospects for genotype-guided dosing unsettled to date.

GWAS of drugs for key side effects is the other side of the remarkable progress that has been made in this field. The Centers for Disease Control reports that almost 7 percent of hospitalization in the United States each year are related to adverse drug reactions.[48] Take the treatment for hepatitis C: it can cause a hemolytic anemia in about 15 percent of patients. GWAS has revealed variants in the ITPA gene that substantially account for the risk or protection from this side effect.[49] Statins, which are used to treat high cholesterol and are the most commonly prescribed drug class in the world, have the primary side effect of muscle inflammation. Common variants in the gene SLCO1B1, which is involved in the liver uptake of statins, are critical— patients with two copies of the variant carry an over twentyfold higher risk of developing severe muscle inflammation.[50] GWAS revealed an allele influencing worrisome reactions to the antibiotic flucloxacillin (Floxapen), which can cause liver toxicity. A variant known as HLA-B*5701 carries an eightyfold risk of liver injury.[51] In the statin and antibiotic GWAS, only eighty-five and fifty-one cases, respectively, were needed to cinch the findings![52]

Likewise, GWAS revealed the underpinnings of bad reactions to Carbamazepine (Tegretol), a drug used frequently for many neurologic conditions such as trigeminal neuralgia, epilepsy, diabetic neuropathy, and migraine headaches. The primary side effect of concern with this drug is a serious allergic reaction that can range from a skin rash to a life-threatening necrosis of skin throughout the body. In 2011, the risk allele for this side effect in individuals of European ancestry was discovered by GWAS with just twenty-three patients, and the HLA allele variant (HLA refers to the major histocompatability complex harboring much of the immune system genomic apparatus) carried over a twentyfold greater risk for severe skin splitting, or dehiscence, known as toxic epidermal necrolysis. Routine genotype screening in Taiwan (testing a different HLA risk allele in individuals of Asian ancestry) for all patients getting Tegretol prescriptions has been shown to dramatically reduce the risk (to zero of more than 4,400 individuals treated).[53]

Another drug, the powerful anti-inflammatory cyclo-oxygenase-2 (cox-2) inhibitor drug lumiracoxib (in the same drug class of Vioxx and Celebrex), which had been marketed as Prexige in many countries, had to be taken off the market because of rare but severe liver toxicity. A GWAS showed that an HLA gene variant (HLA-B*5701) carries a fivefold risk of liver injury with this drug.[54] Such insight could even pave the way to bring this drug back—a "rescue"—with genotyping performed in advance to screen out individuals with high risk of liver damage.

Despite these successes, most drugs have not had a GWAS investigation. Even so, more than 25 percent of commonly used prescription medications have some genetic information that can be useful for guidance.[55] Many drugs used to treat cancer fall into this category: abacavir (Ziagen) (interacts with the same HLA allele as lumiracoxib, HLA-B*5701), 5-flourouracil (Efudex), Irinotecan (Campostar), azathioprine (Imuran), and 6-Mercaptopurine.[56] Other notable examples of drugs with strong genetic data (albeit not via GWAS) to guide selection or dosage include beta-blockers for heart failure; cisplatin, which can have the side effect of hearing loss in children; tamoxifen, which is used in treatment of breast cancer; metformin, which is prescribed for diabetes; and succinylcholine (Anectine), which is used to relax muscles during anesthesia.[57]

The stark contrast between the success of GWAS in predicting drug reactions, and the relative lack of success in identifying disease susceptibility, seems to be a result of the actions of natural selection. Whereas there has been inexorable evolution of humankind in response to many diseases over hundreds of thousands of years, the individual's exposure to drugs can be considered the "new, new thing." There hasn't been a chance for selection pressure in the genome for adaptation to drugs. Finding key gene variants that influence the response to a medication can be likened to hitting the side of a barn—an easy target. This dichotomy between disease-susceptibility genes and drug effects of genes, however noteworthy, should not imply that we'll never get better at predicting disease susceptibility. After all, our knowl-

edge of the variations in genes that drive both disease and side-effect susceptibility is far from complete. The obvious next step would be to go beyond peeking at genomes and look at every base, or at the very least every
base of particular regions of the genome. Getting more granular, deep into
the DNA sequence, will move the field forward.

PIVOTING TO SEQUENCING

We know now that the "common variant, common disease" theory does not
largely explain heritability of complex traits and diseases, so genomics has
turned to much rarer variants, digging down to the 0.1 percent and even
lower allele frequency levels. Much work had indicated that such lower frequency variants would have a higher rate of penetrance. For example, for
"good" high-density lipoprotein (HDL) levels in the blood, it has been shown
that multiple rare variations in several genes, in aggregate, largely explain
the relatively common trait of low HDL. Rare variants with high penetrance
have been found in severe obesity, Type I diabetes, schizophrenia, and many
autoimmune diseases. Of note, some of these variations are not in SNPs but
instead are structural—either deletions or copy number variations (CNVs)
(see Figure 5.1, p. 81). This points to one of the other issues in missing heritability. While SNPs are the most common form of human genomic variation, indels, CNVs, and other structural variations are critical, too, and have
not been adequately emphasized. SNPs in a GWAS can serve as a marker
for some of the structural variations, particularly CNVs, but they are quite
incomplete. The full map and disclosure of these structural variations has
to rely on whole-genome sequencing of thousands of individuals with the
condition (phenotype) of interest.

A good comprehensive sequencing investigation should be, like GWAS,
hypothesis-free. There are two other hypothesis-free genomic approaches:
exome sequencing and whole-genome sequencing.

EXOME SEQUENCING

Exome sequencing refers to sequencing the 1.5 percent of the genome that
actually codes for proteins. This process enables us to detect whether there
are functional variants—so called because they affect the structure and

action of the encoded protein—that contribute to disease. Given its size, it is much easier to work with the exome than the whole genome in order to sort out the needles from the haystack; the former might have tens of thousands of variants, whereas the whole thing can have half a million or more. The trick is finding the one or ones that are functional. The best means of determining whether a genomic variation is functional is to breed mice with the variant (the so-called orthologous mutation, or the equivalent of the human genomic change in the mouse genome) and see whether the mice recapitulate the disease phenotype. But lesser forms of support or proof are becoming acceptable, such as the *in silico* computer-based prediction of whether a variant in a coding element would significantly change the protein structure or binding properties (e.g., by disrupting an enzyme's catalytic site— essentially the business end of the stick). For the rest of the genome outside of the exons, this task is remarkably more complicated—we don't yet have the tools to predict likely functionality of genomic regulatory changes.

The relatively low cost and quick completion of exome sequences, as compared to whole-genome sequences, has recently made the exome the hot, go-to method for cracking diseases. As a result, the previously unknown root causes of a variety of rare, Mendelian diseases were determined via exome sequencing. Several cancer types, such as ovarian clear cell and uveal melanomas, had key mutations determined via exome sequencing. The discovery efforts extended to unexplained mental retardation, severe brain malformations, and even adaptation to high altitude among Tibetans.[58] Thus, there was a remarkable body of progress in just the first year of exome sequencing, with genomic science groups all over the world in hot pursuit. A second phase of the genomics gold rush was officially under way.

The exome is not without shortcomings, however. This approach is not truly "hypothesis-free," since for it to work it requires that important variations exist in the exome. To really solve the problem of inherited disease, the full monty would have to be pursued—whole-genome sequencing on a large scale was the inevitable next step.

WHOLE-GENOME SEQUENCING

As a colleague of mine and I wrote in 2007, "Ultimately, when whole genome sequencing is practical and affordable, it will be increasingly difficult for the genomic basis of health and disease to be left undetected."[59] We aren't at the practical and affordable ideal yet, but the progress that has been made in the past few years has far exceeded virtually anyone's expectations.

The initial efforts in the 1970s were essentially manual, done by tagging bases with radioactive isotopes. Gel-based systems arrived in the 1980s and were automated by 1990; even at that point, however, they could only sequence fewer than 10,000 base pairs per day at a cost of $10 per base. By the mid-1990s capillary sequencing took root, and the method gradually increased the output from 15,000 base pairs per day to over one million, and the cost dropped to about $1 per base called.[60] What has put exome and whole-genome sequencing in reach of people like Nicholas Volker's doctors are the "massively parallel sequencing" machines, which rely on simultaneously reading hundreds of thousands of short stretches of sequence. This has raised output from twenty million per day (on the 454 Life Sciences platform) in 2005 to twenty-five billion (Illumina HiSeq and Life Technologies SOLiD 4) in 2010. In just five years, the speed increased a thousandfold while the price per base fell from $0.01 to $0.000001.[61] The rate of improvement has left Moore's law for computing in the dust.[62]

Considering the required time and cost for some of the landmark projects helps to add perspective. The first human genome sequence, which actually represented a hodgepodge of multiple individuals, took thirteen years and $2.7 billion. In 2007, Craig Venter's genome took four years and about $100 million. The Watson genome in 2008 took only four months and just $1.5 million.[63] By November 2008, multiple human genomes were sequenced in one to two weeks for less than $100,000. In 2009 Stephen Quake, a professor at Stanford, sequenced his own genome in a week for less than $50,000.[64] By 2010, one of the leading genome science companies, Illumina, offered whole-genome sequencing for $28,000, and the newcomer Complete Genomics announced that they would soon perform whole-genome sequencing for $5,000. By the end of 2009 they published in *Science* multiple human genome sequences and declared "the high accuracy, affordable cost of $4,400 for sequencing consumables, and scalability of this platform enable complete human genome sequencing for the detection of rare variant in large-scale genetic studies."[65] By the end of 2011, Complete Genomics was sequencing approximately 1,000 whole human genomes per month.

This high throughput is amazing, but it is only part of what's necessary to make whole-genome sequencing clinically important. The other side of the coin is accuracy, and indeed, that is what has kept the Archon X prize of $10 million—to be given to the first team to sequence a hundred human genomes in ten days—from being awarded.[66] The accuracy of sequencing platforms is dependent on the depth of coverage, depth being a measure of how many times an average base is read during sequencing. If the average is forty times, this is commonly regarded as a deep sequence, although you should bear in mind that, because it's an average, some bases are going to be read a hundred times, and others only ten. At some point there appears to be a "saturation" point at which further depth does not improve accuracy to any measurable or substantial degree.

The other key metric about sequencing is the length of the read. This was not much of a problem in the capillary sequencing era, when the classic Applied Biosystems machine had read lengths approaching 1,000 bases. Massively parallel sequencing started with read lengths of less than 40 bases, and these days it gets up to a few hundred. Unfortunately, short read lengths are unhelpful and can even be misleading when looking for structural variations such as indels, CNVs, inversions, and the like. Remember that we are trying to pick up rare and very low frequency genomic variants, occurring at a frequency of 1 percent or far less in the population. If you are looking for a variant that occurs in only 0.1 percent of the population, and the accuracy of the sequencing is 99 percent, the 1 percent false reads of 3 billion base pairs will yield three million false positive variants. In such a case, the number of correctly identified rare variants would also be about three million, meaning that 50 percent of all positives would be false positives. Sorting these out would be a mess. Accordingly, everything that can be done to improve depth of coverage and read length, and promote as close to 100 percent accuracy as possible, is required to make sequencing and the downstream interpretation of the data practical. Despite the extraordinary reductions in cost, we aren't quite there yet. But that doesn't mean the competition to get there isn't fierce.

The level of competition was exemplified by a scene I witnessed at an annual meeting about genomic sequencing held in February 2008 on Florida's Marco Island. Every genomic sequencing company gets fired up for this meeting, and they hold nothing back to imprint themselves and their products on the memories of attendees, of which there are more than a thousand. A different sequencing company—Illumina, Life Technologies, Pacific Biosciences, or Helicos—sponsored each meal of the conference. Pacific Biosciences, a company that until that point had been operating in stealth mode, was treating the event as a "coming out" party, with an evening of fireworks and celebration on the beach. Even my hotel room key had the name and logo of one of the sequencing companies on it.

I participated in a panel, sponsored by Pacific Biosciences, on how much sequencing capacity would be needed to make the biggest advances in medical genomics. As usual, the emphasis was on more done in less time; at the meeting, the company's representatives spoke about sequencing a whole human genome in less than fifteen minutes at some point in the future. (Three years later, however, they had not yet sequenced one whole human genome.) I think the panel was missing a fundamental issue, and it occurred to me while I was there. Everyone was saying the usual things about accuracy, coverage, read length, throughput, and cost. But what, I wondered, about the people and patients who would be sequenced? In order for us to make any substantive advance on missing heritability, we need to whole-genome sequence large numbers—thousands—of people with a particular condition and compare the findings with suitable controls without the condition. What's more, we would need to make sure that the controls, to truly be controls, would *never* develop the condition. Otherwise, they wouldn't really be controls.

The fact of the matter was that no one was talking about that kind of study. The 1000 Genome project, an international, government-funded collaboration, was just getting under way. It sequences randomly selected, anonymous subjects whose health status is unknown, predominantly at a low depth of coverage. There was no plan at that time, or even a few years later, to tackle the challenge I envisioned, with just a few exceptions, such as a large sequencing project of thousands of individuals with Type 2 diabetes. A big reason is expense: even if one could go with the Complete Genomics or Illumina price of a whole sequence for less than $5,000 by 2011, any such study would still have $10 million or more in sequencing costs.[67]

As Kevin Davies wrote in his 2010 book, *The $1000 Genome*, the price of sequencing keeps falling.[68] But even if we were to reach a $1,000 genome, that would not capture all the costs. Having the sequence is not the whole story, and the rate-limiting step is the in-depth interpretation, which nowadays is estimated to cost several hundred thousand dollars. If we get thirty-times coverage of each person's genome, we would have ninety billion bases to sort through. These, of course, would be in little slices, which would then need to be assembled and compared with the human reference genome and annotated with respect to all of the known functional variants. Here is where the "quants" (the new breed of math whizzes) take over: major analytics, computational biology, and informatics are used to crank out the meaningful findings, moving from raw sequence data to real information. The typical human genome sequence yields around three million sequence variants compared with the reference genome. Approximately 100,000 SNPs are deemed "novel" (first discovered in the individual being sequenced), and of these some 15,000 to 20,000 occur in exons. As more and more individuals are whole-genome sequenced, the lower frequency

SNPs will be found and deposited in dbSNP, and the numbers of novel variants will progressively go down.

Clinically annotating the human genome is another big step. Stephen Quake's paper reporting the findings of his rapid self-sequencing adventure had three authors, and it reportedly took a week to obtain the raw data. But when his genome was annotated, it took thirty-one authors several hundred hours to scour through the genomic variants and all of the related literature to figure out which ones were clinically meaningful. His risks of diabetes, coronary artery disease, and obesity were noted, but of particular interest were the sixty-three predicted pharmacogenetic interactions, which included inability to metabolize Plavix, needing a low dose of warfarin, and unresponsiveness to beta-blockers and routine anti-diabetic drugs. Another rare variant was that he was a carrier for cystic fibrosis, of which he was previously unaware and might have been useful information in conjunction with similar data for his wife, since this is a Mendelian recessive trait requiring two copies of the rare variant. (Also of note was that Stephen had a cousin who had died suddenly at age nineteen, cause unknown. In 2011, it was announced that the cousin's DNA would be sequenced in the first "molecular autopsy.")[69]

So even though there is much we can learn from sequencing whole genomes, it has been—and remains—an expensive proposition. Hence its clinical applications have been limited. Thus far the strategy of whole-genome sequencing (WGS) has predominantly been applied to cancer. By sequencing paired tumor and germ-line DNA, the ability to find driver mutations has been demonstrated for a number of cancers. The first case was that of a leukemia patient, and a small subset of eight genes thought to drive the cancer was identified. To find them, the researchers sequenced ninety-eight billion base pairs from the leukemic cells (the genome being sequenced thirty-three times) and forty-two billion base pairs from germ-line (skin tissue was used, sequenced fourteen times).[70] Several solid tumors, including lung, breast, and pancreatic cancer, have since been sequenced.[71] In small-cell lung cancer, WGS identified signatures related to tobacco exposure.[72] WGS has been applied to a family of four individuals with a rare Mendelian trait known as Miller syndrome,[73] and it provided some incremental findings beyond exome sequencing, which had been previously applied for this same trait (a second gene mutation was identified). It has been used for finding the genetic defect in Charcot-Marie-Tooth in a particular family and applied to discordant identical twins with multiple sclerosis (one affected, one not).[74] In the latter study, surprisingly no explanation was found for this neurologic condition—yet another indicator of how complex the human genome has proven to be.

SEQUENCING TO SAVE A LIFE

The case of Nicholas Volker, touched on in the opening of this chapter, demonstrates the real potential of defining life codes. In medicine we often use the terms "idiopathic" or "cryptogenic" when we really mean "we don't know." Such was the case with this child, who would develop a fistulous connection from his intestine to the skin of his abdomen whenever he ate. This led to more than a hundred surgeries to repair the bizarre channels, a clinical condition that had not been seen before. Ultimately, when all hope was about to be given up, and Nicholas had been enduring protracted hospitalization, intermittently septic, and living in a hyperbaric chamber at times, his pediatrician at the Medical College of Wisconsin asked for Nicholas's genome to be sequenced.

The results were completely unexpected—a mutation in the gene XIAP, known to be pivotal in the activity of the immune system, was responsible. Happily, there was a way to fix it—an umbilical cord blood transplant to replace the stem cells responsible for producing white blood cells, one of the cornerstones of our immune response. Such an operation hadn't even been under consideration. This genomic knowledge and the decision it induced wound up transforming Nicholas from near death to a healthy five-year-old boy. Now at the Medical College of Wisconsin more than forty children in the queue are being considered for sequencing to define, at the molecular DNA level, why they are sick. A committee of physicians, geneticists, genetic counselors, and ethicists meets on a regular basis to consider whether a child qualifies for sequencing after conventional medical measures have been exhausted. And that same committee decides the priority of the patients, determining who will be sequenced next.

The Volker case was groundbreaking, and for the first time, it is possible to imagine a future medicine that has no use for the terms "idiopathic" or "cryptogenic." Many people go from one medical center to another because their condition has not been diagnosed, effective therapy cannot be found, and they are suffering. The fact that the committee in Milwaukee was able to secure reimbursement for sequencing in select cases is particularly intriguing, since one can certainly make a case that sequencing early in life could prove to be cost-effective over the long term: it would preempt extensive and expensive medical evaluations, not to mention totally ineffective treatments.

In 2011, whole-genome sequencing of fraternal teenage twins who had a serious movement disorder (dystonia) led to the precise molecular diagnosis and highly effective therapy.[75] The case of these twins, whose parents were healthy, is particularly instructive and representative because their causative mutation was due to a "compound heterozygote," meaning that mutations in two different bases in the same gene, one from the mother, one from the father, came together to induce the movement disorder. This is a common explanation for why a new condition appears without it ever having shown up in the prior generations, and it will certainly be a common finding in the sequencing era going forward.

A family in Ogden, Utah, had five children in two generations die from a mysterious accelerated aging condition that had never been seen before, but through sequencing this has now been pinpointed to a gene mutation on the X chromosome.[76] For this family, the Ogden Syndrome, as it is now being referred to, could be prevented by in vitro fertilization with selection of the embryo that does not carry this mutation.

The root cause of the rare Proteus syndrome, which is believed to be the basis of the disfigurement of Joseph Merrick, known as the Elephant Man, has now finally been identified as a gene mutation in AKT1, which is also a mutated gene in many cancers.[77]

Although encouraging and intuitively positive, we need to see many more cases like this before it can be considered a new path to demystifying serious, unknown medical conditions. Currently, the ability to sequence is way out in front of our ability to interpret the data. Annotating human genomes to accurately process and interpret the enormous data derived from each individual's DNA, separating the noise (changes in bases or structure that is irrelevant) from the signal (the culprit, functional variation) remains a formidable challenge. This will become less daunting when the genomes of hundreds of thousands of people, representing the full gamut of phenotypes, are sequenced and fully annotated.

THE RACE TO AMP UP SEQUENCING

The sequencing race is multidimensional. It is being waged across countries and continents, across the major companies and technology platforms, and throughout the academic genomic science community. The Beijing Genomics Institute headquartered in Shenzhen, China, has purchased more than 100 of the latest generation sequencing machines (predominantly HiSeq) and has projected completing 20,000 human genomes by the end of 2011.[78] North America and Europe, particularly the United States and the United Kingdom, have over 1,000 of the HiSeq or SOLiD sequencers and project about 10,000 human genomes will have been sequenced by the end of 2011. Complete Genomics, which like Pacific Biosciences became a publicly traded company in 2010 (as compared with Helicos, which was delisted in the same year), is currently sequencing thousands of human genomes each month and projects to be able to provide the life codes for a million people in five years. Their unique "mail order" WGS model is an interesting twist that preempts academic or life science industry laboratories from having to purchase the very expensive sequencers and proprietary reagents, along with the significant costs of specialized personnel. This model could make raw sequence determination a commodity, but time will tell if this is a financially viable model and whether the genome product will be widely accepted by the most accomplished academic sequencing centers.

Another new competitor on the scene in 2010 was Ion Torrent, which was acquired by Life Technologies and offers a smaller "desktop" sequencing platform that is not intended for WGS but might be especially applicable for medical sequencing, such as rapid sequencing of a bacterial pathogen to be able to predict whether a particular antibiotic would be effective (like MRSA, methicillin-resistant staphylococcus aureus, which can be a notoriously deadly infection). Ion Torrent's desktop platform was the first to be based on semiconductors, perhaps representing the most important shift of platform for sequencing. In 2011, the first human genome to be sequenced by transistors was Gordon Moore, one of the most prominent pioneers of semiconductors. Although it took about 1,000 Ion Torrent chips, the number of transistors required to sequence a whole human genome is expected to drop exponentially in the next couple of years. [79]

This represents the ultimate convergence between transistors and sequencers—DNA transistors. In December 2010, *Scientific American* published their "World Changing Ideas" top-ten list; highlighted on the cover was the emergence of DNA transistors as "innovations for a brighter future."[80] Unlike the cumbersome reagent costs and optical instruments needed to read fluorescent tags, this process is a radical redesign of genome sequencing: "Whereas existing dishwasher-size sequences require expensive chemical reagents to analyze genes that have been sliced into thousands of small fragments, the so-called DNA transistor takes an almost naively simple approach." Chris Toumazou, director of the Institute of Biomedical Engineering at Imperial College, London, is a principal advocate for using semiconductor technology to achieve human genome sequencing. He expects semiconductor technologies could enable the sequencing of a "full genome on a chip within a matter of minutes."[81] An already existing prototype of a portable sequencer is the size of a cell phone. If this results in a successful, functional technology, it would take the sequencing and medical genomics world by storm.

Because of all of these platforms, it has been projected that 250,000 human genomes will be fully sequenced by the end of 2012, 1 million by 2013, and 5 million by 2014. Clifford Reid, the CEO of Complete Genomics, projected in 2010 that within five years his company alone would sequence 1 million human genomes. So in the next five years, there ought to be more than enough WGS to digitize the life codes for most human conditions.[82]

PROGRESS IN CANCER GENOMICS

Cancer is a genomic disease. It can't develop without a change in the DNA sequence of the cancer cell genome. While the germ-line DNA may have susceptibility genes that predispose an individual to cancer, somatic cells have developed mutations that have overridden the normal DNA house-keeping repair processes. There is a remarkable spectrum of changes that can occur in the genomes of cancer tissue, ranging from point mutations to the structural variations I have been reviewing in this chapter (insertions, deletions, CNVs). One structural variant is particularly prominent and, in many examples, fundamental: chromosomal rearrangements.[83] These rearrangements involve broken DNA strands that have been combined somewhere else in the genome, either on the same chromosome (intra-chromosomal) or on another chromosome (interchromosomal).

Often these chromosomal rearrangements can have an activating role, whereby a "fusion gene" is created, combining two previously separate genes into a hybrid that in some cases markedly promotes growth and replication, fulfilling the "oncogene" functional category—a gene capable of inducing cancer ("onco" denotes tumor or cancer). Whereas fusion genes were first found in the "liquid" tumors—leukemias and lymphomas—they are now being recognized as occurring frequently in solid tumors, such as prostate and some types of lung cancer (adenocarcinoma). Besides the rearrange-ments, other types of genomic alterations in cancer include new DNA in-tegration, particularly from a virus, which can cause certain cancers; those involving mutations in the 17,000 bases of the mitochondrial DNA; and epigenomic changes, in side chains of DNA and the histones that package the DNA (discussed later in this chapter).[84]

Not to be discounted, environmental exposures such as tobacco or ultraviolet light can induce somatic mutations through many of these mechanisms and are an important part of the story. In parallel to the diverse spectrum in the kinds of mutations possible, there is a wide range in the numbers of mutations. A cancer might involve no rearrangements, or it might involve hundreds; likewise, there can be fewer than 1,000 point mutations or more than 100,000. In an individual cancer, it is thought there might be as many as 20 "driver" mutations responsible for taking the cells off track.[85]

Driver mutations are the functional force behind cancer and can confer resistance associated with recurrence of tumor growth. Out of 22,000 protein-coding genes, about 350 to 400 have recurrently shown up with somatic mutations driving the growth of a tumor. They are disproportionately from specific gene families, and one of the most important is the protein kinase.

By determining the importance of many of these protein kinase genes (such as epidermal growth factor receptor-EGFR, KRAS, and BRAF in particular tumors), there have been some striking advances in cancer therapy. Traditional chemotherapy relies on multiple toxic drugs (with or without radiation) that kill cells indiscriminately; the new therapies are much more specific. One of the most dramatic examples is the specific point mutation (V600E) in the gene BRAF, a driver mutation that is found in about 60 percent of patients with malignant melanoma. Despite multiple chemotherapies and radiation, nearly all patients with malignant melanoma die within one year of diagnosis. But now with an orally active BRAF mutation-directed drug that specifically binds the mutated protein, more than 80 percent of patients respond with rapid tumor shrinkage in only two weeks. The highly specific nature of this benefit is emphasized by the observation that patients not carrying the BRAF mutation actually get worse with the drug.[86] This is a prototypic example of individualized treatment of a particularly lethal form of cancer.

There are many other mutations with biologically based drug coupling that have led to enhanced efficacy of treatment. Two well-known drugs are Gleevec for chronic myelogenous leukemia, which targets a fusion gene, and Herceptin for breast cancer, which targets the HER2 estrogen receptor, but there are many more recent examples. The drug Erbitux (cetuximab), which was tied to the major media exposé involving Martha Stewart and the CEO of ImClone, Sam Waskal, is commonly used in colon cancers.

However, it is not effective if a patient carries a KRAS mutation. For a sub-type of lung cancer known as non–small-cell, the drug Iressa (gefitinib) has proven to be remarkably useful when there are activating mutations in EGFR. In the same type of lung cancer, patients who have tumors lacking a mutation in the gene ERCC1 have a marked benefit with cisplatinum chemotherapy. The drug Sorafenib inhibits the cancer gene RAF and has been shown to improve outcomes in patients with cancers involving this driver mutation—particular individuals with kidney, liver, lung, and thyroid cancer. For tumors that have an ALK gene fusion, which occurs in certain lymphomas and non–small-cell lung cancer, the experimental drug Crizo-tinib has been demonstrated to have a high response rate. In certain types of brain cancer, glioblastoma, the drug Temodar (temozolomide) is effective for when the MGMT gene is methylated. The mutated gene PIK3CA (phos-phoinositide 3-kinase, PI3K) has been shown to be present in a variety of tumors, such as ovarian, colon, certain brain tumors, and others, and a num-ber of drugs have been designed to target this cancer gene and are in the midst of clinical testing.[87]

With the ongoing efforts in cancer mutation screening and sequencing, the discovery of targeted cancer therapies is clearly accelerating. We are learning that the same driver mutation can be found in a diverse set of can-cers, such as BRAF in not only melanoma but also colon cancer and thyroid cancer. In one of the first documented cases of sequencing a tumor and tai-loring its therapy, Dr. Marco Marra of the University of British Columbia in Vancouver had a patient in his eighties with tongue cancer that had metastasized to his lung.[88] The sequence of the tumor found a driver mu-tation in the cancer gene RET (abbreviation for "rearranged during trans-fection"), which led to a successful outcome using a targeted drug that would not have been considered. So connecting the dots between a mutation and a drug may prove to be more useful than guiding therapy by the type of cancer (e.g., colon or lung), which turns out to be too imprecise.

The controversial codiscoverer of the DNA double helix, James Watson, an octogenarian, recently stated that cancer can be cured in his lifetime—at least with some qualifications. He said, "I would define cancer cured as instead of only 100,000 being saved by what we do today, only 100,000 people die. We shift the balance." With sequencing, he believes we'll know all of the genetic causes of major cancers in the next few years.[89]

Although the cancer genome can be remarkably chaotic and is much more complex to accurately sequence and analyze than germ-line DNA, it is clear that sequencing tumor DNA could be the first step in treating a cancer patient. Many have projected that this will likely become commonplace at some point of the future—digitizing the cancer in any individual who is affected. With the same issues of practicality and affordability as WGS, cancer sequencing holds considerable promise. In the next few years, thousands of cancers will be sequenced, and it will ultimately be determined whether a large panel of driver mutations is a useful tool as compared with exome or WGS sequencing.

BEYOND SEQUENCE: RNA, PROTEIN, METABOLITES, EPIGENOMICS

DNA sequence has been the main topic of this chapter and the field in general, but the scientific efforts and progress go well beyond it and touch every aspect of what we collectively term the "omics." Included in this general term are transcriptomics (the RNA transcripts from genes or noncoding RNAs), proteomics (all the proteins that are translated), metabolomics (all the metabolites), and epigenomics.

International scientific efforts are mapping each of these. One example is the Human Protein Atlas, led by researchers in Sweden, India, China, and South Korea, who are defining the human proteome. This, by itself, is a herculean task, since even though there are only about 22,000 genes, there is not a 1:1 relationship of genes to proteins. Proteins undergo alternative splicing and cells can change proteins after they are produced (so called posttranslational modifications), so they can appear in any given individual in many forms, and it is estimated that the human proteome comprises more than one million distinct proteins.[90]

Sequencing the transcriptome, the set of all RNA molecules, provides a digital readout and has gained popularity over its precedent, gene expression profiling, providing insight on what the genome is doing—its activity—at a given moment in time in a particular tissue. Gene expression profiling is used clinically for prognosis in breast cancer, as a means of monitoring rejection after transplantation, and has recently been introduced as a way to diagnose the presence of coronary artery disease.[91] (A study led by our team at Scripps in coronary artery disease gene expression work was recognized as one of the top ten medical breakthroughs by *Time* in 2010.)[92] In each of these examples, large studies were performed using cancer tissue or, in the case of transplantation and coronary disease, using white blood cells to determine the relevant set of gene transcripts that would track with the clinical condition of interest. Whole-transcriptome sequencing, known as

RNA-Seq, has already proven especially helpful for identifying fusion genes in cancer.

Metabolomics is the study of all of the metabolites or cellular end products. It relies heavily on the technique of mass spectrometry, which is used to both identify and quantify the unique chemical fingerprint that represents a particular metabolite. In conjunction with the other omic tools, such as a gene mutation in a coding region, or an abnormal protein, this can determine whether a novel metabolite or pattern is associated with a disease or condition of interest.

The field of epigenomics has really exploded in recent years.[93] The term "epigenetic" has evolved to encompass heritability not related to DNA sequence. There are many examples of how environmental conditions can influence heritability through the generations: people who smoke cigarettes in their youth affect the age of puberty in their children's children; mothers exposed to famine have specific imprinting on the side chains of the insulin growth factor 2 gene (IGF2) six decades later.[94] This transgenerational impact of environmental conditions helps to make sense of and coalesce the long-standing nature versus nurture debate.

Accordingly, there is no such thing as "identical" twins. Even twins emanating from the same embryo—monozygotic—with the same DNA sequence have different epigenomic marks, chiefly influenced by differences in the maternal in utero or perinatal environment. These marks can be manifest on the elements that package the DNA in a number of ways (Figure 5.4): by attaching methyl groups to the side chains of genes (particularly at C or cytosine bases), by changing the chromatin pattern (open or closed), or by affecting the histones (attaching an acetyl or methyl group).

Along with multiple international "big science" programs, a large U.S. government-funded collaborative initiative known as the Roadmap Epigenome Project is dedicated to unraveling the epigenome's complexity. In October 2009, Joseph Ecker and his colleagues from the Salk Institute in La Jolla published the first human methylome,[95] using both skin and stem cells. Their achievement was recognized by *Time* as a top ten scientific discovery for that year.[96] While epigenomic modifications that often result in turning off or "silencing" effects on gene expression were first emphasized in cancer, more recent studies have highlighted their contribution in many other diseases, including mild cognitive impairment and Alzheimer's, behavioral disorders such as schizophrenia, and metabolic diseases of obesity and diabetes.[97] Many studies of Type 2 diabetes, especially, have highlighted epigenomic effects. These include the role of open chromatin for the principal susceptibility gene TCF7L2; the importance of parent-of-origin se-

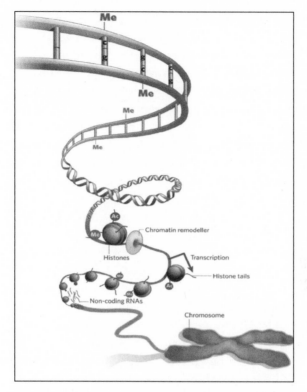

FIGURE 5.4: Schematic of DNA to show methylation of the cytosine side chains, the histone and chromatin packing of DNA, and other features, including noncoding RNAs.

quence variants in passing down risk alleles (whereby, for example, only the SNP of paternal or maternal origin is critical for transmitting susceptibility); and methylation of key genes in muscle biopsies of diabetics.[98] In experiments with rats, fathers with chronic high-fat diets have daughters who develop diabetes; the process is thought to be mediated by a change in the methylation pattern of several genes.[99] The *Economist*, which doesn't miss much in the meaningful progress of genomics, had an article on the origin of diabetes. "Don't blame your genes," the headline said. "They may be simply be getting bad instructions—from you." The illustration was a picture of a woman eating an ice cream sundae, with the caption "piling on the methyl groups."[100]

There is one more "ome" that deserves more than mention as we wrap up the omic landscape—the microbiome. This refers to the species of bacteria, fungi, and viruses that live in each of us. There are efforts throughout the world to map the microbiome: the MetaGUT project in China, the Human Gastric Microbiome in Singapore, the Human MetaGenome Consortium in Japan, MicroObes in France, the Australian Urogenital Microbiome Consortium, the Canadian Microbiome Initiative, and the U.S. Human Microbiome Project.[101]

In our intestine alone bacterial species contain more than a hundred times as many genes as our own human genome. In fact, there are about a hundred trillion microbes in our gut, as compared with ten trillion cells in the human body. Many microbiomes have been linked to diseases—such as the gut microbiome and obesity or heart disease, the airway microbiome and asthma[102]—and this whole area is just beginning to take off because of the power of current ultra-high-throughput sequencing platforms. A recent editorial in the *Journal of Clinical Investigation*, "Which Species Are in Your Feces," pointed out the remarkable value of high-throughput sequencing of the intestinal microbiome to determine antibiotic-resistant, high-risk populations.[103] Some researchers have even resorted to fecal content transplants from healthy patients to those who harbor an extremely difficult bacterium to treat—*Clostridium difficile*—as a potential probiotic therapy.[104] In 2011, a major advance in the human gut microbiome was made with the discovery of distinct "enterotypes" from hundreds of individuals. There were three different patterns of microbiomes of the gut, characterized by dominance or overrepresentation of a specific bacteria species: *Bacteroides*, *Prevoella*, or *Ruminococcus*.[105] The specific enterotype may influence the risk of colon cancer, obesity, and metabolic disorders; the response to many medications; what we should eat; and the optimum antibiotic to use in case of an infection. It's a big new wrinkle in the new science of individuality.

PERSONAL "CONSUMER" GENOMICS

We now progress to a central topic—how will knowledge of my genome prevent diseases and keep me healthy? When are the data ready for prime time?

The field of personal genomics has been mired in controversy, which got particularly sonorous in late 2007 when two companies—DeCode Genetics and 23andMe—commercialized genome-wide scans for the public, followed soon thereafter by Navigenics. These companies offered the same genotyping chips that were being used in the GWAS studies, assaying 500,000 SNPs, and they provided a readout of risk for complex traits and diseases. The kits were available to order over the Internet and the data provided over the Web. The high cost attached to the service—initially ranging from $995 for DeCode and 23andMe to $2,500 for Navigenics—included updates on a frequent basis as new data became available. Since GWAS results were being published in major scientific journals virtually every week, this was a key feature. The expensive Navigenics package included a telephone session with a genetics counselor to review and interpret the results.

In late 2007, I was one of the first subjects to test the Navigenics service, before it was commercially released. The security involved was extraordinary. Once I received an email notification that my results were ready, I had to use multiple passwords and verify my identity using my cell phone to finally get into the site and get my results. Since I'm a cardiologist with no family history of heart attacks, the first thing that caught my eye was this: "Heart Attack—Average Lifetime Risk 52 percent, Your Risk 101 percent." I was pretty stunned, not least because I had no idea how my risk of heart attack could be greater than 100 percent! I remember calling my wife that evening and suggesting that I might not make it home for dinner. I later notified Navigenics, and they confirmed that it was a mistake. It was a good thing they were not already marketing the test yet to general consumers!

Figure 5.5 shows my updated Navigenics output for disease susceptibility for twenty-five conditions. At the top of each column is the lifetime risk category, ranging from less than 1 percent on the far left to more than 50 percent on the far right. The darkened boxes represent conditions for which my risk exceeds that of the general population. I have increased risk of osteoarthritis, at 56 percent in my lifetime compared with 40 percent in the general population. In fact, I already have this problem, so that was accurately forecasted, but doesn't everyone get osteoarthritis if they live long enough? Going back to the heart attack risk, it looks a lot better now than the first pass, but my risk is 30 percent, or 12 percentage points, higher than that of the general population of men. Notably, I have more than a threefold increased risk of developing Alzheimer's disease because I carry one copy of the well-documented apoε4 risk allele. If I had two copies of this allele, my risk of developing Alzheimer's would be increased at least tenfold. I am also at heightened risk for brain aneurysm (although the absolute risk for me is still less than 1 percent), abdominal aorta aneurysm (less than 4 percent), and atrial fibrillation (risk of 33 percent, compared with 26 percent for the general population).

It's not immediately clear what the clinical importance of any of this information is. Take colon cancer. Even though both my maternal grandparents and my paternal aunt died of colon cancer, and I have been undergoing colonoscopies every five years since age thirty-five, my report indicates

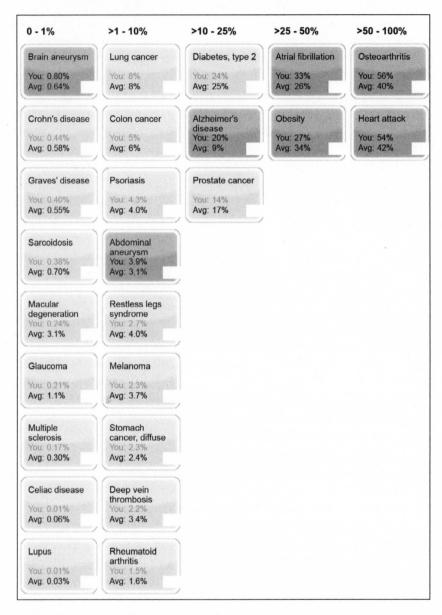

0 - 1%	>1 - 10%	>10 - 25%	>25 - 50%	>50 - 100%
Brain aneurysm You: 0.80% Avg: 0.64%	**Lung cancer** You: 8% Avg: 8%	**Diabetes, type 2** You: 24% Avg: 25%	**Atrial fibrillation** You: 33% Avg: 26%	**Osteoarthritis** You: 56% Avg: 40%
Crohn's disease You: 0.44% Avg: 0.58%	**Colon cancer** You: 5% Avg: 6%	**Alzheimer's disease** You: 20% Avg: 9%	**Obesity** You: 27% Avg: 34%	**Heart attack** You: 54% Avg: 42%
Graves' disease You: 0.40% Avg: 0.55%	**Psoriasis** You: 4.3% Avg: 4.0%	**Prostate cancer** You: 14% Avg: 17%		
Sarcoidosis You: 0.38% Avg: 0.70%	**Abdominal aneurysm** You: 3.9% Avg: 3.1%			
Macular degeneration You: 0.24% Avg: 3.1%	**Restless legs syndrome** You: 2.7% Avg: 4.0%			
Glaucoma You: 0.21% Avg: 1.1%	**Melanoma** You: 2.3% Avg: 3.7%			
Multiple sclerosis You: 0.17% Avg: 0.30%	**Stomach cancer, diffuse** You: 2.3% Avg: 2.4%			
Celiac disease You: 0.01% Avg: 0.06%	**Deep vein thrombosis** You: 2.2% Avg: 3.4%			
Lupus You: 0.01% Avg: 0.03%	**Rheumatoid arthritis** You: 1.5% Avg: 1.6%			

FIGURE 5.5: My Navigenics genome-wide scan results. Each column pertains to the risk level of the disease in the population, ranging from less than 1 percent to greater than 50 percent. My risk for each of the twenty-five conditions is compared to the general population.

that my risk of colon cancer was not increased compared with the average (5 percent versus 6 percent). Does that mean I am not at risk of colon cancer and can stop having colonoscopies every five years? I wish, but unfortunately, the answer is a definite *no*. The scan only tests for common SNP variants, and—as we have seen—most of the heritability of common variants is grossly incomplete. One cannot make accurate individual predictive risk assessments based on the results of large populations—association doesn't equate with prediction. That is, just because I don't have the zip codes associated with colon cancer doesn't mean I will not develop this condition. On the other hand, without question I have a risk of getting Alzheimer's disease, but it isn't clear there is anything I can do about prevention. So what can I do with this information?

It turns out I can do many things. If I had a doubling or more risk of melanoma, I would certainly be much more concerned about sun exposure. Knowing I was at high risk of diabetes would give me extra incentive to maintain my weight and exercise maximally. For blood clots in the leg (deep-vein thrombosis), I would be much more inclined to get up and walk around on long flights and consider taking low-dose aspirin as a preventive strategy. Knowing I have a moderately increased risk of heart disease and atrial fibrillation, if I get chest discomfort with exertion or develop a very rapid heart rate, I've been forewarned that I am predisposed to these conditions. Instead of denying that the chest discomfort might represent lack of blood supply to my heart muscle, I would seek medical attention. Similarly, if I developed a fast and irregular heart rate, I would know what this likely signals and get the appropriate therapy.

Here is another account from a participant in our Scripps Genomic Health Initiative:

> As an employee of Sempra, I participated in this study in May and received my results last week. I was amazed at the informative detail and thoroughness of the report. I was "floored" when I read my Navigenics results which found that my highest elevated risk condition (96 to 98 percent) was colon cancer.
>
> The reason I found this so fascinating is that I have just been diagnosed with colon cancer. This was found during a routine colonoscopy two weeks ago (I am 51 years old and was supposed to have it done last year). Had I procrastinated and not had the colonoscopy procedure done this month, I would have definitely made an appointment to have the procedure done after reading my results. I am a believer in this study! I have shared my results with other family members and

intend to share this report and other risk conditions with my doctor. Thank you for allowing me to participate in this important health study.

Jeff Gulcher, the chief scientific officer of DeCode Genetics, had a genome-wide scan (known as DeCode Me) performed by his company and found he had a doubled risk of prostate cancer. Although he was not yet fifty, when PSA screening might be considered, he had the screening done and learned that his PSA level was significantly elevated. Ultimately, he had a prostate biopsy, which showed cancer, and underwent a radical prostatectomy, which he believed saved his life.[106] Here's another anecdote, which a Stanford professor emailed to me after a lecture I gave on genomics at medical grand rounds in late 2008 (I have withheld his name for privacy reasons):

> Subject: My personal genomics story
> I decided to volunteer for the Scripps Genomic Health Initiative after hearing Eric Topol give medical grand rounds at Stanford. My analysis was mostly reassuring, but showed two areas of increased risk. One was for prostate cancer, which was not a surprise since my father died of prostate cancer. The second was celiac disease (due to HLA-DQ2.5 and other markers), which was a surprise. Although in retrospect, I had some subtle signs and symptoms which could be attributed to celiac disease: poor digestion of fatty foods, low serum cholesterol, LDL, and HDL, a mysterious skin rash (which might be dermatitis herpetiformis) and recurrent aphthous ulcers. So I followed-up the Navigenics report with a serologic test for celiac disease, which was positive, and an upper endoscopy, which was also positive, for moderately severe celiac disease. A dexa bone density study showed that my bone density was around 1.5 standard deviations below the mean for sex and age-matched controls, which is probably due to chronic vitamin D deficiency from celiac disease. A gastroenterologist has recommended a gluten-free diet, which is thought to be effective at reducing the many possible complications of celiac disease: diabetes, thyroid disease, liver disease, small bowel lymphoma, other cancer. In addition, I am taking calcium and Vitamin D, and my first-degree relatives are also being tested for celiac disease. It is amazing to me, at the age of 52 years, and being a physician, that my diagnosis and treatment was possible only because of your DNA test. Thank you for that.

Opinion about whether such genetic risk information should be available to anyone who seeks it has been split. Among much of the medical establishment, however, the backlash was strong. *Nature Genetics* had the first shot entitled "Risky Business" in December 2007,[107] which was a quick response to the first commercial release, and by January 2008, the *New England Journal of Medicine* began a relentless series of editorials and perspectives, the first of which was entitled "Letting the Genome Out of the Bottle: Will We Get Our Wish?" coauthored by the journal's editor, Jeffrey Drazen.[108] Dr. Terry Manolio, who worked directly for many years for Francis Collins (the leader of the public Human Genome Project) at the NIH, concluded her review of GWAS with this statement: "Patients inquiring about genome wide association testing should be advised that at present the results of such testing have no value in predicting risk and are not clinically directive."[109]

On the other hand, the journal *Nature* took a more measured outlook, arguing that scientists needed to figure out how to talk about personal genomics with the public.[110] By November 2008, *Nature* dedicated a whole issue on personal genomics; its cover featured a human genome inside a container labeled "Break Glass When Ready" and "Your Life in Your Hands: Instructions for the Personal Genome Age." One of its articles, "Misdirected Precaution," pointed out that personal genome tests are "blurring the boundary between experts and lay people." And its authors stated, "We welcome a shift from genetic protectionism to a situation in which individuals become experts on, and active governors of, their genomes."[111]

Collins has been one of the few advocates in the medical establishment for giving the public access to genomics. In a 2008 segment of "The DNA Age" series on personal genomics by Amy Harmon, discussing the slow uptake of these tests, Collins, then director of the Human Genome Research Institute and now director of the NIH, said, "It's pretty clear that the public is afraid of taking advantage of genetic testing," and "if that continues, the future of medicine that we would all like to see happen stands the chance of being dead on arrival."[112] In his book *The Language of Life*, Collins writes about how learning his own genome scan results changed his behavior. His risk of Type 2 diabetes was high, he said, so he lost twenty pounds and got into a regular exercise routine with a personal trainer, which might otherwise not have happened.[113]

Technologists have taken an approach more like that of Collins than that of Manolio. Sergey Brin, the cofounder of Google, is married to Anne Wojcicki, who started 23andMe. Sergey's mother developed Parkinson's, as did his great aunt. Sergey had a genome-wide scan and found that he carried the variant in the LRRK2 (leucine-rich repeat kinase) gene, which carried a high risk—about 70 percent—of developing Parkinson's disease.

And his mother, who also had the 23andMe genome scan, had the same variant. Sergey started a blog, and his first post was titled "LRRK2." He wrote: "I know early in life something I am substantially predisposed to. I now have the opportunity to adjust my life to reduce those odds (e.g. there is evidence that exercise may be protective against Parkinson's). I also have the opportunity to perform and support research into this disease long before it may affect me. And, regardless of my own health, it can help my family members as well as others. . . . I feel fortunate to be in this position."[114] His perspective is reminiscent of the view of Charles Sabine, the former NBC correspondent who learned he had the gene for Huntington's disease and said, "Knowing my genetic condition both empowers and motivates me. . . . Nothing incentivizes more than knowing your genetic code."[115]

Other journalists have been positive as well. Amy Harmon, who later won a Pulitzer prize for her coverage, wrote a front-page article for the *New York Times* in November 2007: "Learning My Genome, Learning About Myself." Thomas Goetz, executive editor of *Wired*, had a cover article, "Your Life: Decoded. A New $1,000 DNA Test Can Tell You How You'll Live—and Die. Welcome to the Age of the Genome."[116] *Time* named retail genome scanning the top invention of 2008.[117] Goetz also wrote a feature in *Wired* about Sergey Brin, which reviewed Brin's experience with personal genomics. Goetz argued that having access to DNA results can make some think that "it holds dark, implacable secrets" or "toxic knowledge." This wasn't the first time the medical establishment had thought such a thing, however. He reminded us that "in 1961, 90 percent of physicians wouldn't tell their patients if they had cancer."[118]

That coverage, though, proved to be most of the few bright spots in the history of personal genomics thus far, and it's worth recalling that genome-wide scans have been under attack. Nevertheless, this form of direct-to-consumer genetic testing uses research-grade genotyping chips and should be differentiated from "snake oil" DTC genetic tests that have no scientific basis, such as analyzing one's genes for what foods one should eat or avoid, finding out whether you are "compatible" with your lover, or predicting whether a child will become a world-class athlete. But there are many legitimate concerns that I separate into five general categories: (1) fear; (2) interpretability, utility, and errors; (3) privacy and security; (4) lack of regulation; and (5) access and cost.

To address some of these issues for which there was no or limited anecdotal data, the Scripps Genomic Health Initiative performed a large study of more than 3,600 individuals, who underwent the Navigenics genome-wide scan at a markedly discounted charge ($200 instead of $2,500). The study assessed the impact on diet and exercise, psychological effects, and medical screening and diagnoses. In follow up at approximately six months, for more than 2,000 participants who submitted their data, there was no evidence, as assessed in well-validated tests of psychological impact, of depression, distress, or heightened anxiety. Unhappily, there was also no clear evidence of lifestyle improvement or of undergoing appropriate screening tests related to conditions of risk, but there was strong support for intent to have such tests. For individuals with an increased risk of colon cancer there was a significant increase in intent to undergo colonoscopy, as well as a similar increase in intent to undergo mammography for breast cancer risk, PSA testing for prostate cancer risk, and eye examination for those at heightened risk for macular degeneration.[119]

The colonoscopy finding was interesting. My wife, who participated in the study, found she had a doubled risk of colon cancer. But at age fifty-five, five years past when she should have undergone her first colonoscopy, she had previously avoided the procedure. She underwent it and was relieved to find out it was normal. At least two participants in the study similarly had their first colonoscopy and had polyps removed; one also was found to have cancer and felt gratified that it was detected early. Of note, more than half of the population over age fifty does not undergo colonoscopy as recommended. But this alternative to mass screening, directed by increased risk of common gene variants, may someday prove to be a worthwhile means of directed, individualized screening, particularly when more of the heritability for each condition can be precisely defined.

The overall findings of the study were both sobering and provided some reassurance. Changing behavior to improve lifestyle—to lose weight, exercise more, and eat healthy foods—is one of the greatest challenges in health care, and there have not been any major success strategies to date. Unlike with Collins and Brin, results of a genome-wide scan did not induce salutary lifestyle changes in a large population.[120] On the other hand, the "toxic knowledge" concern could be put aside. There was solid evidence pointing away from any detrimental psychological impact. This goes along with an important study on the apoε4 allele in families with Alzheimer's disease, which concluded, "The disclosure of apoε genotyping results to adult children of patients with Alzheimer's disease did not result in significant short-tem psychological risks."[121] Our study, which included apoε genotyping along with over twenty other conditions, yielded similar findings. That doesn't mean that everyone can deal equally well with his or her DNA results, but it strongly suggests that willing participants are well equipped to do so. It certainly counters the assertion of the officials of California's

Department of Public Health that these tests were "scaring a lot of people to death."[122]

The questions about utility are more grounded. From an article by my colleagues in *Nature* in 2009, it became clear that the results across the three companies could be conflicting.[123] The main reason had nothing to do with the genotyping, which was remarkably accurate, but with the lack of consensus on which studies to use and how the calculations were made for reporting the risk of the various conditions. There were no standards for what level of evidence was necessary to incorporate a genetic marker or what magnitude to assign its effect. An individual who has tests from all three companies might have an increased risk for a disease in one and protection from the same disease in another. I have had all three of these genome scans, along with a fourth test by Pathway Genomics, and have seen some of these discrepancies myself. By 2010, there were demands resonating from both government regulators and the science community for developing uniform standards for reporting.[124]

The biggest vulnerability of present-day genome-wide scans cannot be avoided: as GWAS peeks at the genome, most of the risk will be missed. Hopes are being pinned on whole-genome sequencing to be the means of providing the critical missing information.

Some individuals have had whole-genome sequencing and written about the experience, but the revelations have been far from striking. Craig Venter in 2008 published "A Life Decoded: My Genome, My Life," which contained tidbits of his genomic sequence data, such as an elevated risk of heart attack from a gene variant that results in slow metabolism of caffeine.[125]

In 2009, Steven Pinker, a highly regarded Harvard psychologist, published "My Genome, My Self" in the *New York Times Magazine* in which he pointed out, from exome sequencing, that he had a genetic predisposition to baldness but couldn't have more hair on his head, and already knew he had one copy of a rare variant for a serious disease known as familial autonomic dysautonomia. Pinker's reflection on his apoε allele was noteworthy. He knew that James Watson, in his own scan, had omitted that sequence data when published in 2008 in *Science*. As Pinker put it: "All of us already live with the knowledge that we have the fatal genetic condition called mortality, and most of us cope using some combination of denial, resignation and religion. Still, I figured that my current burden of existential dread is just about right, so I followed Watson's lead and asked for a line-item veto of my APOε gene information." As an aside, Pinker is an advisor to Counsyl, a company with a genomic test for screening couples who are planning to conceive a baby or have a history of infertility or multiple miscarriages.

Counsyl screens one hundred rare recessive mutations for carrier state for $350. Pinker, with the rare mutation for familial autonomic dysautonomia, had his wife screened; she carried this allele too. He said, "We met too late in life to have children, but if we had met a few years earlier we would have been playing roulette."[126] More recently, with next-generation sequencing it has become technically feasible to screen more than 500 recessive conditions (out of a known 1,139 recessive Mendelian traits), which has the potential to supersede Counsyl's more limited platform for screening couples who are planning to conceive a child.[127]

Esther Dyson, the venture capitalist and one of the first ten participants in the Personal Genome Project, wrote a *Wall Street Journal* op-ed entitled "Full Disclosure" in which she explained why she's posting her genome and medical records on the Internet. She quoted a colleague: "You would no more take a drug without knowing the relevant data from your genome than you would get a blood transfusion without knowing your blood type."[128] Misha Angrist, a genetics researcher at Duke and former genetic counselor, published a book in 2010, *Here Is a Human Being*, about his experience in participating in the same Personal Genome Project as Pinker and Dyson and getting his whole-genome sequence data.[129] While long on the adventure and checkered with humor, it was notably short on meaningful genomic data. Glenn Close presented some of her sequence results at the Society for Neuroscience's annual meeting in San Diego in late 2010. She has a history of mental illness in her family that she vaguely described as a "neuroscience family" and had other family members contribute to the presentation.[130] Even though she was the first identified woman (rather than an anonymous one) whose diploid genome was sequenced, there were no specific gene findings to note. Even the notorious Ozzy Osbourne had his genome sequenced; the data were presented at the TEDMED 2010 meeting. Surprisingly, it had no insight to offer on his long history of substance abuse.[131] While these are just a sampling of individuals in the nascent WGS phase, the results show us that a lot more work needs to be done to make the six billion life codes medically fruitful.

Here I'll return to the controversy over genome-wide scans to offer some predictions of what it will be like when WGS is widely available. Even though participants in the Personal Genome Project were required to have their sequence data posted online, the issue of privacy is a central concern. Some of the worries about employer or health insurer misuse of the results were diminished by passage of the Genetics Information Nondiscrimination Act (GINA) in 2008.[132] The law is already seeing action in court: The first lawsuit under GINA was filed in 2010 by a woman who sued her employer for terminating her job after she had confirmed very high risk of breast cancer with the BRCA2 mutation and underwent a double mastectomy. The law has clear limits, however. Insurers can still request genetic information before covering procedures, and while GINA litigation aims to prevent abuse of genetic data by health insurers or employers, it does not protect against inappropriate handling by life insurers or long-term disability insurers and does not apply to military coverage or to companies with fewer than fifty employees.[133] These residual issues may and should be addressed in the future, but legitimate concerns about privacy of life code data remain.

Further concerns about privacy are indexed to the potential of misuse of the data by the consumer genomics companies. For example, what if a consumer genomics company marketed the data to pharmaceutical companies? What happens to the data if the consumer genomics company goes bankrupt, which is not a far-fetched scenario? A de-identified, anonymized genome-wide scan can still be used to identify a particular person. So there should be protection for privacy in any database that contains genome-wide scan data, let alone WGS data. Nevertheless, personal genomic companies have performed high quality research with their database and been able to directly connect to a large number of individuals with a particular condition, such as through the discovery of novel gene variants for Parkinson's disease by 23andMe.[134]

Government regulation of consumer genomics companies has been a centerpiece (and the semblance of a circus) in their short history. Back in 2008, the states of California and New York sent "cease and desist" letters to the genome scan companies.[135] State officials were concerned that the laboratories that generated the results were not certified as CLIA (Clinical Laboratory Improvement Amendments) and that the tests were being performed without a physician's order All three companies developed work-around plans in California and remained operational but were unable to market the tests in New York.[136]

In 2010, the regulation issues escalated to the federal level. In May it was announced that 7,500 Walgreens drugstores throughout the United States would soon sell Pathway Genomics's saliva kit for disease susceptibility and pharmacogenomics.[137] While the tests produced by all four companies had been widely available via the Internet for three years, the announcement of wide-scale availability in drugstores (which was cancelled by Walgreens within two days) appeared to "cross the line" and set off a cascade of investigations and hearings by the FDA, the Government Accountability Office (GAO), and the Congressional House Committee on Energy and Commerce. The FDA's Alberto Gutierrez said, "We don't think physicians are going to be able to interpret the results," and "genetic tests are medical devices and must be regulated."[138] The GAO undertook a "sting" operation with its staff posing as consumers who bought genetic tests and detailed significant inconsistencies, misleading test results, and deceptive marketing practices in its report.[139]

All four personal genomics companies are struggling. In the cumulative four years since the first commercial launch, about 100,000 consumers have bought the tests.[140] The prices have dropped dramatically, to between $200 to $400, and the scope of the scans has narrowed to approximately 50,000 to 100,000 SNPs (down from 500,000 to 1 million SNPs) to save costs. The price is clearly a significant factor for consumers. During our Scripps Genomic Health Initiative study, the Sempra energy company in San Diego offered to defray the costs for participation for 1,000 of its employees. There was a veritable stampede of Sempra staff to put their saliva in cups! This correlates to the response I get when I ask groups of lecture attendees, "How many would get a genome scan if it was free?" More than 90 percent typically raise their hands.

Although these companies have been under siege and may not survive, several important impacts have forever changed the landscape of genomics for consumers. One is the realization that there will likely never be a "right time"—after we have passed some imaginary tipping point giving us critical, highly actionable, and perfectly accurate information—for it to be available to the public. The logical conclusion is that the tests should be made available. What's more, the fact that they have been available has meant that the democratization of DNA is real.[141] Consumers now realize that they have the right to obtain data on their DNA. As a blogger wrote in response to the Walgreens flap, "To say that this information has to be routed through your doctor is a little like the Middle Ages, when only priests were allowed (or able) to read the Bible. Gutenberg came along with the printing press even though few people were able to read. This triggered a literacy/literature spiral that had incredible benefits for civilization, even if it reduced the power of the priestly class."[142]

The American Medical Association (AMA) sees things differently. In a pointed letter to the FDA in 2011, the AMA wrote: "We urge the Panel . . . that genetic testing, except under the most limited circumstances, should be carried out under the personal supervision of a qualified health professional."[143] The FDA has indicated it is likely to accept the AMA recommendations, which will clearly limit consumer direct access to their DNA information. But this arrangement ultimately appears untenable, and eventually there will need to be full democratization of DNA for medicine to be transformed. Of course, health professionals can be consulted as needed, but it is the individual who should have the decision authority and capacity to drive the process.

The physician and entrepreneur Hugh Rienhoff, who has spent years attempting to decipher his daughter's unexplained cardiovascular genetic defect and formed the online community MyDaughtersDNA.org, had this to say: "Doctors are not going to drive genetics into clinical practice. It's going to be consumers. . . . The user interface, whether software or whatever, will be embraced first by consumers, so it has to be pitched at that level, and that's about the level doctors are at. Cardiologists do not know dog shit about genetics."[144]

As a cardiologist I would tend to agree with Hugh's point here, although I would not express it in these graphic terms. As I discuss in Chapter 9, physicians are not prepared for genomic medicine. In my 2007 *Wall Street Journal* op-ed, "What You Can Learn from a Gene Scan," shortly after consumer genomics became a reality, I forecasted that "when a consumer arrives in his or her doctor's office to get help in interpreting the genomic data, the doctor is likely to respond, 'What's a SNP?'"[145] Four years later that hasn't changed, and we have only 1,500 medical geneticists in the United States and fewer than 2,000 certified genetic counselors in this country for 310 million people.[146]

The necessary and appropriate democratization of DNA data extends beyond what the companies willingly supply. Each individual has the right to request his entire data set, particularly worthwhile when scans include 500,000 to 1 million SNPs. One can look up any SNP with data on the SNPedia website or run it through George Church's downloadable software, known as Trait-o-matic, to yield considerable information on conditions that are not generally reported. In 2011, the iPad app Genome Wowser became available to graphically review the findings of any individual's genomic variants. Being able to explore one's genome by manipulating the tablet screen is a particularly dynamic experience.

One other contribution of the consumer genomics companies may be the "sweet spot"—pharmacogenomics. Remarkably, the data for all four companies were completely concordant across many drugs—Plavix, warfarin, statins, and many more. Pathway Genomics initiated a pharmacogenomics panel for $79 as part of the drugstore roll out, and all four companies have now incorporated such genotyping. Having had all four tests, I learned that I have a marked sensitivity to warfarin and a very high risk of developing bone-marrow suppression from the anti-cancer drug Irinotecan (Camptosar). I also learned that I am a poor metabolizer of caffeine; my risk of heart attacks would rise if I drank more than three to four cups of coffee per day.[147] This information might be considered a bargain, since any single drug genotyping costs over $200 when processed by national labs like LabCorp or Quest Diagnostics. When I ask large groups, "How many would like to have their pharmacogenomics panel?" everyone raises his or her hand. Someday we will all have this information; it will be fairly comprehensive and markedly improve the precision of safety and efficacy for prescription medications.

That someday will likely involve whole-genome sequencing, given its inexorable march to becoming affordable and the software that will inevitably be developed to provide exceptional annotation, with constant updating. The first decade since the human genome sequence was drafted will, in retrospect, be viewed as a long warm-up to making a difference in day-to-day medical practice. Already we can see the impact that sequencing can make in cancer treatment, definition of individual gene-drug interactions, and unraveling idiopathic conditions. The foundation for genomic medicine has been laid. The revolution is ongoing: even though it has taken longer than initially projected, we are moving irrevocably forward in the second postsequence decade. Routine molecular biologic digitization of humankind is just around the corner.

ANATOMY

From Imaging to Printing Organs

The exquisite depictions of anatomy and function generated by modern imaging technologies have blinded many physicians to the limitations and potential harms of radiologic diagnosis.
—Bruce Hillman and Jeff Goldsmith, 2010[1]

THE DATE WAS December 27, 2009. That evening, I had just gotten my hands on the first Vscan—a pocket-sized digital imaging device that provides high-resolution ultrasound imaging—in the United States. Until this point, the only way to get an ultrasound of the heart, an echocardiogram, was to send patients to a lab, where they'd be studied with a $300,000 machine the size of a refrigerator. This was an exciting and liberating event. Naturally my first step was to image my own heart.

To get the images is quite simple. A transducer, which has the shape and size of an electric toothbrush with no brush, is placed on the chest after some gel is put on its tip to help transmit the ultrasonic energy (see Figure 6.1). The transducer is moved around the chest to find a good "window" to acquire digital movie loops of the heart in multiple standard views, each revealing something different about the heart's structure and function. Combined, the movies give us information about the heart muscle function, the thickness of the heart's walls, the status of the four heart valves, the size of the four chambers, how the segment of the aorta near the heart looks, and whether there is any fluid in the pericardium (the sac around the heart).

Placing the transducer on my chest, I quickly got a crystal clear image of the main pumping chamber—the left ventricle—and the mitral valve. I was not surprised that my heart muscle function looked fine. But then I put the color flow on, which uses ultrasound to track the blood flow, and it showed that my mitral valve was leaking badly—so badly, in fact, that I had just become a potential candidate for open-heart surgery to repair the valve! I finished the rest of the ultrasound exam of my heart, and everything else was OK, with the minor exception of a moderate leak from my aortic valve. (At least that one wasn't severe enough to warrant surgery.) The whole scan had taken less than five minutes, and even most of that was taken up by the shock of seeing, and repeatedly examining, my leaky mitral valve.

It just didn't make sense. I had been feeling well and exercising vigorously almost every day. I knew, however, it's possible to have a slow, insidious, progressive leak without showing any symptoms. So I got out my stethoscope to see what I could hear. I listened to my heart in various positions, and I could hear some leak from the valve, but it sure didn't seem like much—maybe 1+, on the cardiologists' scale of 1 to 4, but not the 3+ the Vscan showed. It was peculiar, but the Vscan was tracking the blood flow, and I was looking right at it—lots of leak and enough to be requiring a consultation with a heart surgeon.

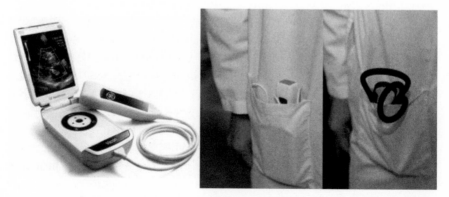

FIGURE 6.1: Picture of the pocket-sized mobile echocardiogram unit known as the Vscan and comparison of a stethoscope and the Vscan in the pocket of my white coat.

It was too late in the day to go to an echocardiography lab for the forty-five-minute, gold-standard examination. I set up an appointment for the next day, went home, and had a bad night. I already knew, genomically, that I was at a high risk of a heart attack, and this discovery left me with a lot to think about: I thought about who would do the surgery (I hoped for a repair of the valve rather than a replacement); how I would have to take off a few weeks or a month from work to get this fixed; all the patients I had looked after with leaky mitral valves, especially in men, who tend to have the most severe cases, and even in a few cardiologists I had trained. I hoped I would tolerate the operation well.

My only hope was that maybe this was a mistake. After almost twenty-four hours of restlessness and high anxiety, my formal echo showed only 1+ mitral valve leak. I escaped the heart surgery. I then informed GE, the manufacturer of the Vscan, about what had happened, and after they reviewed the problem, they found there was a bug in the color-flow software. Once it was reloaded, the software showed only a minimal leak. It was a great relief and the second time I'd learned not to be the first person to try out new medical tests—not to mention another illustration of the potential hazards of treating a scan rather than the patient.

I had gotten the Vscan to demonstrate it at the January 2010 Consumer Electronics Show in Las Vegas. This is one of the largest trade shows in the country, with over 100,000 attendees, and features almost every digital gadget known to humankind. My mission was to demonstrate some exciting advances in wireless and digital medicine. With the Vscan not even yet released in the United States (it became commercially available for $7,900 in February 2010), it was an ideal technology to present. There were several thousand people at the session during which Paul Jacobs, the CEO of Qualcomm, introduced me. Feeling a bit like Dr. Gizmodo, I brought out the Vscan from my old black medical bag—the one I had since medical school in the 1970s but hadn't used in decades. And then the thing didn't work. The demo demons were clearly at work. Paul called out to the audience, "Is there anyone from GE here?" Less than a minute later, even though it felt like fifteen, I got the Vscan to image my heart in front of the audience. The ultrasound of my heart muscle contracting, with my valves moving to and fro, was projected on the big screen. The audience started clapping, and I thought it was pretty funny—they had no idea what an echo image ought to look like, and I sure wasn't going to tell them about my leaky valve.

Pocket-size, high-resolution ultrasound is one of the most significant advances in medical imaging in decades and is replacing the stethoscope, which has been around since 1816.[2] I now use it to examine every patient

I see in clinic, and it usually preempts the need for another appointment for a formal echocardiogram study. It not only saves time but a lot of money (a combined technical and professional charge of about $1,500 per echo). With over twenty million echocardiograms done per year in the United States, there's certainly room to improve efficiency.[3] A further benefit, and one I had not anticipated, is that I can discuss the image and finding with the patient in real time. When a patient goes for a formal study, the ultra-sonographer does not communicate the results to the patient. They are reviewed by an attending physician, who does so off-line, and the results are then typically conveyed the next day or many days later (if at all).

Modern imaging technologies can be divided into two general categories: those that rely on some sort of ionizing radiation (either a form of light, such as X-rays, or particles, such as electrons) to generate an image and those that do not. Ultrasound, as the name suggests, relies on sound, which makes it generally very safe.[4] It is not the most accurate imaging technology, as the acquisition and interpretation of the images is influenced by the person doing it, and it has its limits: ultrasound cannot be used to image through bone (e.g., for the brain) or through gas or extensive fat tissue. But it can be considered the go-to medical imaging technology for a wide range of applications.

The only other imaging technology that does not rely on ionized radiation is magnetic resonance imaging (MRI). MRI combines the use of radiofrequency waves and magnetic fields to produce extraordinary images—highly detailed and in three dimensions—of all parts of the body.[5] Injecting dyes into a patient can delineate blood vessels. Unfortunately it's quite unpleasant, especially compared to an ultrasound: patients must lie still for a prolonged period in a big, noisy machine, and for someone with claustrophobia, it represents a big challenge even with sedative medications. (There are "open" magnets to help minimize claustrophobia.) Likewise, the process requires a specialized facility with sophisticated multimillion-dollar equipment, with the magnets often cooled by liquid helium to nearly –270 degrees Celsius.[6] Thus MRI will never become a portable, real-time, and wireless digital device.

IONIZING RADIATION

Ionizing radiation technologies include projection radiography, such as X-ray and mammograms, fluoroscopy, CT scans, and nuclear scans. These methods, unlike sonographic or magnetic methods, involve some risk: the interaction of radiation and atoms in the body can produce free radicals and can induce DNA damage.[7] Furthermore, they can uncover what are known as incidental findings—essentially abnormalities that, although they are real, do not require further tests or procedures but nevertheless often do lead to more tests or procedures. In spite of these shortcomings, the ability to image virtually any part of the human body and obtain high-resolution pictures is extraordinary and continues to reshape current and future medical practice. There are even software applications for integrating images taken from multiple modalities in a given patient to provide an extremely high-resolution view inside the body.[8] Below I focus my discussion on the use of imaging in three major areas: heart, brain, and cancer.

The amount of ionized radiation is typically measured in millisievert units or mSv. Table 6.1 shows typical imaging exposure and compares each to the exposure in a routine chest X-ray reference standard.[9] There is some radiation exposure—about 3.6 mSv—from natural background sources, a figure that has not changed appreciably in the past thirty years. But the dose of radiation from medical imaging has increased more than sixfold. Indeed, the use of ionizing radiation for medical imaging has skyrocketed beyond any outlandish projection. In 1980, only 15 percent of the population's exposure to radiation was derived from medical imaging; by 2010 it had become 50 percent. In 1980 there were fewer than three million CT scans in the United States; in 2010 that number had grown to over eighty million. Each year 10 percent of the population in the United States undergoes a CT scan, and the use of CT is still growing by more than 10 percent per year.[10] As with magnetic resonance imaging, CT scanners are more than twice as common in the United States per capita than in any other country in the world.[11] There are about twenty million nuclear medicine procedures and a similar number of angiograms and fluoroscopic procedures done each year in the United States.

An annual exposure of radiation in excess of 20 mSv is considered high and correlated with an increased risk of cancer. A dose greater than 50mSv is considered "very high." It is now estimated that approximately 2 percent of all cancers in the United States may be related to the use of ionizing radiation.[12] Evidence that radiation exposure is linked in a linear relationship with cancer, and that there is no threshold below which exposure does not matter, has been reinforced by a fifteen-country study of 407,391 nuclear industry workers. In this cohort, there was a significant association between dose of radiation and cancer-related death as well as death in general; the rates of death were excessive at lifetime (not annual) cumulative doses as

Procedure	mSv Adult Dose	Equivalent No. of Chest X-rays
Airport whole body backscatter scan	0.002	0.1–0.2
Dental X-rays	0.005–0.01	0.25–0.5
Chest X-ray	0.02–0.1	1
Mammography	0.4	20
CT scan of head	2	100
CT angiogram of heart	16	800
Nuclear lung scan	0.2	10
Nuclear heart scan	41	2000
Angiogram of brain	5	250
Angiogram of heart	6	400
Coronary stent procedure	15	750

TABLE 6.1: The amount of ionizing radiation exposure in millisievert (mSv) units for various medical imaging procedures, with comparison to airport screening and dental X-rays. Adapted from *White Paper: Initiative to Reduce Unnecessary Radiation Exposure from Medical Imaging*, Center for Devices and Radiological Health, U.S. Food and Drug Administration, February 1, 2010, and R. Fazel, "Exposure to Low-Dose Ionizing Radiation from Medical Imaging Procedures," *New England Journal of Medicine* 361 (2009): 849–57.

low as 5 to 50 mSv. A recent Canadian study of over 82,000 patients in Quebec followed for more than ten years showed that the risk of developing cancer within five years increased 3 percent for every 10 mSv of radiation exposure.[13] Accordingly, there is really *no* safe annual dose of ionizing radiation, and the cumulative dose is clearly an important metric of exposure and risk of cancer. This is especially a concern among children, who have experienced a sharp increase in radiation exposure through the liberal use of CT scans in emergency rooms.[14] One would anticipate that there is a strong genomic predisposition to DNA damage and cancer, whereby some individuals are protected and others are highly susceptible, but any variations in genes that account for this risk have not yet been identified.[15]

There is also as much as tenfold variation in the radiation dose from ostensibly identical procedures. In 2010, a *New York Times* front-page investigative feature article, "The Mark of an Overdose: The Radiation Boom—A Test Turns Dangerous," found that more than four hundred patients at eight hospital facilities in California and Alabama suffered massive radiation overdose from CT scans of their heads.[16] Multiple individuals who were interviewed suffered acute radiation toxicity with symptoms including hair loss, headaches, memory loss, and confusion. The biggest danger for these individuals is the subsequent risk of cancer. Compounding the overdose mistake is the fact that some 20 to 40 percent of CT scans are medically unnecessary.[17] Double CT scans of the chest (one scan with contrast dye, one without), with each scan equal to 350 chest X-rays, are being performed in over 30 percent of U.S. hospitals without any justification for the second scan.[18]

In 2009, a study of a large patient population of 952,420 adults between the ages of eighteen and sixty-four sought to quantify the radiation exposure from different procedures.[19] The largest contribution was heart-perfusion imaging, with an average dose of 15.6 mSv and accounting for 22 percent of the radiation exposure from all procedures. This was followed by CT scans of the abdomen, pelvis, and chest, with an average dose of 6 to 8 mSv, which were responsible for 18 percent, 12 percent, and 8 percent, respectively, of the radiation dose of all procedures.[20] This study also found that 2 percent of the individuals were getting an annual dose of radiation that exceeded 20 mSv. These patients were often subjected to multiple imaging studies as the result of findings that were unrelated to the indication for the scan but that might require further evaluation, although they usually proved not to be problems.

The Mayo Clinic published a study of imaging-related incidental findings in 2010. Of 1,426 imaging examinations, almost 40 percent had at least one incidental finding. CT scans of the abdomen and pelvis had the highest likelihood of this problem, almost twentyfold higher than ultrasonography.[21] CT scans of the lung and MRIs of the brain had a twelvefold risk of incidental findings. Most of the incidental findings were nodules, such as in the lung, which resulted in serial CT scans and lung specialist consultation. In very few individuals, with a three-year clinical follow-up, was there evidence of clear benefit for chasing down these incidental findings.[22]

The benefits and risks of CT scans are exemplified by one of the largest trials ever undertaken by the National Institutes of Health—the National Lung Screening Trial (NSLT). This trial cost over $250 million, conducted at thirty-three medical centers from 2002 to 2010, and enrolled 53,500 individuals, ages fifty-five to seventy-four, who were either former or current smokers. Instead of the usual lung CT, the trial used the helical low-dose CT that carries much lower radiation exposure, closer to that of a mammogram than that of a regular lung CT. The aim was to reduce the toll of lung cancer, which accounted for almost 160,000 deaths in the United States in 2010—more than colon, pancreatic, breast, and prostate combined. It has long been thought that lung cancer is diagnosed too late in its progression and that the late pickup, in part, accounts for the 85 percent death rate.[23]

By random assignment, half of the individuals in the study had lung CT screening; the other half had chest X-rays. These imaging studies were repeated twice during the next two years, and the patients were followed for five years. There was a 20 percent reduction in lung cancer deaths, but since the incidence of cancer was low in general, that works out to one person saved per three hundred screenings. Furthermore, 25 percent of the individuals had incidental findings, often prompting more procedures, and nearly all were benign. Since forty-six million Americans smoke, the researchers extrapolated that thousands of lives would be saved. But mass lung CT scanning would cost billions of dollars each year.[24]

HEART IMAGING

Coronary artery disease is the most common heart disorder, so medical imaging of the heart is focused on detecting blockages in the arteries that supply the heart muscle. A treadmill stress test with multiple EKG readings is the "plain vanilla" assessment. Since this is not considered a very reliable test, imaging of the heart is done in conjunction with the stress test and EKG—either echo or nuclear. Both of these are indirect measurements, since they do not indicate whether there are any blockages in the arteries but instead show whether parts of the heart muscle are not contracting properly (in the case of echo) or if the radioactive tracer is symmetrically getting to the different regions of the heart (nuclear perfusion study).

Approximately 30 percent of the overall radiation exposure to the U.S. population related to medical imaging is attributed to heart imaging.[25] The chief culprits are nuclear perfusion scans. Most patients with established coronary artery disease have an exercise test with nuclear perfusion scan on an annual basis. There are well over ten million performed each year, increasing at a rate of more than 6 percent per year. This test alone accounts for more than 10 percent of the entire cumulative effective dose to the population from all medical imaging sources.[26] In 2010 Columbia Medical Center in New York City published a study of 1,097 patients, each of whom had at least one heart nuclear perfusion scan over a ten-year period. It found that the average patient had fifteen ionizing radiation imaging procedures, and more than 30 percent had cumulative exposure greater than 100 mSv. Multiple heart nuclear perfusion scans were performed in 38 percent of the patients with an average cumulative radiation exposure of 138 mSv.[27] Another study of 64,071 patients with heart attack found the average dose of radiation was 15 mSv as a result of having, on average, more than four ionizing radiation imaging procedures.[28]

That's a lot of radiation, so finding some safer method would seem imperative. Unfortunately, not every patient is especially "echogenic," making ultrasound images difficult to interpret. Fortunately, there are other tools available—coronary angiography and CT angiography.

Coronary angiography is considered the gold standard, the direct way to assess blockages. A tiny catheter is inserted in the artery in the arm or the leg and threaded to the area in the aorta where the coronary arteries originate, and contrast dye is injected to provide the roadmap. This procedure is not without risk: the very rare but notable risks of stroke, damage to an artery, and damage to the heart itself. What's worse is that most people don't need to take the risk. In a large national study of 663 hospitals and 398,978 patients, only 38 percent of those undergoing the procedure had significant blockages![29] Still, when actually called for, the procedure can often include implanting a stent to restore the normal heart blood supply.

The other alternative to directly detecting blockages is the CT scan angiogram. This test got legs in 2004, and at the time its boosters promised it would set off a revolution in medicine, or at least cardiology.[30] Over the course of the next few years, the test evolved from sixteen to sixty-four detectors to increase image resolution, but a large multicenter comparison of the test with conventional coronary angiography did not show it was as accurate, especially for patients who did not have significant blockages.[31] Both forms of angiography use contrast dye, which carries its own risk of kidney toxicity and allergy, and if the CT scan is positive for a blockage, it would require a conventional coronary angiogram to address the narrowing (assuming the patient was suitable for a stent instead of bypass surgery). The radiation exposure for CT heart angiography is substantial and was recently demonstrated by a group of fifteen hospitals in Michigan that this can be reduced from an average of 21 mSv to 10 mSv without degradation of image quality.[32] This is still almost double the radiation dose of a conventional angiogram, but a major step in the right direction.

In cardiology, we do not have a problem in detecting significant coronary blockages, although we do far too many tests and use much too much radiation to get that accomplished. The challenge to preventing deaths and heart attacks is that we do not have a way to detect the individuals who have a minor blockage that might suddenly crack or rupture. This will likely require molecular diagnostics, such as detecting the arterial cells that are sloughing off into the circulating blood from the diseased and inflamed artery, which typically precedes a heart attack, or detecting some nucleic acid signature reflecting this vulnerability. Active research is ongoing in this area, and the good news is that, if ultimately successful, it will not likely require any radiation exposure.

BRAIN IMAGING

One of the most vital areas of biomedical research is to prevent or optimize treatment for Alzheimer's disease, the leading cause of dementia of the aged and one that so far has defied all attempts at prevention. In 2010, 35 million individuals had Alzheimer's disease worldwide, with 5.5 million people diagnosed in the United States. Unless something happens to stop this epidemic, those figures will triple by 2050.[33]

The unequivocal diagnosis of Alzheimer's has required the clinical symptoms of memory loss and diminished ability to carry out daily activities plus the classic brain autopsy findings of two sick proteins: plaques of beta-amyloid protein and neurofibrillary tangles in the temporal lobe filled with tau protein. The diagnosis is difficult to make, since dementia is frequent among the elderly, and at least 20 percent is unrelated to Alzheimer's disease.[34]

In 2010, the field underwent a big shakeup for the first time in twenty-five years, partly related to advances in imaging and biomarkers.[35] For the first time there was consensus reached by the National Institute of Aging (part of the NIH) and the Alzheimer's Association that the disease goes through three major phases: a preclinical phase, when there is accumulation of beta-amyloid but no symptoms or clinical manifestations, lasting ten or even twenty years; a phase of mild cognitive impairment (subdivided into early and late), lasting one to four years, characterized by memory lapses and later by poor decision making; and finally progressive dementia, with inability to carry out activities of daily life.[36] During the preclinical phases and mild cognitive impairment phases, beta-amyloid deposits in regions of the brain that form new memories, and this induces damage to synapses (the connecting space between brain cells).[37] Later, approximately one to five years before the diagnosis, tau is building up inside the brain cells of the temporal lobes, and extensive phosphate groups attach to this sick protein, forming tangles and causing further damage. Ultimately, areas such as the hippocampus (the key memory area and the place where beta-amyloid markedly accumulates) and the cortex (important for higher-level cognitive function) atrophy, as a result of the substantial death of brain cells.[38]

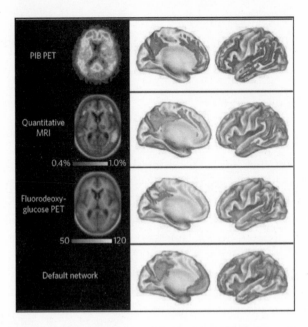

FIGURE 6.2: Different imaging modalities to show the characteristic findings in Alzheimer's disease. At top is the Pittsburgh B Compound (PIB) PET imaging, with corresponding parts of the brain schematically depicted. Similarly, for quantitative MRI and fluorodeoxyglucose PET, the affected areas of the brain are shown. The "default network" is the part of the brain involved in mind wandering (daydreaming) and in the individual's sense of self, which is especially affected in Alzheimer's. Source: R. Perrin, "Multimodal Techniques for Diagnosis and Prognosis of Alzheimer's Disease," *Nature* 461 (2009): 916–22.

Part of the acceptance of this path to Alzheimer's was attributed to the use of brain imaging and chiefly two different modalities: magnetic resonance imaging and nuclear positron emission tomography (PET) scanning. There are two different MRI scans for Alzheimer's—structural (also known as volumetric) and functional. Structural is used to quantify the atrophy of the brain. The functional MRI enables us to see which parts of the brain are activated by specific tasks. The PET nuclear imaging, by using glucose tagged with a radioactive isotope, reveals activity too; low uptake of the sugar reflects inactive or dead cells.[39]

The "new new" imaging thing in 2010 was a concept incubating for several years—the PET scan with Pittsburgh Compound B (PIB) (see Figure 6.2), which relies on a radioactive fluoride compound.[40] There was convincing evidence that PIB could track the actual beta-amyloid plaque—for the first time, the medical field had the ability to light up the parts of the

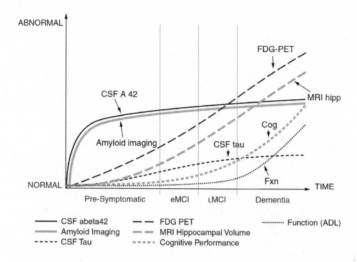

FIGURE 6.3: The pattern of various biomarkers over time in the stages of Alzheimer's disease from presymptomatic to early and late mild cognitive impairment (eMCI and lMCI), displaying the trends in imaging, cerebrospinal fluid, activities of daily life (ADL), and cognitive performance (Cog). Source: *Future Opportunities to Leverage the Alzheimer's Disease Neuroimaging Initiative* (Washington, DC: National Academies Press, 2010).

brain that had accumulated this sick protein.[41] This nuclear test (now called Amyvid) was recommended for approval by an expert panel convened by the FDA in 2011 based on data in 152 patients, but still awaits final commercial approval.

An alternative means of detecting the process was through performing a lumbar puncture to collect cerebrospinal fluid and measuring both the tau and the beta-amyloid proteins.[42] A high tau to amyloid ratio was found to be strongly suggestive of risk of developing Alzheimer's. The graph shown in Figure 6.3, part of the Institute of Medicine's 2010 report *Future Opportunities to Leverage the Alzheimer's Disease Neuroimaging Initiative*, pulls most of these concepts together.[43] In the preclinical phase PIB imaging can show the beta-amyloid deposition in the brain, and the cerebrospinal spinal fluid (CSF) will demonstrate presence of the amyloid protein species known as Abeta-42, which is especially prone to aggregation and damage. During mild cognitive impairment from early (eMCI) to late (lMCI), not only are these markers evident and progressing, but tau can be detected in the spinal fluid, MRI lights up the abnormal hippocampus, and the PET scan shows that regions of the brain are not metabolizing glucose normally. As MCI is progressing, cognitive impairment accelerates, and the deterioration of the individual's daily life functional capacity is fully manifest.[44]

While all of this tracking ability for early detection or risk of Alzheimer's is academically interesting, and will be helpful to develop effective drug or vaccine strategies, individuals will likely be reluctant to undergo spinal taps or multiple scanning procedures. There is considerable work being done to use a blood marker that would identify individuals with increased risk. Back in 2007, the company Satoris published intriguing findings using an eighteen-protein test that was nearly 90 percent accurate in detecting the likelihood of progressing from mild cognitive impairment to Alzheimer's.[45] The National Institute of Aging reported the correlation of blood levels of a protein known as Apoε with the amount of amyloid plaque in the brain, and a group of investigators led by the University of California, San Francisco, showed the correlation between a beta-amyloid blood plasma level and subsequent cognitive decline.[46] Another recently developed blood test approach leverages the presence of auto-antibodies.[47] The Alzheimer Disease Neuroimaging Initiative (ADNI) Consortium and the pharmaceutical industry are doing ongoing work in this area.[48] But as Sanjay Pimplikar wrote in a *New York Times* op-ed, "brain scans and spinal taps may do patients little good."[49]

To date, there are no effective therapies and a veritable graveyard of drugs (numbering more than twenty) that have blatantly failed.[50] The only commercially approved drugs are the so-called cholinergic drugs (donepezil, rivistigmine, galantamine), and their effect is minimal. The number of theories for the root cause of Alzheimer's is telling by itself—is it indexed to the sick beta-amyloid or tau proteins (production, accumulation, or clearance), or the damage to or intrinsic health of the synapses or the mitochondria? Part of the problem here may be that "Alzheimer's disease" may look similar in scans and via biomarkers, but for any given individual there may be a specific gene and biological pathway that are responsible. This is "the many roads to Rome" concept. For some it may involve beta-amyloid; for others, tau protein accumulation.

There are over a hundred drugs in development for prevention of Alzheimer's, reflecting that this condition is one of the highest priorities of the pharmaceutical industry.[51] In fact, except cancer therapies, no field in medicine has more novel drugs being pursued. The programs can be viewed as elegant, targeted approaches at nearly every level that has been theorized to play a role: blocking beta-amyloid production, preventing beta-amyloid aggregation, revving up beta-amyloid clearance, increasing brain resistance to beta-amyloid, inhibiting tau protein, improving the function of synapses, and preventing mitochondrial dysfunction.[52] Most of the drugs have had success in mice models that attempt to simulate the disease, but the failure of so many similarly validated drugs makes one wonder how good a model a mouse is. Nevertheless, the newfound ability to use imaging and biomarkers to define individuals at risk, and to intervene very early in its progress, or what in medicine is commonly referred to as "the natural history," may ultimately lead to effective prevention.

MAPPING THE BRAIN AND CONTROLLING THE MIND

Functional MRI (fMRI) has had an extraordinary ability to read a person's mind. A major feature of the technology is the ability to interview the subject, show images or movies, or monitor task performance, among many other stimuli or provocations, while performing the test. A recent study by Cal Tech tested the ability of individuals to deal with multiple competitive stimuli using images of Marilyn Monroe, Josh Brolin, Michael Jackson, and Venus Williams. The patients in the study had intractable epilepsy, so they also had intracranial electrodes placed in addition to the MRI scans at the time of surgery for possible resection of the part of the brain that is the culprit. The study showed the remarkable ability for the individuals to control their temporal lobe stimulation by focusing on a single image. Functional MRI has been used to determine whether an individual is lying, although the tests have not yet been permitted as evidence in the courtroom.[53]

Functional MRI has also been used to determine brain maps consistent with political activism, introspection, courage, confidence, feelings of romance toward and deep attachment to a significant other, the impact of learning to read on children and on those who are illiterate, the positive effect of placebo treatment, and a long list of human behaviors.[54] Medically, it has been applied to characterize autism, schizophrenia, depression, and attention deficit disorder. Mapping the brain with MRI has set the foundation for image-guided neurosurgical interventional procedures. The Japanese, in particular, have been using near-infrared spectroscopy (NIRS) as an alternative to fMRI for tracking blood flow by leveraging the property of hemoglobin to absorb much more light than other tissue constituents. In Japan it has been used to diagnose bipolar disorder and schizophrenia, along with other psychiatric disorders.[55]

One of the most extraordinary things I have ever seen in medicine is the effect of deep brain stimulation (DBS) in a patient with severe Parkinson's disease. In the late 1990s, I referred a patient to Dr. Ali Rezai, a neurosurgeon who headed up the program of neuromodulation at Cleveland Clinic. The patient had such severe tremors and involuntary movement that he was unable to walk or carry out most activities of daily living. After an extensive clinical evaluation and brain imaging, and mapping during the surgery with electrodes placed on the brain tissue, doctors placed a pacemaker lead in the patient's brain region known as the globus pallidus. A generator, placed in the neck, was connected to the lead and controlled whether the electrical activation was on or off. Following the surgery, the patient not only had a dramatic reduction of requirement for medications that he was not tolerating or responding to but was able to resume his prior activities, which were extensive and included playing golf. Not so many patients have such a dramatic impact from a brain pacemaker, and there are multiple potential complications, but it is striking to watch a patient turn on or off the generator and go from seriously physically impaired to appearing virtually nondisabled and fully ambulatory. Bringing this image to mind reinforces a relatively newfound and extraordinary capacity for a medical device implant to take control of one's brain.

Deep brain stimulation is currently used for some other indications, including severe essential tremor, Tourette's syndrome, and severe cases of obsessive-compulsive disorder, and it is being explored for a wide range of neurologic and neuropsychiatric conditions, including bipolar disorder, posttraumatic stress, epilepsy, depression, autism, and schizophrenia. In these investigational pursuits, the brain is mapped before surgery with an fMRI and during the operation with intracranial electrodes to define what specific region, if any, may benefit from pacemaker activation.

Dr. Rezai told me about another particularly memorable case. A patient had such severe depression that he was deemed catatonic, with no ability to show emotion and barely any physical activity. He had been this way for years, despite multiple drug treatments and the use of electroconvulsive therapy. He underwent the surgery while being only mildly sedated, which is typical for deep brain surgery, since the brain has no sensory receptors. Brain activation was mapped with electrodes on its surface. Dr. Rezai activated one region of the brain, and the patient smiled, probably for the first time in many years, and further mapping indicated it was an ideal spot for the pacemaker lead. All went well during the surgery, and the patient afterwards had an entirely different physical appearance and friendly behavior, interacting with the hospital staff and his spouse. He was discharged on a Friday. Over the weekend, however, Dr. Rezai got an urgent call from the patient's wife, who begged him to turn off the pacemaker. The patient, who hadn't wanted to have sex for fifteen years, now wanted to have sex continuously. Fortunately Dr. Rezai was able to titrate the electrical activity, and over time this unanticipated problem settled down.

Perhaps the most exceptional mind control story came in 2007, when Dr. Rezai and colleagues from Cornell University used brain mapping and deep brain stimulation in a thirty-eight-year-old man who, following severe traumatic brain injury, had been in a minimally conscious state, completely unable to communicate or respond, for more than six years. An fMRI showed that a large language network was preserved, so a deep brain stimulation was pursued in the hope it might restore consciousness. Indeed, within forty-eight hours of surgery, the patient became aroused, turned his head from side to side, had sustained eye opening, and over the weeks ahead became capable of naming objects, feeding himself, and being progressively interactive.[56]

Beyond brain pacemakers to stimulate the mind, there has been much interest in "brain training" or "brain calisthenics" with cognitive stimulation, with various tools like crossword puzzles and Sudoku; interactive computer software such as CogniFit, Posit Science, and Happy Neuron; video games like Nintendo Brain Age or Sega Brain Assist; and brain gyms. The theory, promoted particularly by companies with proprietary brain fitness products, is that training will promote brain plasticity—the ability to form new synapses. So far, this has not been convincingly proven, and a recent study of over 11,000 participants conducted in Cambridge using a six-week online stimulation program found no evidence of improvement in general cognitive functioning.[57]

CANCER IMAGING

Besides the use of imaging to screen and detect early cancer, such as the massive lung cancer study using spiral CT scanning previously discussed, or MRI to back up mammography for more accurate detection of breast cancer, the major use of imaging in cancer is to characterize the tumor and the extent of metastasis ("staging"), and then to track response to treatment.[58] Imaging that demonstrates shrinkage of the tumor for at least a six-month period is encouraging, but even that has not been shown to correlate with overall, long-term clinical benefit for patients.

In the early stage of testing a new drug, a marked response by imaging may be a solid indicator to fulfill the "proof of concept" that the therapy has considerable promise. In the case of malignant melanoma, which is usually fatal within one year of diagnosis, almost 60 percent of individuals carry a specific point mutation in their tumor in the gene BRAF (known as V600E).[59] By 2011, more than three hundred fifty patients with malignant melanoma received an oral drug directed to the V600E mutation in BRAF, and 81 percent had a striking response within two weeks as manifest by PET scan (see Figure 6.4).[60] There was a 63 percent increase in survival compared with conventional therapy (dacarbazine) at one year. Interestingly, those individuals without the BRAF mutation who received this highly targeted drug actually got worse. In the United States the use of imaging for following cancer treatment has exploded in recent years. Researchers at Duke reported on over 100,000 patients diagnosed and treated between 1999 to 2006 with a diverse array of cancers—breast, prostate, lymphoma, leukemia, colon, and lung. The annual increase in PET was from 36 to 54 percent, for nuclear bone scans 6 to 20 percent, for MRI 4 to 12 percent, and for ultrasound 1 to 7 percent. This imaging accounts for only 6 percent of the total annual costs—in excess of $250 billion—of cancer care, but it remains unclear whether the increase in imaging has led to better outcomes for patients.[61]

In certain situations the use of sophisticated imaging has been shown to make a difference. In a randomized trial of a particular type of lung cancer (non–small-cell), the combined use of PET with CT tomography was compared to conventional staging alone. The accuracy of the PET-CT approach was clear, with an increase in prevention of unnecessary lung surgery, which requires opening the chest. Combining ultrasound and MRI has been evaluated for improving the accuracy of breast cancer detection. The technique known as magnetic resonance elastography uses low-frequency sound waves at 60 hertz passed through plastic tubes while the patient is simultaneously undergoing the MRI. The variations in how the tissue moves by ultrasound, picked up by MRI, may help differentiate cancerous versus normal tissues. An alternative and much more practical approach for breast cancer imaging involves the use of a handheld near-infrared scanner that quantifies how light energy scatters through the tissue. Another handheld device that uses Raman spectroscopy, a technique that detects vibrational states of molecules, has been found useful for distinguishing melanoma from benign moles.[62]

PRINTING ORGANS

At the 2011 TED (Technology, Entertainment, Design) conference in Long Beach, California, Dr. Anthony Atala, the director of the Institute of Regenerative Medicine Institute at Wake Forest University, talked about "printing" a human kidney.[63] He discussed how the information from an individual's three-dimensional CT scan was sufficient to design and print

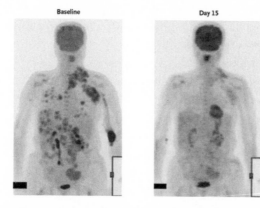

FIGURE 6.4: PET scan of an individual with malignant melanoma before and two weeks after taking an oral medication directed against
the BRAF V600E mutation, showing marked resolution of the metastasis, indicated by considerably less uptake of the isotope (darkened
areas throughout the chest, abdomen, and left arm). Source: G. Bollag,
"Clinical Efficacy of a RAF Inhibitor Needs Broad Target Blockade in
BRAF-Mutant Melanoma," *Nature* 467 (2010): 596–99.

a three-dimensional, working kidney for that patient. He talked about how
it takes "about seven hours to print a kidney" and how it is like baking a
cake, and then showed the audience a "kidney structure" that was made by
the high-tech printer. As part of the presentation, patient Luke Masella,
who had been born with spina bifida and ten years previously had received
a bladder tissue engineered by Atala, discussed how he was now leading a
regular life as a college student. The audience gave him a standing ovation.
The media reports were quite dramatic: "Surgeon Prints New Kidney on
Stage," "Need a Kidney? Just Hit Print," and "Next-Step in 3-D Printing:
Your Kidneys."[64] The reports confused the bladder tissue engineering with
a kidney being printed for Luke ten years earlier. For a brief period, it seemed
that we were ready to forego organ transplants, since all we needed to do
would be to print organs. Then Wake Forest University issued a press release
to correct the misconception: the kidney that Atala showed was just a mold
(we can't just blame the media—he certainly did not make that clear during
the presentation), without any blood vessels or internal structures, even
though it looked like a kidney from the outside. Nevertheless, the whole
incident brings to mind exciting futuristic possibilities.

Atala led a team in 2006 that published a report in the *Lancet* on their
experience of growing bladders in culture from a biopsy and then implanting

the organs into seven patients, ranging from four to nineteen years old, all of whom had been affected by spina bifida.[65] (Luke Masella was one of those patients.) To grow a whole bladder from the limited tissue derived from a biopsy, and then implant that back into individuals whose bladders were not functioning, was surely a biomedical triumph. It remains unclear whether it will become widespread. More recently an entire synthetic retina was generated from embryonic stem cells in mice. The bioengineered bladder and eye are possible because given the chance in a three-dimensional culture, cells are capable of self-organizing into highly specialized tissue. Another integration of 3-D printing and self-organizing cells was recently accomplished with a successful lab-made trachea transplant for a thirty-six-year-old patient with cancer of the windpipe. A plastic trachea was produced from 3-D images of the patient's own trachea, and stem cells derived from the patient's bone marrow were introduced onto this scaffold before the transplant.[66]

Printing organs—enabled in part by the exquisite three-dimensional images we can make of any organ—is another story. In 2011, the *Economist* had on its cover "Print Me a Stradivarius" and included a feature on the progress being made in three-dimensional printing, as manufacturers use the technology to make landing gear for airplanes, jewelry, boots, lampshades, racing car parts, parts of a violin, customized mobile phones, and solid-state batteries.[67] A 3-D printer is somewhat like an inkjet printer, and it works by printing successive layers according to a computer-aided design. The design itself can be extracted from 3-D CT scanning, which captures its images layer by layer. Already dental crowns and titanium medical implants for bone are now being produced via 3-D printers. The company Organovo, based in San Diego, is making solid progress in printing blood vessels. As one of their scientists was quoted in a *Wired* feature article, "Sir, Your Liver Is Ready: Behind the Scenes of Bioprinting," "Right now we're really good at printing blood vessels. We printed ten this week. We're still learning how to best condition them to be good, strong blood vessels."[68] Certainly the requirement of having an intact blood vessel supply for any organ makes this step a vital building block.

Parts of the story of organ printing are beginning to come together, but despite the excitement induced by the kidney mold at TED, bioprinting of organs has to be seen as several years away from reality. Nevertheless, the exceptional progress of 3-D medical imaging is what makes it all the more likely to be eventually possible. And the ability to go from digitizing an organ to someday printing a new one reinforces the concept of where digitizing human beings can take us.

ELECTRONIC HEALTH RECORDS AND HEALTH INFORMATION TECHNOLOGY

American health care is one of the last great industries to remain largely undisrupted by the information technology revolution of the past few decades.
—Vijay Vaitheeswaran, *The Economist*[1]

Information is the lifeblood of modern medicine, and health information is destined to be its circulatory system.
—David Blumenthal, *New England Journal of Medicine*[2]

EVENTS LIKE WHEN President John Kennedy was assassinated in Dallas, Texas, in 1963 or when the airplanes crashed into the World Trade Center towers on 9/11 become index, vivid, lifelong memories. Some, like these, are shared by almost everyone who was alive for the event. Some are more personal. For me, one such memory of significance occurred on November 29, 1999. The Institute of Medicine, a division of the National Academy of

Sciences, representing our most prestigious scientists and physician researchers, published the report *To Err Is Human*, which proclaimed that "at least 44,000 people, and perhaps as many as 98,000 people, die in hospitals each year as a result of medical errors that could have been prevented," but which arose because "faulty systems, processes, and conditions" led people either to make mistakes or to fail to prevent them. Beyond the human toll, these errors cost between $17 billion and $29 billion.[3] The report created a media frenzy that sent shock waves throughout the public. Medical errors are far more common and serious than had generally been perceived, and certainly far greater than what I had ever estimated. The errors killed more people than highway accidents and breast cancer combined, and more than six times greater than the deaths from AIDS at the time, which is particularly noteworthy since many people believed then that AIDS was by far the leading cause of death. This alarming exposé of medical errors raised awareness of how poorly and chaotically patient data is recorded and how it is notoriously inaccessible—perhaps representing the root cause of the problem. Until this point, most physicians had considered medical record keeping exceptionally mundane and inconsequential.

An alarming *Washington Post* editorial on this subject, "A Medical Enron," declared that "these various errors reflect the arrogance of the medical priesthood" and "thousands will continue to die needlessly with no one held to account."[4] A *New York Times* article on the day the report was released described a possible remedy: the report called for a new Center for Patient Safety and the minimum goal of reducing medical errors by 50 percent in the following five years.[5] There was a surprising omission in the plans, however: the computerization of medical records.

But the next Institute of Medicine (IOM) report on this subject—*Crossing the Quality Chasm: A New Health System for the 21st Century*, published in March 2001—zeroed in on health information technology: "Health care organizations, hospitals, and physician groups typically operate as separate 'silos,' acting without the benefit of complete information about the patient's condition." With medication errors especially prominent as a leading cause of hospital related deaths and harms, the IOM advisory committee pointed out the need for embracing automated computer systems for ordering medications, monitoring for proper dosing, averting mistaken prescriptions, and allergic reactions. It trumpeted the need for a frank change of the medical environment summarized by a couple of key sentences: "The committee calls for a nationwide infrastructure to support health care delivery, consumer health, quality measurement and improvement, public accountability, clinical and health services research and clinical education. This commitment should lead to the elimination of most handwritten clinical data by the end of the decade."[6]

Now a decade later, there is little to show for this shout-out. There is minimal hard evidence for either the reduction of medical errors or the adoption of electronic medical records. Forty-two percent of Americans report that someone in their family has been a victim of a medical error.[7] In a report in late 2010 in the *New England Journal of Medicine*, ten well-regarded North Carolina hospitals described how, having actively pursued quality improvement measures to reduce errors, they tracked 2,341 admissions between 2002 and 2007 to see whether there was evidence of improvement during that time. The data showed none. There were more than 60 patient injuries per 1,000 patient days both at the beginning and at the end of the study. Roughly two-thirds were deemed preventable. Overall, some form of harm to patients accompanied 25 percent of hospital admissions.[8] In 2011, the journal *Health Affairs* published a special issue, "Still Crossing the Chasm of Quality," with multiple articles on the persistent and serious problems related to medical errors. One group calculated the yearly cost of the errors in the United States was more than $17 billion, another found that the actual number of errors could be ten times greater than the estimates, and an expert in this topic from Johns Hopkins, Peter Pronovost, wrote, "For the past decade, health care quality has largely sought quick fixes and run from science; the results are evident."[9]

Other reports have been grimmer. In 2002, Dr. Barbara Starfield published her perspective in the *Journal of the American Medical Association*. She claimed hospitals killed roughly 225,000 patients per year: 113,000 by medication errors and 80,000 by hospital-acquired (nosocomial) infections, many attributable to unnecessary surgeries and procedures.[10] Later, in 2007 in the *British Medical Journal*, the Netherlands reported that approximately 30,000 patients each year suffer avoidable harm in the hospital, leading to about 1,700 deaths.[11] These and many other publications reinforced the notion that the IOM was not hyping the problem. What's more, these reports discussed only hospitals; many serious errors occur during office or clinic visits or while phoning in prescriptions or filling them at pharmacies.

The ultimate solution to this enormous problem is electronic record keeping. Of more than 3,000 American hospitals surveyed in 2009, only 1.5 percent had fully electronic health records and health information technology (HIT) systems, and these were largely confined to large teaching hospitals in big cities.[12] Only 4 percent of clinics and physician's offices were using fully electronic medical records.[13] The situation is not so bad elsewhere; in fact, the United States is one of the slowest countries to adopt

HIT in the developed world. Countries like Denmark have fully integrated hospital and clinic HIT, connecting virtually every citizen, physician, clinic, and hospital. Even in India, the Apollo hospital chain is a pacesetter for adoption and streamlining advanced electronic health records.[14] Among countries ranked for health information technology capabilities, the United States ranked eighth after New Zealand, Australia, United Kingdom, Italy, Netherlands, Sweden, and Germany.[15]

UNBUILDING A TOWER OF BABEL

Digitizing the words and files from the hospital chart or the doctor's office notes is the core of the electronic health (or medical) record. Ideally, the EHR, as it is more frequently being referred to, would be a comprehensive file that includes all laboratory data and reports from procedures, operations, diagnostic tests, hospital discharges, and office visits with all physicians and health care practitioners. The remarkable fragmentation of the American health care system makes this difficult. The average person in the United States age sixty-five or older receives care from seven physicians spread through four organizations each year. Even if some or many of those groups use EHRs, only in select circumstances are they collated. Even if they tried, collation might be difficult, as hundreds of companies, hospitals, and physician practices have developed proprietary, unconnected software for EHRs. Taken together, we get fragmentation to an exponential level. You might just call it chaos, or a Tower of Babel. The buzzword goal to achieve is "interoperability." "Inoperable" is perhaps the best description of our current, incompatible systems. Nevertheless, pilot projects in the United States, involving most of the large EHR companies, were launched in 2011 to foster an open-source model, which might break down some of the interoperability barriers.[16]

An alternative or complementary strategy to EHRs is the electronic personal health record (PHR), in which all of the data is coalesced and made available to the patient via a flash drive or a login to cloud-based information or both.[17] The advantage of the PHR is that the patient is the mechanism for collecting each piece of data as he or she travels from doctor to doctor, test to test, and one health care system to another. It sounds attractive and would well serve the individual who is on vacation in a remote place and has an unanticipated medical problem. Getting the data to populate the PHR, as I will discuss, represents a formidable challenge.

Truly powerful heath information technology (HIT) is more than just records, however.[18] For example, it might alert a physician writing a prescription for a medication allergy. Or it might provide decision support to inform the physician that a patient needs a vaccination or that an imaging test that was just ordered is not supported by the latest evidence from the medical literature. HIT also includes radio-frequency identification tags or bar codes to track medications given in the hospital setting.[19] Without these capabilities, EHRs alone would not be enough to reduce medication errors. The aggregate database of HIT would allow for patients to check their own laboratory or test results by logging into the system. And far beyond this, HIT could be used for population surveillance, such as detecting the earliest signs of an impending epidemic or the adverse effects of a newly released prescription medication. Aggregating HIT systems throughout the country—and in developed countries, or even worldwide—would make it very effective indeed.

In 2006, in the *Annals of Internal Medicine*, a group of researchers from Los Angeles combed 257 studies to see just how effective electronic patient data could be in avoiding those 100,000—or 225,000, or more—deaths by medical care. HIT was shown to improve compliance to guidelines, decrease medication errors, and improve monitoring for such preventive tactics as flu vaccination. However, they concluded "a disproportionate amount of literature on the benefits that have been realized from a small set of early-adopter institutions that implemented internally developed health information technological systems."[20] In fact, one-fourth of the studies came from just four academic teaching hospitals, and only 9 of the 257 studies used commercially developed EHRs. In 2009, a group of University of Minnesota researchers followed four years of Medicare data to determine whether EHRs had any improvement in patient safety, and they found minimal evidence of support—such as two infections fewer each a year at an average hospital. They concluded, "Health IT's true value remains uncertain."[21]

In late 2010, the evidence that EHRs reduce errors was summarized as follows: "EHR users overwhelmingly report improvement in the quality of care they provide. On the other hand, despite experts' optimism, there is currently no evidence that the use of EHRs reduces diagnostic errors."[22] Yet another study in 2011 on outpatient care concluded there was "no consistent association between better quality of care and electronic medical records."[23] Indeed, one of the problems is connecting the dots. Just collecting data, without processing it into actionable information and providing vital feedback to physicians, nurses, and patients, is not enough. Automating a broken process won't provide the fix. As Steve Lohr asserted in a recent review of the topic in the *New York Times*, "what is also beyond doubt is that the promise of digital records will be unfulfilled if doctors refuse to adopt them, because they regard the technology as cumbersome, time-consuming and possibly dangerous."[24]

Another important dimension of the problem, which will be approached by enhanced digitization of humans in the future, was articulated by my colleague Paul Yock, a cardiologist at Stanford, as quoted in the *Economist*: "The dirty little secret about medicine is that we physicians make decisions all the time based on woefully incomplete information."[25] This brings us back to the fundamental tenet of Chapter 2—incomplete information. Comprehensive digital medical records for each individual, it is hoped, can provide critical scaffolding to alleviate this deficiency.

EHRs may not be saving large numbers of lives yet, but they do already provide significant improvements to medicine. Think of the illegible handwriting of doctors, which, many have joked, is a requirement for graduating from medical school. Eradicating that alone would be a worthwhile digital objective, but when you add real-time abilities to capture, store, exchange, access, and analyze vital medical information—theoretically, from anywhere in the world—it adds up to be irresistible. We can give patients the record of their visit immediately, with recommendations and medications spelled out. HITs can, as I outlined above, provide guidance or evidence for medical decision making, or even make the correct diagnosis in the first place.[26] They can certainly reduce costs, improve productivity and quality of care, and reduce malpractice liability. They can also eliminate the unnecessary duplication of blood tests, X-rays, and other diagnostics; indeed, at least 10 percent of such tests are estimated to be redundant, which amounts to billions of dollars of waste each year.[27]

The American think tank the RAND Corporation projected that there could ultimately be $77 billion a year saved from efficiency if 90 percent of doctors and hospitals adopted HIT.[28] With the anticipated reduction of medication errors and other harms, RAND forecasted the savings figure could easily double. A follow-up report from the IOM on medication errors provided some quantitative context of the problem: "The use of medications is ubiquitous. In any given week, more than four of five U.S. adults take at least one medication and almost a third take at least five different medications." According to the IOM, there are 1.5 million people injured by medication errors each year, and hospital related medication errors alone result in a cost of over $3.5 billion per year.[29] The Society of Actuaries estimated the overall cost of medical errors, for which medications represent just one component. Their study found that there were 6.3 million medical injuries in the United States in 2008 and that the average cost per error was $13,000.[30] More than ten million excess days of missed work and short-term disability were part of the calculated $19.5 billion nationwide cost for 2008. The discrepancy between the financial estimates of the RAND Corporation and the Society of Actuaries is noted, but both groups come up with very big numbers for economic impact.

Of course, this won't all necessarily go without a hitch—the potential problems are neatly captured by a terrific parody website of EHRs known as Extormity. Extormity's tagline is "expensive, exasperating, exhausting." Here are some notable quotes on the Extormity site: "Our slow and painful change process significantly interrupts patient volumes and revenues, and this cumbersome transformation can only be appreciated in hindsight and with the aid of prescription medication," and "Extormity service and support is available on an hourly fee basis, with a minimum initial commitment of 225 hours at an hourly rate generated by a confusing algorithm." Under the "Perpetual Investment" tab:

> Operating the Extormity Bundle requires a phalanx of servers, which of course need to be replicated for redundancy. Fortunately, Extormity acts as a value-added reseller of these servers, which we pre-load with operating software. This allows us to mark-up the cost of the servers and charge for server configuration. In addition, the server software carries with it steep annual license fees. In addition to investments in servers and software, these servers generally have additional physical, communication, power and environmental requirements. Planning for this additional infrastructure can be provided by the Extormity Strategic Consulting unit, with implementation provided by the Extormity Solutions and Services Business Unit. These Extormity business units operate in silos, ensuring that you receive and pay for duplicated services. Of course, the Extormity EMR Software Suite must be integrated with other systems. Extormity software development engineers create custom, one-of-a-kind interfaces for each and every system or piece of equipment. In the event that we have already built an interface with a particular vendor, we employ a 'reinvent the wheel' framework so that each of our customers receives a solution customized for their special needs, no matter how similar to work we have already performed for another client.[31]

Who said electronic health records and HIT couldn't be entertaining?

MODEL EHR HEALTH SYSTEMS

Whenever EHRs and HIT are advocated, two large model systems in the United States are invariably cited—the Department of Veterans Affairs' Veterans Health Administration (VHA) and the medical firm Kaiser Permanente.

I have to laugh when I hear that the VA has come up with the model HIT and EHRs for the United States. When I was training in internal medicine from 1979 to 1982 at the University of California, San Francisco, we rotated through Fort Miley VA Hospital for one-third of the three-year experience. Fort Miley is one of the most beautiful hospital settings in the United States, located in the northwest corner of San Francisco overlooking the rugged Pacific Coast and with a glorious view of the Golden Gate Bridge. The patients typically arrived by public transit, and they seemingly always brought along a suitcase, anticipating their hospital admission. The hospital was a much better place to stay than where most were living: luxurious quarters with food that was at the very least edible, room service and meals in bed with great TV reception, a spectacular view from the predominantly semiprivate and some private large rooms, and the chance to feel better—all for free.

Otherwise, the place had problems. In all the time I spent at Fort Miley seeing hundreds of patients, I can only barely remember ever getting the charts at the right time—before or even while evaluating the patient. The protocol then was you had to order the chart at least a week in advance. The chance it would actually show up was maybe 50 percent. Each morning and afternoon, several carts would be wheeled out with dozens of thick charts, and you would go on a treasure hunt to see if any of your assigned patient charts was one of the lucky ones to arrive. (As badly as this system worked, it was useless if a patient showed up in the emergency room unexpectedly.) The absence of the charts was exacerbated by the complexity of the patients, who typically had five to eight significant chronic diseases and at least fifteen medications. The average patient I saw in that era was diabetic and obese; had chronic obstructive pulmonary disease from cigarettes, congestive heart failure from coronary artery disease (also from smoking, in part), gastritis and ulcer disease, atrial fibrillation, and intermittent claudication from peripheral vascular disease; and abused alcohol. Imagine trying to come up with a sound plan starting from scratch and having a short time to process the minimal information that could be provided directly from the patient, whose memory was frequently impaired, in absence of a medical chart. Not so infrequently the patients had also severe cirrhosis and a corresponding "liver flap" of his hands that goes with cognitive impairment (known as hepatic encephalopathy), but without a chart you would have to guess at the condition. If you were ever lucky enough to get the chart, it was hard to decipher the handwritten notes of dubious legibility along with a chaotic, nonchronological collection of clinic visits and prior hospitalizations. And now this is the leading HIT system in the United States. Perhaps it's the sign that there is hope for EHRs and HIT in this country after all!

The VHA is the largest integrated health system in the United States and one of the earliest adopters of EHRs and HIT. Reports from the VHA that reinforce its value include using EHRs to systematically track the need for and administer pneumococcal vaccines: the rate of vaccination doubled with a halving of the hospitalizations for pneumonia. This was projected to prevent the deaths of approximately 6,000 veterans who had emphysema, saving $40 million, and the initiative became a national benchmark. The other frequently cited evidence of the efficacy of the VHA is the medication error rate of 7 per 1 million prescriptions; in the United States overall, the rate is more than 7,000 times higher.[32] My colleagues who work at the VHA continually point out the ease of getting access to any patient's EHR. There are some problems; the ability to search the VHA database is known to be suboptimal, and the system is closed to outside facilities and practitioners. Within the VHA, however, the system works exceptionally well.

The other model is Kaiser Permanente, a very large health system comprising nine million individuals, 14,000 physicians, 431 medical offices, and thirty-six hospitals in nine states and the District of Columbia. Kaiser spent $4 billion in 2003 to develop the largest civilian installation of EHRs. Beyond providing integrated EHRs for all of its hospital and office visits, it has used the aggregated data to propel quality of care initiatives, such as the assured use of a preventive medication for specific conditions or providing feedback on how practitioners are performing against each other for productivity or patient satisfaction metrics. The Kaiser HIT system alerts its doctors and nurses on screening practices, reminds them to schedule follow-up appointments when they're necessary, provides some decision support and relevant guideline information for many conditions and diagnoses, and can monitor side effects of medications. Kaiser's patient database served as one of the first means to recognize the higher than expected incidence of heart attacks that occurred as a result of Vioxx. One of the most impressive features of its HIT is to enable more than 3 million of its patients to access their data and fully communicate with their physicians via secure email, leading to a 26 percent decrease in office visits over the past four years in a recent study. On any given day, more than 100,000 patients access their data. All Kaiser patients can tap into their laboratory data, which is an efficient way to transmit the information, because all too often physicians in practice do not have the time to follow through in getting the results back to patients in a timely fashion or at all. But like the VHA, it does not communicate well with non-Kaiser facilities or heath care practitioners; it is an effective but closed loop. There are efforts to fix that; in San Diego, for example, the VHA and Kaiser are exploring methods to combine their databases and make information transferable between their systems.[33]

CHALLENGES WITH EHRS

In contradistinction to the virtues of EHRs and these successful, albeit rarified models, there are many limitations and challenges. The absence of interoperability is a huge problem, and because physicians and hospitals are too often not connected to start with, and frequently at odds, collaboration in developing EHRs and HIT is difficult. The most serious concerns, however, lie at the level of the individual physician-patient encounter.[34] Rather than looking the patient in the eye, the physician is looking at a screen and typing in the data. Most doctors are uncomfortable typing and are slow and prone to making errors. Instead of the customary pre-EHR era, in which the doctor freely narrated particular thoughts about the patient's symptoms, condition, or treatment, everything is point and click. Dr. Danielle Ofri recently wrote about this challenge, including 1,000-character limits to the typed assessments of patients and an untoward influence on how doctors think: "The system encourages fragmented documentation, with different aspects of a patient's condition secreted in unconnected fields, so it's much harder to keep a global synthesis of the patient in mind."[35]

So in the mind of the physician, a lot of insight is missing as a result of less direct communication with the patient and less ability to freely express one's thoughts. Symmetrically, the patient feels less direct contact and is often disturbed by the doctor's pecking on the keyboard and looking at a screen. The sense of not being heard or understood is often prompted by this distracted, electronically fettered encounter. Such feelings are exacerbated with the brief time that the doctor and patient come together, typically less than ten minutes, and frequently this occurs after more than an hour of waiting.

One result of this bilateral dissatisfaction has been hiring scribes to enter the information in an EHR as the doctor and patient have a real visit. The reaction of using EHRs in one health system in Portland, Oregon, is telling. The 140 physicians, nurses, and physician assistants there faced a crisis in morale because they were spending three hours after a shift working on EHR data entry. They hired three scribes to help them and radically improved the situation. The president of the group said, "It allows our doctors to see more patients, physician satisfaction is up a lot, and patients are much more satisfied because the doctors can spend time on bedside interaction. Now, if you don't have a scribe it feels like you're showing up without a stethoscope."[36] The fast pace of seeing patients in the emergency department, along with the requirement to enter EHR data, has also made the use of scribes there particularly popular.

The scribes are mostly young people, often premed or prenursing students, who get paid only $8 to $10 per hour.[37] There are already three big companies that train and coordinate the hiring of scribes: Scribe America in California, which has fifty programs; PhysAssist in Texas with forty; and Emergency Medicine Scribe Systems in California with thirty. So much for bringing costs down and efficiency up—EHRs have managed to spawn a new profession.

This compensatory mechanism is just one small feature that introduces the extraordinary expense of building EHRs and highly functional HIT systems. For hospitals, it is estimated to cost at least $100,000 per bed, and this led to Jonathan Bush, CEO at Athena Health, an HIT company, and cousin to ex-president George W. Bush, to declare, "Hospitals will enter a financial crisis on the scale of the subprime mortgage crisis."[38] The cost estimates for converting outpatient practices to EHR vary considerably, and part of this relates to the legacy paper records. There is no magic on-off switch to digital records; to do this right it involves digitizing records of each patient going back many years. With these high cost considerations, many groups have seen little to no incentive to go digital.

That changed, in 2009, when the Obama administration came up with HITECH, the Health Information Technology for Economic and Clinical Health Act. This was part of the economic stimulus bill and allocated potentially over $100,000 for each doctor ($44,000 through Medicare and $63,750 through Medicaid) and between $2 million to $10 million for each hospital to become "meaningful users" of EHRs.[39] This program represents an investment of more than $36 billion over ten years to accelerate adoption of EHRs and HIT.[40] As David Blumenthal, the initial leader of HITECH put it, "this funding will provide support to achieve liftoff for the creation

of a nationwide system of EHRs."[41] Note the national integrated character-
ization that is intended, even though the bare bones of just getting offices
and hospitals off the ground is where we sit today. Although President
Obama said in a speech in January 2009, before taking office, that "we will
make the immediate investments necessary to ensure that within five years
all of America's medical records are computerized," the goal will likely prove
elusive. Only 10 percent of hospitals and 20 percent of physicians were
using these systems in 2011. Meaningful use has now been defined by
HITECH; there are very specific, conservative criteria to fulfill, such as having more than 50 percent of patients with vital signs data, smoking status,
and demographic data (e.g., sex, date of birth, ancestry) and more than 80
percent with a complete and accurate medication and allergy list. Furthermore, more than 50 percent of patients must be able to get a copy of their
EHR within three business days.[42] The criteria from the HIT side were even
easier to meet (for example, performing at least one test of data submission
to a public health agency or demonstrating capability of electronically sharing some information among providers).

Meaningful use looks easy to me, but many health systems called the
HITECH requirements unrealistic. Notably, Intermountain Healthcare
based in Salt Lake City, Utah, stated that it "could not meet 36 of the 48
meaningful use requirements." The reason this is particularly important is
that Intermountain, like the VHA and Kaiser, is considered a model system
of HIT and has been frequently praised by President Obama. Dr. Thomas
Lee, president of Partners HealthCare physician network, which includes
some of the prestigious Harvard teaching hospitals in Boston, concurred
with Intermountain's position.[43] Nevertheless, they need to do it; in addition
to financial support, the law also includes penalties, including decreased reimbursement from the government if meaningful use criteria are not met
by 2015.

The initial period after installation of EHRs and HIT systems often
sees an increase in errors. In 2010, the FDA received more than two hundred fifty reports of HIT bugs, flaws, or crashes, with multiple deaths and
injuries as a result.[44] Examples included incorrect data on allergies and
blood pressure, and it is widely thought that EHR errors have been grossly
underreported. That assertion is substantiated by many isolated reports
at a variety of health systems. For example, at Geisinger Health System
in Danville, Pennsylvania, $35 million was spent in 2005 to purchase and
install the Epic EHR system, but incompatibility between the pharmacy
database and the Epic system led to several major medication errors per
week.[45] Children's National Medical Center in Washington, DC, had an

eightfold increase in dosage errors for high-risk medications after spending $30 million to purchase the Cerner EHR system; it switched back to paper records until the glitches could be worked out.[46] At the University of Pittsburgh Children's Hospital there was a doubling of patient deaths in the five-month period following installation of a computerized order-entry system, which was mainly related to marked delays in administration of the prescribed drugs and attributable to the then-new Cerner software.[47] In 2011, a group from the RAND Corporation published a study in the *American Journal of Managed Care* on the process of health systems going electronic. The researchers told the *Wall Street Journal* that "trying to introduce an EHR system to an already complex health-care workplace causes a myriad of unintended consequences in terms of workflow and communication."[48]

On the other hand, integrating EHRs and a bar code of all medications has recently been shown to reduce medication errors substantially. At Harvard's Brigham and Women's Hospital, with over 14,000 bar-coded medications and more than 3,000 physician orders transcribed, there was a 41 percent reduction in error rate and a 51 percent reduction of adverse drug events compared to traditional, non–bar coding of medication prescription and administration to patients. Interestingly, 39 percent of the errors were at the level of physician ordering, and 12 percent were from transcribing the physician orders. Dispensing by the pharmacist accounted for 11 percent of the errors, and nurse administration of the drug was responsible for the remaining 38 percent.[49] The encouraging findings from this study suggest that when an EHR is integrated with other electronic tagging and surveillance, a marked improvement in quality may be achievable.

Cloud computing could be ideal for HIT with relatively inexpensive, seemingly limitless places to store, process, and maintain patient and medical information. Nevertheless, the reluctance to embrace EHRs is matched by a reluctance to place the data in the cloud. A 2011 *Economist* article on "Heads in the Cloud" highlighted the "reflexive conservatism and technophobia of medical folk" and that the medical industry can be seen as anti-innovative. Still, there hasn't been total rejection of the idea. One big chain of rehabilitation centers has incorporated both cloud computing and connected mobile devices to transmit all health data. Some U.S. health systems are embracing the use of private cloud computing to lower costs of HIT. Rather than spending $80 million on a new data center, the University of Pittsburgh health system, representing twenty hospitals, 50,000 employees, 4000 physicians, and over $8 billion annual revenue, has moved most of its computing operations to a private cloud.[50]

PRIVACY AND SECURITY OF THE DATA

Beyond financial considerations and lack of confidence that going electronic will fix the problem with medical errors, the issues of privacy and security represent significant concerns. At least 80 percent of Americans fear that their health data could be stolen, used fraudulently, or abused for marketing purposes. The possibility that hackers could get access to the medical data for any individual looms large, especially when there have already been examples of unintended breaches of data security from a stolen laptop computer with personal health data on 300,000 individuals and a burglarized hard drive with data from almost 200,000 people. There is also the worry that employers or health insurers would get access to the data and use it in a discriminative or exploitative fashion.[51]

The basis for comparison is paper records, which are much less apt to be lost to a large-scale breach of security or an electronic "phishing" scheme.[52] But a single individual's medical record is relatively easy to access, and there is no way to tell if the record has been read or copied. An EHR's metadata—the data of the data, in the form of tags or descriptors for different data elements—serves as a permanent electronic footprint that can track when and where a record has been accessed or modified.[53] It can also enable partial transmittal of the information in an EHR. For example, if you are brought to an emergency room, should the hospital have complete access to your medical information? If you have a history of mental illness, should that data be accessible? Are the data for certain conditions like diabetes fair game but other information on previous cancer to be withheld? If other physicians are consulted in your care, should they have complete or partial access to your EHR? If you go for a second opinion, do you want that physician to know the first opinion?

Besides customization of the data flow, the tagged data elements and metadata properties are essential to building optimal privacy and security. The patient information data elements are encrypted or unreadable, when stored or transmitted, and ideally not ever stored on the same computer system as the one that holds the encryption key. The metadata and the patient data itself are inseparable and are typically protected by a digital signature. To maximize security, accessing the data can require two-factor authentication, with not only a password but also a smartcard, physical credential, or biometrics, such as a fingerprint. Predetermined customized authorization by the patient and extensive audit processes are further safeguards that are part and parcel of secure HIT systems.[54]

Ironically, although EHRs are felt to diminish medical errors and reduce the liability of malpractice, the metadata feature can also be seen to facilitate malpractice claims.[55] The recording of all electronic transmissions from the input of medication or other orders to time stamps of activity creates discoverable evidence and can be used to establish a doctor's culpability. If the EHR was modified at a later and inappropriate time after the treatment was rendered, this can be tagged and used against the physician. Such examples have already been tested in the courts and have demonstrated the game-changing potential of metadata in substantiating physician and hospital malpractice.[56] The other looming and intriguing feature of malpractice and EHRs as this field evolves is the anticipation that a hospital or doctor who has not adopted digital records or HIT will be liable for deviating from the standard of care.

OPEN NOTES TO PATIENTS

Although EHRs will give patients easy access to their full medical records, the legal right to access has only been assured in the United States since 1996, with the passage of the Health Insurance Portability and Accountability Act (HIPAA).[57] Nevertheless, very few patients have actually seen their records. The obstacles to access include a charge for copying the records, the ability to review records only when a doctor is present, and inordinate delays following patient requests. The long-standing lack of access to the patient's medical information reflects the traditional paternalistic views of medical professionals; moreover, until recent years, when a patient requested a copy of his or her medical records, the first thing a physician would think of was a lawsuit. Under HIPAA, the only legal exception to the right for full access is in the case of mental illness, if a psychiatrist believes that access to the medical record could be detrimental to the patient's condition.

Without question, there are marked potential advantages to full patient access to records. Not only are patients more engaged in their care with enhanced understanding of their condition, but such involvement may foster improved compliance, self-care, patient-doctor communication, and prevention of medical errors by having another pair of eyes (in this case the most important person of all) involved. As Morris Collen, a Kaiser Permanente physician—now in his nineties—and the father of HIT, with pioneering efforts dating back to the 1960s, points out, "the patient has lived with his medical problem and often knows it better than the doctor."[58]

But there are also several potential problems. One example is interpretation of medical jargon, such as the frequent use of "SOB," for shortness of breath, which could be taken by the patient to mean something altogether different in reading the statement "the patient appears SOB." And then there's NERD, which actually means "no evidence of recurrent disease." Or patients may take offense by descriptors of their appearance with common terms such as "obese" or the insinuation that the patient is "somatisizing" or is a "hypochondriac." Review of the notes may actually engender more confusion and anxiety for the patient and increased time for the physician to explain the note contents and clear up any misunderstandings. The impact of full access to medical records for patients is being prospectively studied by an initiative called "Open Notes," involving 25,000 patients in four health systems with well-developed EHR and HIT systems, funded by the Robert Wood Johnson Foundation. In the study, over one hundred primary care physicians prompt their patients to review their records through secure electronic patient portals.[59]

The potential drawbacks of EHRs—usability, cost, security, privacy, medico-legal liability—are also counterbalanced by the big picture, or macro effect, of ultimately creating a national health network that can be used for several meaningful purposes. For a new medication or device that is being released for commercial use, a coordinated national system using de-identified patient data can detect and quantify unanticipated or low-frequency side effects. Already a Swedish national registry has detected blood-clotting problems with drug-coated stents and problems with liver toxicity after the release of a new drug intended to prevent clotting.[60] In both cases the frequency of the side effect was less than 1 percent, but the large sample size of individuals in the population database made the detection possible.

The nationwide HIT would be ideal for the earliest possible detection of an impending epidemic of the flu or another pathogen, or the emergence of antibiotic resistance to a bacterial strain.[61] Such public heath electronic monitoring transforms the population to a clinical research platform. Instead of the random sampling, voluntary reporting, and incomplete methods that are currently used by government agencies like the Centers for Disease Control and the FDA, there is potential for comprehensive assessment and the capability of studying the effect of different treatments on large segments of the population.

ESCAPING THE SILO WITH
PERSONAL HEALTH RECORDS

Dr. Michael Harrison, a physician on the faculty at the University of California, San Francisco, has proposed a Twenty-Eighth Amendment to the U.S. Constitution, which reads as follows: "Like Life, Liberty, and the Pursuit of Happiness, the Pursuit of Health is an unalienable right of every living human being. Each of us has a right to information about our own body, access to knowledge about conditions that affect our health, and the responsibility to use it in the pursuit of health."[62] And an organization called the Markle Foundation takes a similar tack, arguing that the "primary beneficiary" of improving HIT must be the patient.[63] PHRs are one of the foundation's major goals, providing the means for each individual to control his or her own information over an entire lifetime from all providers with secure accessibility from any place at any time.

Patient-centered health records provide a potentially powerful tool for improving patient care as well as their ability to control their own health information. One study has shown significantly better use of preventive health services like flu shots, mammography, and colonoscopy.[64] They could have other powerful effects as well. For one, they might provide an escape from the problem posed by the myriad different EHR systems in use at different hospitals and practices. A new U.S. government program, known as Blue Button, enables every individual in the Medicare and VHA databases to download his or her data for personal records (although the quality and utility of the Blue Button–derived PHR remains unclear).[65]

Nevertheless, the embrace of PHRs is notably weak and in stark contrast to the enthusiasm and clear rationale for their potential benefit. While the Markle Foundation and American Medical Association survey indicated that only 44 percent of physicians are willing to use the patient's PHR as part of their medical effort, it may be even more surprising that only 2.7 percent of individuals have established a PHR. To date there have been only three major PHR products developed in the country—Microsoft's Health-Vault, Google Health, and WebMD's Health Manager—and all are struggling, even though they are free applications. In mid-2011, Google Health was shelved. Google Health was good for printing out or emailing a complete or partial record to share, but it didn't have tools to evaluate one's health. HealthVault has self-evaluation tools, and you can invite someone to view all or part of your records, but printing out parts of the PHR can be challenging. The WebMD product has the health evaluation features and is fine for printing, but it doesn't allow exporting the file in either of the two main industry formats.[66]

The real problem with the PHRs of today is not the presence or absence of particular features. That can be readily fixed. The central issue is populating all of the data fields, which is more than an arduous task. It requires dozens of hours of effort only after requesting and successfully retrieving all of one's medical records from a multitude of sources and providers. When we offered HealthVault as part of our research program on consumer genomics, out of more than 3,000 participants in the study, only 30 collated their data within a PHR. This problem will likely only get resolved when the importation process is automated, such as through a mechanism derived from hospital and physician EHRs. Since part of the HITECH definition of "meaningful use" includes transfer of data to patients, there has been some groundwork and financial incentive toward this worthy objective. A large survey conducted by United Health in 2011 suggested that 77 percent of individuals want a PHR to track medications, test results, and their medical data, and are not worried about cost or privacy issues. So it's chiefly a matter of getting a system that is user-friendly, interoperable, and hassle free.[67] With so many groups and institutions working on this goal, the prospects are encouraging.

THE EHRS AND PHRS OF THE FUTURE

Now that we are concluding our "tour" through the four principal digital arenas—genomics, wireless biosensors, imaging, and HIT—it is fitting to picture what the digital record of an individual will look like in the years ahead. Simply put, everything should be in it. This will inevitably require cloud-based storage: the data sets will be huge, at terabytes of data per individual, to encompass a record that starts prenatally and extends to the end of an individual's life. All of the actual imaging data should be incorporated to allow physicians to view the files directly rather than having to only rely on the report. The size of the typical files have been quantified by the Federal Communications Commission (see Figure 7.1). It will include one's DNA sequence along with all major forms of digital imaging.[68] It must also include data for all radiation exposure, as the risk is cumulative over one's entire life. Collectively, with a national or, even better, international public health monitoring capability, we will be able to determine such environmental effects of radiation exposure on increased risk of cancer, and

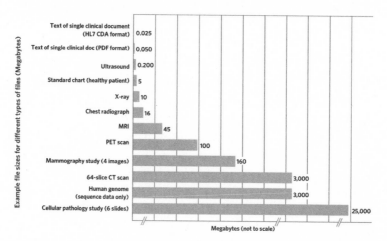

FIGURE 7.1: The size of various files and pieces of data that would be a part of any electronic and personal health record of the future.

particular genomic variations that predispose particular individuals to ill effects of radiation exposure at even low doses.

There are concerns about incorporating genomic data, since the GINA bill (discussed in Chapter 5) does not protect abuse of these data for life insurance or long-term disability coverage.[69] However, using tagged data elements should provide the means to keep everything together but partitioned to control access to DNA sequence and other "-omic" data as required.[70] This is for the protection not only of the individual but of any relatives. It is worth pointing out that no matter how granular and comprehensive the EHR and PHR are developed for an individual, this does not provide a guarantee of avoidance of medical errors or infallibility of the HIT system. But having all the essential and relevant health data from digitizing a baby, child, or adult, and having it be fully accessible to the individual or the parents, will undoubtedly foster better care—at both the individual and the population levels.

While some may consider the topic of electronic medical records prosaic, it should now be abundantly clear that their ultimate adoption and full interoperability will prove fundamental to the future of medicine. Only via full electronic convergence can all the tools of digital medicine be in sync and immediately useful. With the torrent of individualized data flow that is coming from whole-genome sequencing, remote physiologic monitoring, and medical imaging, electronic information storage and processing will become more essential than even envisioned today.

THE CONVERGENCE OF
HUMAN DATA CAPTURE

IN 2007, THE IPHONE, a radically innovative technology, took the digital world by storm. A slick mobile communication device that combined rapid access to the Internet with an impressive new personal computer operating system was a great achievement in its own right. An even greater innovation was the combination of its native capabilities with its open development platform, which led to hundreds of thousands of apps specifically designed for the device. It was this coalescence, or convergence, of distinct technological capabilities that has made the iPhone so formidable, so revolutionary.

Up to this point, I have been treating medical innovation in a manner analogous to a catalog that treats the operating system and applications discretely. Here, I start combining them, for that is where the real power lies. Even within the four modalities—wireless physiological monitoring, genomics, anatomical imaging, and electronic data storage—I have covered so far, there are hundreds of different permutations and combinations that demonstrate the power of technological convergence. I won't describe them all, but even by examining a limited number, I will still be able to demonstrate the great potential of digital medical convergence.

COMBINING WIRELESS SENSORS AND GENOMICS

Heart Attack

One of the biggest mistakes in cardiological dogma has been the belief that a cholesterol plaque gradually develops in one of the arteries that supply the heart muscle and that a heart attack is the final point of this progression. What a mistake that was! In the 1980s, by performing angiograms in the early hours of heart attacks (medically known as myocardial infarctions, MI), along with insights from autopsies of MI patients, we learned it was the sudden rupture or erosion of the plaque, which in the majority of patients was only minor or moderate, that was the proximate cause of heart attacks. This means that at one moment you could have a 20 percent narrowing of an artery and then—suddenly!—a crack in the artery wall develops, a blood clot forms, and a full-boat heart attack ensues. The "attack" means damage to the heart muscle, and if this is extensive enough, death can result.

While we now have ways to reopen the artery, either by dissolving the blood clot with an agent like the protein t-PA (Chapter 2) or by using rapid balloon angioplasty and stenting, such treatments are essentially fire drills, and it is hard to actually prevent damage to the heart muscle from occurring. The average patient takes two hours to arrive at a hospital after a heart attack begins, and treatment takes another hour to accomplish, so even though a patient with an artery reopened typically does well, there have still been three hours for the MI to damage the heart. What's more, hundreds of thousands of people each year suffer heart attacks and never make it to the hospital at all.

Naturally, then, preventing heart attacks, not treating them, would be a major achievement for cardiology. To this point, there have been no effective or refined strategies besides population medicine: lowering LDL cholesterol, avoidance of smoking, and modification of the various risk factors, such as excess weight, lack of exercise, high blood pressure, and uncontrolled diabetes. We have been doing stress tests for decades, only to have patients suddenly die of heart attacks within days or weeks of "passing" the test (recall the Tim Russert story in Chapter 3). We have had no way to identify the vulnerable individuals.

Now comes the ability to digitize humans and ultimately identify susceptible individuals. Sequencing the genome for risk variants will certainly be one way to do this, and we already know some important genes and genomic regions that have nothing to do with cholesterol but nevertheless indicate meaningful risk of a heart attack. A complementary strategy is the hunt for certain cells or their constituents in the blood. Around the time of a heart attack, a large number of cells from the artery lining are sloughed off into the circulation and can be detected by specialized assays that use antibodies and magnets to isolate the cells.[1] This method has been known for more than a decade. We also know that the syndrome of "unstable angina," a precursor to heart attack, is associated with these cells. By monitoring the presence of the nucleic-acid constituents of such cells in the blood, we may be able to know who is truly vulnerable to a heart attack, not just dichotomously but also at a particular moment in time.

Monitoring would ideally use an implanted nanosensor, smaller than a grain of sand and capable of finding its targets in even one millionth of a liter of blood, communicating with a patient's smart phone.[2] (See Figure 8.1.) Individuals who would get the nanosensors would be those whose genome sequence or other biomarkers had already put them at risk for heart attack. Well before the horse was out of the barn, the nanosensor could alert the individual to seek attention; therapy then would consist of both anti-clotting and anti-inflammatory medications. At some point further in the future, nanosensors will likely have the capacity to release medications on their own in response to high levels of circulating cells or nucleic acids. This closed-loop sensing and dosing model may seem far-fetched, but the technical capabilities exist today. Nano autopilot mode to prevent heart attack crashes will exist someday—sooner than you think.

This technologic capability was recently brought home with a report on a tiny implantable magnetic microsensor to detect heart damage.[3] *Fast Company* published a description of the sensor in its article "Heart Attack or Vicious Burrito? Embedded Sensor Knows."[4] While the sensors didn't pick up the risk of heart attack, and while they were only implanted in rodents, they can be considered a step in the right direction. It is not unusual for patients with a heart attack to think they have a severe case of indigestion, and the output from such a sensor would provide appropriate and immediate guidance. Moreover, some medications used for cancer therapy today, like doxorubicin, induce the death of heart muscle in some patients. In this report, such damage was rapidly and quantitatively detected.[5]

This concept of a sensor to detect a serious internal problem is something you are already familiar with in your car. When you see the "Check Engine" sign light up on your dashboard, you know what to do. Of course, a smart phone beeping that a heart attack (or other condition, as we'll see) is on the way would be far more terrifying. Even worse would be the possibility of a false alarm, so high accuracy for nanosensor monitoring would

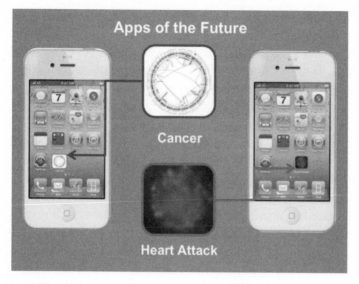

FIGURE 8.1: New apps for the smart phone that rely on an embedded nanosensor to detect a molecular signature in the blood.

be necessary before many of us have nanosensor chips embedded in us. Nanosense but no-nonsense has to be validated. There aren't yet apps for that, but they're coming.

Cancer

We're not very good at detecting and fighting cancer. The mass-screening model, as with mammography or prostate specific antigen (PSA) testing (discussed in Chapter 2), is enormously expensive and leads to an untold number of false positive results and more unnecessary biopsy procedures. Doing serial sensitive scans like PET or CT would likely make this problem worse, both by increasing the false positives and incidental findings and by exposing individuals to ionizing radiation that itself causes cancer. As one sequencing entrepreneur, Luke Nosek, has put it, "While we have 10X better computers and video games every ten years, we do not have 10X better cancer cures, and we do not really understand what causes the major killer of the first world other than the cop-out term 'aging.' We must change this."[6] Genomics offers real opportunities. I have discussed the use of genome sequencing of the tumor and germ line to guide therapy, but not for the prevention of cancer. "Gene peeking"—genome-wide association studies— for finding the genomic signatures of cancer risk will only become more powerful as whole-genome sequencing is undertaken in large numbers of individuals who have had cancer, making prediction less of an art and more of a science.

Other tools are available, too. Most tumors shed cells, known as circulating tumor cells (CTCs), which can be found in the blood.[7] Although we have the means of counting those cells, the assay is not widely used, and then only to monitor patients who have already been diagnosed with cancer. The same principles that apply to heart attacks apply here, however; implanting a nanosensor in at-risk individuals could reveal either CTCs or circulating nucleic acids, the latter being more convenient and likely even more sensitive to small signs of incipient cancer growth. The nanosensor communicates with the individual's smart phone and, if so desired, the primary care physician if a high level of circulating targets is identified. At that point it might be appropriate for the individual to undergo high-resolution imaging to define whether there is any "macro" evidence for cancer. This is biomarker-guided imaging, just like biologically directed therapy.

To date, only one method is approved in the United States, CellSearch, whereby cells that are validated as cancerous are counted from a single tube of blood. But a chip being developed at Massachusetts General Hospital can detect a single circulating tumor cell among more than a billion cells in the blood via a combination of antibodies and magnetic beads. An area of current research is to use these cells as a means of further characterizing the tumor, in a sense as a fluid-phased "biopsy" of the cancer tissue.[8] Isolating the cells sets up the opportunity to do gene expression to see which genes are activated and to sequence the DNA or RNA transcriptome. These techniques may ultimately provide a much more granular and insightful means of understanding the status of the individual's cancer at any point in time and provide meaningful guidance for targeted therapy.

In 2010, a group of researchers at Johns Hopkins University introduced a remarkably clever concept to track cancer using circulating DNA, which they called PARE—"personalized analysis of rearranged ends."[9] As Chapter 5 discussed, cancer is a genomic disease typically characterized by intra- and interchromosomal rearrangements of the somatic genome. These rearrangements lead to multiple, specific DNA fusions that have been readily identified through cancer genome sequencing. The success of the drug Gleevec, which targets an active fusion gene in patients with chronic myelogenous leukemia, is anchored to the same principle, but in this example the fusion gene represents the driver of the cancer, and it is a malignancy of blood cells that can be readily sampled.[10] With PARE, the object is to find DNA fusions in solid tumors—not because the fusions are likely to be driving the cancer or because they will provide guidance for therapy, but simply as biomarkers. In the first report of PARE, sequencing the tumor of four patients with colon cancer and two with breast cancer identified an average of nine DNA fusions each.[11] These fusions, which were patient- and tumor-specific, could then be assessed in the circulating tumor DNA to track the cancer.

The results showed the extraordinary ability to use the blood DNA fusions in each individual to monitor the results of surgical resection of the primary tumor or a site of metastasis and the response to chemotherapy. While this technique relies on extensive sequencing, computational analysis, and identification of DNA fusions, it may ultimately be scalable and be far more sensitive than current macro imaging modalities. In order for the method to be successfully adopted, it will need to be performed quickly, inexpensively, and reliably, and we'll need to establish whether DNA fusions are stable or whether further mutations might delete old fusions and create new ones. It is also encouraging to see that a common fusion gene associated with prostate cancer can be detected in the urine and may prove useful to predict the risk of prostate cancer risk and reduce unnecessary prostate biopsies (a topic discussed in Chapter 2).[12]

Nanosensors can also serve as a cancer breathalyzer.[13] At the Technicon Israel Institute in Haifa, Israel, sensors using gold nanoparticles have been used to detect distinct breath chemical signatures for various cancers, including lung, breast, and prostate cancer. Everyone's breath has certain organic chemical constituents, such as benzenes, alkenes, and alcohol; the engineer leading this effort, Hossam Haick, suggests that the breath signature will reveal if cells begin producing an abnormal array of organic chemicals, which is thought to happen before the cells become cancerous.[14] While quite provocative, this clearly needs considerably more validation.

Lastly, the use of fecal material holds promise for early detection of colon cancer. In a study of 1,100 patients, a search for four altered genes in the DNA in cells embedded in stool revealed 85 percent of the cancers found through standard colonoscopies. This may seem like a bit of a miss, but consider that Pap smears only detect 50 percent of cases of cervical cancers. An alternative strategy for detection of colon cancer via a blood epigenomic marker known as Septin 9 also appears to be promising, with very high sensitivity and specificity rates.[15]

Transplant Rejection

With any organ transplant, there is a risk that the recipient's body will reject it. Unfortunately, the standard approach for monitoring rejection is typically invasive, such as performing serial biopsies of heart tissues after a heart transplant. While these procedures are considered safe, they are traumatic: in the case of a heart transplant, a special catheter with a bioptome (small scissors to collect tissue) is inserted into the jugular vein of the neck to reach the heart. A better means—just validated—is to monitor what genes are being expressed in the blood, as rejection has a specific expression profile. A far more attractive and sensitive method, however, is to seek the DNA of the donor in the recipient's blood; high levels would indicate rejection of the tissue.[16] This would be well suited to detect via an embedded nanosensor, capitalizing on the abnormal appearance of the donor's DNA in the recipient's bloodstream.

Type 1 Diabetes

Type 1 diabetes is one of eighty different autoimmune disorders that affects humans, but it is one of the most important, especially in children. Typically it occurs around age eight, but in recent years, increasing cases have occurred earlier in the patients' lives, sometimes even before age five. The large number of genome-wide association studies have confirmed the autoimmune basis of this disease, with almost every gene of nearly thirty identified having something to do with the immune system. Slow, progressive destruction of the pancreatic beta-cells occurs years before the onset of diabetes, as shown in Figure 8.2: the child showed antibodies for pancreatic beta-cells at age seven but did not manifest the disease until age fourteen.[17]

Ongoing sequencing studies investigating Type 1 diabetics have identified rare gene variants that carry either exceptional risk or protection from this condition, helping us to zoom in on the root cause of the disease in certain individuals. The genes include interferon induced with helicase C domain 1 (IF1H1), sialic acid acetylesterase (SIAE), and the interferon regulatory factor 7 driven inflammatory network (IDIN).[18] With babies screened for these gene variants, along with many others yet to be discovered, it will be possible to know not only which infants have a significant risk of developing diabetes but also what gene and specific immune system defect is operative. By matching intervention to the defect—say, with specific vaccines or courses of immunotherapy—there is an excellent chance that the condition could be prevented. Although trials of prevention using islet cell auto-antibodies have uniformly failed in the past, they were only started after the children already had high levels of the antibodies in their blood, which means that cell destruction was well under way; furthermore, in none of those trials was anything done to determine the specific immunological

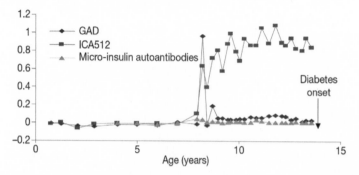

FIGURE 8.2: Development of different antibodies to the pancreas beta-islet cells detectable in the blood years before the diagnosis of diabetes in a child. Source: J. Bluestone, "Genetics, Pathogenesis and Clinical Interventions in Type 1 Diabetes," *Nature* 464 (2010): 1293–300.

pathway that was causing the diabetes. Trials with inhibitors of an immune system protein known as CD3, such as with the antibody tepluzimab, are currently ongoing to modulate the immune system and spare the destruction of the precious islet cells.[19]

There is much improvement to be done to the work undertaken so far. It will be expensive: with diabetes, there are "many roads to Rome"; much as any particular flu vaccine won't stop all strains of influenza, many treatments will be necessary to prevent the many kinds of Type 1 diabetes. But this remarkable feat is waiting to happen and, in the era of digitizing humans, we can expect a large proportion of Type 1 diabetes will someday be deemed fully preventable. At the very least, the same concept of an embedded nanosensor to detect levels of auto-antibodies or other molecular signals of pancreatic islet cells dropping off could be adopted as a solo strategy or used to know precisely when immune system modulation would be indicated.

Asthma

Asthma is the most common reason for prescription drugs in children, and one of the leading causes of death in children.[20] The airways in children are much smaller than in adults, so that any inflammation can have more serious impact. We now know the common gene variants associated with asthma, and we are beginning to identify the low-frequency and very rare gene variants too. We also know that a variety of environmental exposures are exceptionally important for precipitating asthma attacks, and this "exposome" is frequently a specific pattern for an individual.[21]

FIGURE 8.3: Prototype of a handheld genotyping and sequencing device.

The true prevention of asthma attacks relies on a multipronged genomic and wireless sensor approach. By knowing which individuals are susceptible based on DNA sequencing, and using targeted medications that address the specific root cause of asthma for the individuals, far better protection from inflammation of the airways could be achieved. But that will not likely be the complete approach; a useful adjunct would be biosensor monitoring of the key environmental metrics of air quality, pollen count, dust, and mold—coupled with physiologic metrics of oxygen concentration, respiratory rate, heart rate, and forced expiratory volume. Combined with a social network like the Asthmapolis program (see Chapter 4), digitized genomes and wireless sensors provide a potential solution either by preventing the disease from manifesting altogether or, at the least, by protecting individuals from severe attacks that lead to hospitalizations, disability, or death.

WIRELESS GENOMICS

Currently, the hardware for genome sequencing or ultra–high-throughput genotyping is expensive, typically costing more than $500,000; the reagents and the labor involved aren't cheap either. Although a "desktop" sequencer was introduced in 2011, an even bigger step is to enable handheld sequencing in a mobile device, which could be used for medical application of rapid sequencing of a pathogen, like influenza or a bacterium.[22] The prototype pictured in Figure 8.3 uses a microchip transistor to genotype or sequence, without any need for sample preparation or expensive reagents. A cotton swab from inside the cheek or a saliva sample is put into the device at the top end, and the genotype, or ultimately the sequence, can be rapidly generated and sent wirelessly anywhere at low cost. While this is not likely to evolve for whole-genome sequencing, it shows the potential for a fusion of technologies and even the possibility of real convergence between the smart phone per se and DNA sequencing. Already one's sequence data can be displayed and interpreted through the app "Genome Browser," which is available for the iPad.

GENOMICS AND DRUG DEVELOPMENT

Disease in a Dish

Another major convergence is developing from the ability to induce ordinary skin or blood cells to become pluripotent stem cells, which can be done by manipulating just four genes. Once created, these cells can be coaxed to grow into any tissue of interest, whether it be the heart, liver, or brain. (See Figure 8.4.) This technique has already been used to investigate many rare disorders, such as spinal muscular atrophy, Rett syndrome, and familial dysautonomia (the disease in Steven Pinker's family, discussed in Chapter 5) and now is being applied for common diseases like schizophrenia, Alzheimer's, amyotrophic lateral sclerosis, and Parkinson's.[23] In our laboratories at Scripps we are using this technique for individuals who are predisposed to heart attacks, creating artery lining endothelial cells and genomically editing the cells to understand what exactly underlies the process at the molecular level.

This technique should enable us to determine which drugs might be effective for preventing a disease in an individual who is predisposed to a serious illness. As it is not really possible to do a biopsy of an individual's artery or brain, being able to grow these tissues represents a formidable advance. A biotech company, iPerian, has been formed to use this concept for drug discovery. CEO Michael Venuti said, "Although we're not putting the patients themselves into the process, we are introducing them earlier than they have ever been included in the history of drug discovery."[24]

Being able to look at an individual's cells, differentiated to the organ or tissue of interest, in a dish and thereby determine the appropriate therapy or prevention may be considered the extreme form of individualized medicine. Recently, scientists at the Salk Institute were able to use lab-generated nerve cells to determine precisely which drug was effective in correcting neuronal defects found in a series of individuals with schizophrenia.[25] In a

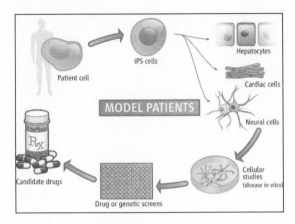

FIGURE 8.4: Schematic showing how the disease-in-a-dish process can be used to recapitulate the condition of interest in the target organ or tissue of choice and determine drug responsiveness. Source: G. Vogel, "Diseases in a Dish Take Off," *Science* 330 (2010): 1172–73.

recent trial, skin cells of patients with the rare form of early onset, familial Parkinson's Disease were obtained through biopsy and transformed into nerve cells; their genomic mutation was edited, and the diseased cell functions were restored to normal.[26] In another study, stem cells derived from a patient with a lethal heart arrhythmia genetic condition were coaxed into heart muscle cells, which fully mimicked the abnormal electrical conduction and served as a basis to test which drugs would be effective in blocking the defect—drugs that would not have otherwise been predicted to work.[27] These examples represent the epitome of digitizing human cells at the molecular level, which is likely to become an increasingly useful tool in the future to find optimal treatments and even cures.

Furthermore, we know that much of the genomics story, as reviewed in Chapter 5, is at the tissue and cellular levels. The regulatory part of the genome is dominant and quite tissue-specific, and this not only involves enhancers, promoters, repressors, and insulators but all of the epigenomics apparatus—methylation of DNA side chains as well as histone and chromatin modification. Just studying the genome of cells from the blood does not give us access to the vital operating instructions at each cell and tissue level. Certainly, insights from the cells may not be fully representative of the tissue from the individual's body, since it only accounts for one compartment of a highly integrated, human systems biology network. Nevertheless, it may forgo the need to breed transgenic mice, which is the customary approach to study genomic variations, and also may be a major

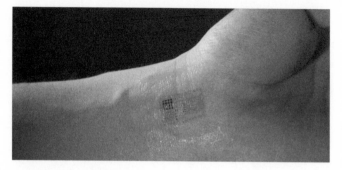

FIGURE 8.5: A chip tattoo.

improvement on the technique, as there have been many questions raised about whether such findings in mice are applicable to humans.

In a similar vein, a mechanically active "organ-on-a-chip" of a lung has been recently created in a microdevice that provides a platform for drug screening.[28] The ability to reconstitute cells, tissue, and organs of an individual to elucidate the molecular defect and determine appropriate remedies is a new extension of digitizing humans in vivo.

Electronic Skin

In 2011, a group of engineers from the University of Illinois published a breakthrough paper on chips that could be directly integrated with the skin, essentially creating a chip tattoo. These remarkably elastic chips were shown to capture an individual's heart rhythm and rate, muscle activity, and brain waves when attached to the appropriate body location (on the chest, near a muscle, or on the forehead, respectively).[29] (See Figure 8.5.)

Fetal Whole Genome Sequencing and Wireless Sensors

Recently, it was shown that the DNA of an unborn baby could be isolated and sequenced from the mother's blood sample.[30] Although the presence of free fetal DNA fragments in maternal blood has been known to exist since 1997, it took another decade and a half to determine whether the whole fetal genome would be represented and for the technological sequencing triumph to be accomplished. (See Figure 8.6.)

Why is this development important? Today, for the diagnosis of major genetic abnormalities of unborn children, we rely on the invasive procedures of amniocentesis or chorionic villus sampling. This procedure requires using

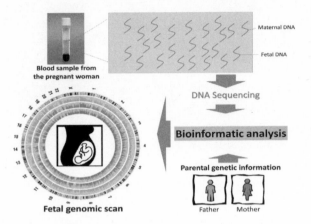

FIGURE 8.6: Schematic of taking a maternal blood sample from an expectant mother and capturing fragments of fetal DNA for sequencing, using the mother and father's DNA to help anchor the process. Source: G. Vogel, "Diseases in a Dish Take Off," *Science* 330 (2010): 1172–73.

a needle, with ultrasound imaging guidance, to extract some amniotic fluid in the sac around the fetus. The procedure is performed commonly, about 250,000 per year in the United States, and almost half of women who are pregnant and over age thirty-five will undergo this procedure. There is a risk of miscarriage in 0.2 to 0.3 percent of cases, along with a small risk of injury to the fetus, infection, and leakage of amniotic fluid.[31] But the newfound capability of sequencing the genome of the fetus through a maternal blood sample, using the mother and father's DNA sequence for anchoring its analysis, makes comprehensive prenatal whole-genome sequencing assessment possible.[32] Ultimately, this should supplant amniocentesis; even more powerful, however, is the ability to anticipate such risks as neonatal blood sugar dysregulation or sudden infant death syndrome. If the DNA indicated such possibilities, an appropriate biosensor could be used to monitor the baby and provide customized treatment.

Even the metabolic conditions (such as phenylketonuria, maple syrup urine disease and many others) screened for at birth, using a heel stick to obtain a droplet of blood from a newborn, typically take many days to weeks for the results to come back.[33] Such diagnoses could be anticipated well before birth by fetal DNA sequencing to assure the proper diet or treatment at the earliest possible time. Even more far-reaching and eminently achievable is the ability to use the sequencing data to modulate the diet or institute treatment while the baby is still *in utero*.

Genomics and Social Networking

Traditional methods of determining the origin of an infectious outbreak include specific genotyping and tracing contact between infected individuals. But a report in 2011 from researchers in British Columbia, the site of an outbreak of forty-one tuberculosis cases, combined two digital technologies and came up with completely different results compared with the classical methods.[34] Instead of simply genotyping specific markers, known as variable number tandem repeats (VNTR), of the tuberculosis strains, they performed whole-genome sequencing. As a result, they found they had two strains of the tuberculosis bacterium involved. The VNTR method had revealed only one. And instead of traditional contact tracing, they did comprehensive social networking analysis. By integrating that with the sequencing data, they were able to accurately identify "super-spreaders"—the individuals responsible for the outbreak. This combination of genomics and social-network analysis is likely to prove the new standard for investigating outbreaks of infectious disease.

This sampling of digital medical convergence is clear evidence that with these technologies, the sum is greater than the parts. It will not be at all unusual in due course for all four of the digital medical tools to be used in every individual. The results will be unprecedented. In 2011, the $10 million Tricorder X-Prize for mobile diagnostics was announced, anticipating this opportunity to change medicine via digital tool convergence. Its name derived from the handheld device used in *Star Trek*, the prize is aimed to "reward the inventor of a single portable device that, without human input, can diagnose an array of diseases with the same level of accuracy as a panel of physicians."[35]

Now that the tools have been laid out for capturing human data, along with the power of combining these tools, we are ready to examine how this will affect the future of medicine. The implications for the different players—doctors and the medical community, life science industry and regulators, and consumers—are remarkably heterogeneous yet unvarying in importance. The convergence and harmonization of these tools, players, and the digital world infrastructure will serve as the means of radical disruption of health care as we know it today.

PART THREE · THE IMPACT OF *HOMO DIGITUS*

DOCTORS WITH PLASTICITY?

Medical education is fundamentally conservative, indoctrinating new generations into the failed ways of the old. For too long we have hugged the shore of safe and acceptable tradition.
—Richard Horton, *The Lancet*[1]

We're using 3,000-year-old tools to deliver health care in the richest country on the planet.
—Jay Parkinson, *Fast Company*[2]

ATHEROSCLEROSIS, REFERRING TO a progressive and degenerative process of artery walls, is typically translated for a lay audience as "hardening of the arteries." We've never needed a similar word to describe the medical community. It came with sclerosis built in. Of all the professions represented on the planet, perhaps none is more resistant to change than physicians. If there were ever a group defined by lacking plasticity, it would first apply to doctors.

The inherent "hardness" of physicians and the medical community suggests they will have a difficult time adapting to the digital world. Before the emergence of the Internet, physicians were high priests, holding all the knowledge and expertise, not to be challenged or questioned by the lowly consumer patient. "Doctor knows best" was the pervasive sentiment, shared by patients and especially physicians.

Changes in the landscape began in the 1990s, and have been building exponentially since, with the emergence of scorecard medicine in magazines. People started to ask, essentially, why they couldn't have data about the skill or reliability of a physician when they wouldn't consider buying a car or TV without consulting a rating. It was only then that the old lack of transparency began to fade, and appreciation for how frequent and profound a problem that medical errors represented became apparent. As consumers grew comfortable with the Internet and health information online became more reliable, it became common for patients to arrive at office visits with a long list of Web-derived questions, or for a family member or designated advocate to show up at a hospital for an interrogation session of the health care team about the course of treatment or even the members' backgrounds. The blogosphere, patient online communities, and social networking sites like Facebook and Twitter amplified the trend. Suddenly the empowered patient—"someone who has figured out that healthcare is no longer best practiced in a paternalistic and beneficent way," as Trisha Torrey, founder of the website *Every Patient's Advocate*, puts it[3]—was no longer an oxymoron or an anomaly. It was par for the course: an aspect of the "shifting nature of trust" toward individuals and away from brands and institutions that Nick Bilton describes in *I Live in the Future*.[4] Medical advice itself was crowdsourced. Alternative treatments grew more popular, and suddenly magazines began offering do-it-yourself advice for avoiding doctors.[5]

Simultaneously, albeit probably coincidentally, visits to doctors across the United States were dropping.[6] And, of course, the most significant legislative reform in health care in several decades, particularly related to access and insurance coverage, was passed, and more than thirty million previously uninsured individuals were assured to get care. The alarms of a growing physician shortage were sounded.

Change was hitting doctors from other sides, too, leading them to sit on the sidelines through practically all those social movements. Reimbursement from insurance companies and government programs was quickly ratcheting down, forcing most practices to have to see more patients more quickly and to work longer hours. Unable to keep up with scientific discoveries occurring at an unprecedented and breakneck pace, physicians instead were preoccupied with RVUs ("relative value units," the basis of Medicare's fee compensation schedule), a preposterous number of forms and "bureaupathic" approvals and denials, and survival tactics to stay afloat and not be absorbed as a salaried employee of a local hospital. Justifiably, the principal goals were simply to take care of their patients, have a real life outside of work, and earn a living to support their families. The situation was captured nicely in a recent *New York Times* front-page feature profiling three generations of physicians. Rather than engage in the kinds of battles the older generation has confronted—maintaining an independent practice in the face of all those factors I have described—many young doctors aren't even trying; instead, they simply join a hospital. The father of a new practicing physician said of this trend, "On the one hand, it bothers me that the generation of doctors that my daughter is in doesn't work as many hours and isn't willing to do the stuff that I did. On the other hand, I'm almost a little jealous."[7]

One of my favorite authors is Atul Gawande, a surgeon who is particularly interested in promoting the quality of health care. In his book *The Checklist Manifesto* he compares medicine to aviation, exploring how the vital ritual of the pilot's checklist can inform the work of surgeons.[8] I think aviation holds a second lesson for medicine. Too many doctors today are trying to fly a biplane into the jet age, the indicators all analog and completely out of whack. It's no wonder they seem apt to crash.

All, however, is not lost. Doctors need to evolve, not just to survive but to thrive in the world of digital medicine. There is a crisis here, but I have a plan that could enable physicians to leverage their remarkable assets and create new opportunities in the exciting environment I've described so far. Nevertheless, as you will see, profound obstacles exist.

EDUCATION

Published in 1910, the famous Flexner report, *Medical Education in the United States and Canada*, a 346-page monograph, summarized Abrahams Flexner's investigation of the 155 medical schools that existed in North America at the time. One of the report's major conclusions was that, with no regard for public welfare, the schools had allowed "an enormous overproduction of uneducated and ill trained practitioners."[9] As a result of Flexner's indictment, the curriculum of medical schools was completely revamped, the standards were radically improved, and the competence of trained physicians and the care patients received were dramatically upgraded.

A century later, among the 150 accredited medical schools in the United States and Canada, albeit not the same schools, the problems are just as deep seated. In *Educating Physicians: A Call for Reform of Medical School and Residency*, Molly Cooke and her coauthors found similarly deep problems as enormous advances in knowledge, technology, and specialization intersected with a "near-chaotic system of health care delivery."[10] It might seem that all the advances I've discussed so far will only make things worse, but, perhaps ironically, the ability to digitize humans should prove a potent means of organizing and upgrading medical care.[11] At present, however, it does not show up in any medical school curriculum. How can this be the case?

Having started a medical school in Cleveland back in 2002, I was able to learn firsthand about some of the challenges involved in changing medical education. The curriculum at each school is fixed and formally reviewed by the Liaison Committee on Medical Education, an entity jointly sponsored by the American Medical Association and the American Association of Medical Colleges. The faculty senate or some other representative body of the school also typically approves it, and since most medical schools are part of a university, the curriculum is approved at that higher level as well. It is extremely complex and onerous to simply think about, let alone actualize, a new, innovative curriculum in an established medical school. Perhaps this led, in part, to Richard Horton, editor of the *Lancet*, to pen a pointed statement about higher education in medicine: "Their questionable admission practices, ossified curricula, out-of-date learning models, invalid assessments, lack of incentives to match health professional to public need, deficits in disease prevention, and the largely absent leadership to put social responsibility at the heart of their educational mission all point to a bankruptcy of vision by our overpaid academic leaders."[12]

Let's drill down a bit into Dr. Horton's opinion. Yes, the admission practices are questionable, since the undue stress on premed classes and grades in organic chemistry have been shown to poorly correlate with the downstream quality of physicians.[13] It's certainly the case that contemporary and critical topics of genomics, wireless sensors, digital imaging, and health information technology barely show up in any medical school. At last assessment, there were two out of 150 medical schools that had more than a brief, cursory curriculum dedicated to genomics of complex traits or pharmacogenomics; traditional genetics of simple but rare Mendelian traits was taught in 62 percent of schools for a total of twenty to forty hours of instruction.[14] The hit on out-of-date learning models is quite appropriate, too: There has been substantial advancement in the science of learning, and we know that the more interactive, participatory, and less didactic a lesson is, the more likely knowledge it contains is assimilated. Real learning occurs when students leave the educational session and discuss and internalize what was talked about. The old model of the professor giving a one-way lecture in the auditorium to a large group of medical students is clearly passé, but you wouldn't know it if you were a medical student in most schools today. On the other hand, you would find that attendance in these antiquated classes is pathetic, and most students simply work on their own. Pedagogy is out; collaborative learning is in. When there are simulators available in medical schools to virtually teach how to do procedures, there is outstanding participation—because the students are actually engaged. The medical schools at both Stanford and the University of California were among the first to give each entering student an iPad, which has on it all of the lectures, slides, course materials, and textbooks. This makes learning much more fun and convenient, not to mention setting the framework for the digital doctors of the future.

There are other learning models out there. Take the Khan Academy. A young entrepreneur and former hedge fund manager, Sal Khan, put together a large series of ten- to fifteen-minute videos on math and science topics like algebra, calculus, and biology and placed them on the Web through YouTube. By the fall of 2010, Khan had made over 1,800 video tutorials, and they were being viewed at least 70,000 times per day, reaching a cumulative 200,000 students. By 2011, more than 2 million students per month were viewing the more than 2,300 videos posted. There isn't any glitz in the videos, but the content is solid, and complex topics are made eminently understandable. While the Khan Academy isn't teaching medicine, the same concept could easily be adapted. Reflecting this point, one video posted in 2011 was on diabetes.[15]

Other examples are the Carnegie Mellon University Open Learning Initiative and Massachusetts Institute of Technology OpenCourseWare (OCW), which use new media to reach large audiences, averaging one million visits each month.[16] OCW completed online publication of over 1,800 courses in thirty-three disciplines in 2007 and has been described as "probably the single most important and cost-effective contribution to the world's knowledge base in the past ten years."[17] Serious Games takes advantage of the engagement that occurs with gaming to drive the learning experience. It doesn't necessarily take a "shock and awe" video website to convey valuable material in a lucid, attractive, and highly digestible manner.[18] Once a talented educator can be identified to make complex material relatively simple and, at least to some degree, entertaining, such educational initiatives can go viral via the Web. And they wouldn't have to reach 250 million viewers on YouTube—as "Charlie Bit My Finger" did—to be a success!

Horton's allusion to the social mission is also justified. Most American medical schools derive a large proportion of their funding from research grants, particularly the National Institutes of Health, and put the highest priority on those faculty with hot research programs. Dr. Fitzhugh Mullan of George Washington University School of Medicine analyzed and published data from 60,000 medical school graduates in 2010 in the *Annals of Internal Medicine*. His findings were bleak: "If we continue to produce more doctors in the system we have now, we won't be able to address the needs, the health outcomes, and certainly the populations that are underserved, dying, and suffering as a result of it."[19] Horton was right, too, when he said medical education doesn't focus on prevention. Of course, this problem is pervasive throughout medicine today. But the tools of digital medicine present new and extraordinary opportunities to address this void—finally!

In June 2010, Stanford University School of Medicine initiated a course called Genomics and Personalized Medicine. Adding this eight-week elective required a twenty-seven-member task force of bioethicists, basic and clinical science faculty, genetic counselors, lawyers, students, and "education officials" (not sure what that really means!), who debated for a full year before ultimately approving the offering.[20] Although the concept was proposed by a Stanford medical student, it was deemed highly controversial; among other things, the course would make Stanford the first medical school in the country to offer students the option of having the school subsidize a commercially available genome-wide scan for its students. Of course such scans had been available to the public since late 2007, but no matter—there was controversy. Of the fifty-four students who took the course, thirty-three opted to have the genotyping, and they were given full access to a genetic counselor.[21] The copay was supposed to indicate the student commitment. At the end of the elective one student concluded, "Having undergone the kind of testing and knowing the anxiety it can provoke and the meaning of the results and the limitations—that is an experience that is useful for physicians. Now we'll understand what kinds of stresses patients will have. I think it's a unique way for physicians to empathize with patients because you went through it firsthand." Well put. And the faculty sponsor, Stuart Kim, said, "You don't wait until it's perfect before you learn it. You learn what you can. You learn the principles of how human genetics work, so that you can best use new genetic discoveries as soon as they are made."[22] His is another well articulated reason to have this learning experience, despite the heated debate of many Stanford faculty who were adamantly against it. At least the brief elective went forward. It's hard to imagine how difficult this must have been to accomplish, and it's even harder to comprehend what a low threshold there is to get labeled as "pioneering" in this space.

Meanwhile, a survey of over 10,000 physicians, conducted by the American Medical Association and Medco, the largest American pharmacy benefit manager, showed that 98 percent are aware that patient genomics influence response to drug therapy, but only 10 percent believe they are adequately informed and comfortable with the use of genetic information to guide treatment in clinical practice.[23] Paradoxically, in a large survey, when more than 3,000 consumers were asked, "Who do you trust with your genetic data?" 90 percent said their physicians.[24] I guess they don't read USA Today, which published an article in 2010, "Most Doctors Are Behind the Learning Curve on Genetic Test."[25] With only 2,000 certified genetic counselors and considerably fewer than 1,000 medical geneticists involved in patient care in the United States, and both of these professional groups predominantly trained in rare, monogenic diseases rather than common, polygenic traits and pharmacogenomics, the number of educated medical personnel who can serve as a resource and earn the public trust is exceptionally small. Why shouldn't all physicians have genome-wide scans, or at least pharmacogenomic screening, and get oriented to and on board with this body of knowledge? At the 2011 Aspen Ideas Festival, Dr. Ezekiel Emanuel, who had been a top White House advisor and is now on the faculty at the University of Pennsylvania, called for a shake-up in the required premed college courses to alleviate the genomic knowledge chasm—by getting rid of organic chemistry and physics and replacing these courses with molecular biology.

The deficiencies in education extend to all areas of digital medicine. Data from the 2009 American Association of Medical Colleges demonstrated that less than 20 percent of medical schools have a required radiology clerkship, despite ubiquitous use of ultrasound, MRI, CT, nuclear, and PET imaging across virtually all medical specialties[26]—ditto for almost a complete void of medical education in electronic health records, health information technology, wireless biosensors, and telemedicine.

Back in the late 1970s, when I entered medical school, it was customary for each student to be given a stethoscope by a pharmaceutical or medical device company. Having one of these icons of medicine in your pocket certainly made the students feel they were on track to becoming real doctors. Eventually that ritual faded away when medical schools deemed such gifts unacceptable. But now with miniature, digital ultrasound imaging with devices like the Vscan, at least one manufacturer—General Electric—is restarting the concept of equipping each medical student with a pocket scanner. The School of Medicine at the University of California, Irvine, purchased fifty-four Sonosite laptop ultrasound devices to teach their medical students comprehensive diagnostic ultrasound throughout the four-year curriculum.[27]

Even if these changes in medical education take off, the impact of such programs will not be felt in the real world of medicine (outside of teaching hospitals with medical schools) for many years. Medical school forms the foundation of medical education, but most of the more than 700,000 physicians in practice in the United States today are beyond its influence. Reaching them requires changing the other categories of medical education. One is GME, or graduate medical education, which includes the internship, residency, and fellowship training programs; the other is CME, or continuing medical education, for doctors in practice, who make up more than 95 percent of physicians.

In most states, renewal of one's medical license to practice is contingent on a significant number of CME credits, usually obtained by attending postgraduate courses of one's choosing. However, more than 90 percent of CME in the United States is supported via the life science industry, and many medical specialty organizations are pushing back against this source of financial sponsorship because of perceived conflict of interest issues.[28] That has left the profession with a desperate unmet need and diminutive funding. So how can we break the education gridlock? In their book *Macrowikinomics*, Don Tapscott and Anthony Williams cite MIT president Charles Vest and the vision of a growing open access movement, the "emergence of a meta-university—a transcendental, accessible, empowering, dynamic, communally constructed framework of open materials and platforms."[29] With successful remote learning experiences like the Khan Academy and MIT's OpenCourseWare, along with leveraging the digital world of online medical communities and YouTube, it seems that this open source model can be accomplished with relatively little capital. If only doctors were keen to use social media networking, then the word could spread, and a real educational buzz might even be ultimately created.

ACCOUNTABILITY

Back in 1994, a colleague and I published an article in the *Annals of Internal Medicine* on scorecard medicine. A new movement was just taking root of releasing medical data to the public, having initiated mandatory collection in 1989. New York State's Department of Public Health, probably the most tightly regulated state for health care, with oversight of a relatively small network of thirty-three hospitals that perform open-heart surgery, started disseminating data via newspapers and brochures distributed at super-markets that presented hospital-specific (and later surgeon-specific) volume of operations and the death rates at thirty days after a bypass operation. The data were adjusted by risk, such as the age of the patient, strength of the patient's heart muscle before surgery, other medical conditions, and key variables known to influence the outcome. The public had sought these data to make intelligent choices about which hospital and which surgeon to request, and now they were getting the data. Hospitals with low volume of operations or poor results or both were exposed. Other states like Cali-fornia and Pennsylvania initiated similar programs: California reported out-come data for treating heart attack, and Pennsylvania included the cost of open-heart surgery. The prestigious University of Pennsylvania took a hit when it was found that, of nearly forty programs in the state, it had the highest cost and the worst outcomes. Soon thereafter the leadership for cardiac surgery was replaced.[30]

The idea was that the scorecard would soon extend to virtually all med-ical specialties and procedures. More than fifteen years later, however, score-card medicine has made relatively few inroads. In September 2010, *Consumer Reports*, the well-regarded independent publication for rating consumer devices and automobiles, published their first rating of heart by-pass surgery groups.[31] They rated 221 groups in forty-two states, capturing just 20 percent of hospitals performing this operation (and no data specific to the surgeons); the vast majority of the 1,100 programs did not allow the information to be published! Nevertheless, the accompanying editorialists in the *New England Journal of Medicine* called it "a watershed event in health care accountability."[32] Another way to describe it would be setting the bar quite low for transparency and accountability. Or even that the medical profession is not accountable to consumers. A Potemkin village of sorts?

Health insurers, too, are trying to initiate the ranking of doctors by qual-ity and cost. The American Medical Association has responded aggressively and negatively, citing a RAND Corporation study in Massachusetts that found that a two-tiered rating based on costs would incorrectly classify an estimated 22 percent of doctors.[33] One response to all the resistance and difficulties getting to the information has been Castlight Health, a San Fran-cisco company launched in 2010 that promises to reveal "the true cost of medical care."[34] The company has raised more than $80 million from investment firms and the Cleveland Clinic; it sells large employers cost data (charging per employee per month) for health care services, and its first client was the Safeway grocery chain, with 200,000 employees.[35] However, although the data for cost are likely accurate, quality and outcome data are not supplied, at least not yet. As with parallel efforts, price transparency has attracted the most interest here.

This isn't to say that no one is watching quality; measures for tracking up to twenty different metrics have been placed in more than 3,000 U.S. hospitals for the past decade. (These metrics include the use of beta-blockers for a heart attack or the administration and discontinuation of preventive antibiotics at the appropriate times to reduce surgical site infections.)[36] It's just that these metrics provide little guidance to consumers. The relatively small list of ways to assess quality, mostly representing checklists of whether something was done or a medicine was given for a particular diagnosis, reflects how difficult these data are to come by and has left the door open for the raft of medical websites I reviewed in Chapter 3.[37]

The next phase of accountability was created and sanctioned by the 2010 Affordable Care Act: accountable care organizations (ACOs).[38] To be launched in 2012, ACOs are conceived to be groups of primary care and multispecialty physicians, with or without hospitals, which form an integrated network to promote quality and affordability of care. The organization must collate data on its quality of care and rely on significant electronic information infrastructure. This leads to some uncertainty as to whether physician groups, which tend not to be well organized and often without adequate HIT support, will be able to organize ACOs, or whether hospitals with stronger assets will dominate these new entities.[39] The reward for ACOs will be financial bonuses from Medicare and private commercial insurers for meeting performance goals. Unfortunately, patients have been left out of the loop on the formulation and discussion of ACOs,[40] but there is reason to think they might help. A randomized study—in which Medicare beneficiaries either were provided with personalized information on cost or were simply directed to a website—found that personalized information led to a striking increase in switching health plans. Accordingly, while ACOs are intended to provide better care at lower costs, the "accountability" is directed to the government and insurers and does not include furnishing provider-specific data on outcomes or costs to patients.[41]

Clearly, digital medicine needs to do far better in transparently providing useful and specific data to consumers. There is no reason to hold back data, either on volume of procedures or on the results. Physicians have a long track record of not being optimally accountable; this even extends to dealing with impaired and incompetent colleagues.[42] In the future with EHRs and a comprehensively digitized world of health care, there is an exceptional opportunity for physicians to lead the accountability charge.

PHYSICIAN DEMOGRAPHICS

With the graying of the American population and 36 percent growth of the number of Americans over age sixty-five in the next decade, a major mismatch between the patient and physician populations has already been percolating. Some one-third of doctors are due to retire. Added to that, with the passage of health care reform, thirty-two million newly insured Americans were brought into the system. The American Association of Medical Colleges reported that in 2010 there were 709,700 physicians in practice and a shortage of 13,700 relative to demand. By 2015 the shortage is projected to be 62,900 doctors, and by 2025 the projections are well above 140,000, or about 19 percent fewer doctors than are needed to provide adequate care of the population.[43] Densely populated states like New York and California will be particularly affected, along with vulnerable and underserved populations throughout the country. Already the problem has surfaced with an inadequate number of pediatricians: in states such as Mississippi, Arkansas, Oklahoma, and Maine, there are more than 3,000 children per pediatrician or family physician.[44] Surprisingly, most of the shortage in the next five years is projected to be among surgeons and specialists rather than internists.

Given the shortage, the rules of demand and supply would suggest that physicians should be getting paid more. That could not be further from reality. Not only has medical school tuition soared, but at the same time the reimbursement for physicians has been plummeting. Another somewhat unexpected factor cropped up in 2010—a reduction in visits to doctors.[45] With the escalating costs of health care, there are higher copays even for Americans who are insured. Large employers are using nurse "care managers" to reduce or preempt the need for doctors. For the past two years, more patients have gone to the Web for assistance than consulting physicians. There was an 11 percent drop in patients in large multispecialty groups over the past few years, and that represented the first time there was any decline in thirty years of tracking such data. Correspondingly, Thomson Reuters reported a 7.6 percent drop in doctor visits and more than a 2 percent reduction in hospital admissions.[46]

The decline in doctor visits might seem like a way out of the problem of the physician shortage, but the decline may well be a transient phenomenon, so there are many ideas coming forth for how to deal with the impending shortage of physicians. In a *Wall Street Journal* op-ed, "How to Care for 30 Million More Patients," Dr. Pete Vanderveen, a pharmacy school dean, offered up the 300,000 pharmacists in the United States as a potential solution. Citing positive experiences in Asheville, North Carolina, and other cities where these two groups of professionals worked collaboratively, he wrote, "What we need is a new health-care delivery model in which the primary-care physician is complemented by a team of professionals and providers."[47]

The consolidation of physicians as salaried hospital employees is a major change, too.[48] In a *New England Journal of Medicine* roundtable discussion of experts on health care delivery and quality of care, one of the experts, Dr. Lawrence Casalino, said, "I think one of the most significant and least publicly noted things that's happened in the health care system in the last 9 years or so is very rapid increase in hospital employment of physicians, not just primary care physicians but specialists, and not just physicians at the end of their career or the beginning of their career but physicians at all stages of their career. It's changing the demographics of physician practice very quickly."[49] This trend reflects the general unwillingness of physicians to maintain their independence and control the fate of their practice. Instead they are succumbing to the pressures of a turbulent time in medicine and hoping that hospitals and health systems can provide a sanctuary of sorts.

DOCTORS AND EMAIL

One of the most common solutions offered to address the instability facing physicians is a higher level of efficiency and productivity, such as digital medicine offers. In that same *New England Journal of Medicine* roundtable, Dr. Casalino said, "I actually think that probably a good 50 or 60 percent, if not more, of visits to primary care physicians, face-to-face visits, don't need to be face-to-face."[50] There are certainly data to support that point of view. A Kaiser Permanente study conducted in Hawaii showed that secure email reduced patient visits by 26 percent. Published in *Health Affairs* in 2010, a subsequent Kaiser Permanente Southern California study of 35,423 people with diabetes, hypertension, or both demonstrated that patient-doctor email messaging significantly improved health indicators for these conditions.[51]

But in spite of these studies, the use of email by physicians is remarkably low. Fewer than 7 percent of physicians surveyed in 2008 were using email with patients routinely. Taken from the other side, for more than 2,000 consumers surveyed in 2010, only 9 percent reported using email to communicate with their doctors.[52]

The low email use rate by physicians is, in part, related to lack of compensation. Notably, email use is more frequent among doctors who are salaried compared with those working on a fee-for-service basis.[53] But the reasons for not using email are more than just financial. To comply with federal privacy rules, doctor-patient email is supposed to be through secure websites, not Microsoft Outlook or Gmail. Unauthorized disclosure of identifying health information, a violation of the federal law HIPAA, can result in fines up to $250,000 and/or imprisonment.[54] The American Medical Association has an email policy it developed with the American Medical Informatics Association that states email should not be used to establish patient relationships—only to supplement "other, more personal, encounters."[55] Many physicians and health systems view email as a liability for malpractice. This might be because the physician did not respond to an email in a reasonable period of time, because the advice took the place of seeing the patient face-to-face and might constitute negligence, or because the text was quickly written and could be misconstrued or even contain errors.[56] On the other hand, it has been acknowledged that doctor-patient email can strengthen a relationship with patients, beyond reducing office visits and improving health.[57] Nevertheless, it is clear that something as simple and ubiquitous as email, an entrenched standard of modern communication, is being suppressed by the medical establishment.

In contrast, an appointment-scheduling mobile app known as ZocDoc has been notably popular among physicians, who are willing to pay $250 per month to fill open appointments. By 2011, 700,000 patients each month in nine U.S. cities were using ZocDoc to locate a nearby physician who will accept their insurance and book an appointment online. While the service is free for patients and provides reviews of the providers, the attractiveness for physicians is the efficiency of filling their schedules, even for last-minute cancellations. The success of this mobile app to date among physicians, especially with its revenue base derived from them, is clearly distinct from their prevailing pattern of electronic avoidance.[58]

DOCTORS AND SOCIAL NETWORKING

Dr. Daniel Sands is the director of medical informatics at Cisco and a physician at Deaconess Medical Center in Boston. Regarding digital communication with patients, he said, "Most doctors haven't embraced Health 2.0 technologies in any significant way. They're still back in Health 1.0, stuck on the question 'should I use e-mail with my patients?'" Even if they do anticipate the benefits of emailing with patients, he says, it's likely physicians know nothing at all about online communities.[59] My own experience agrees with Dr. Sands's analysis. When I give presentations to physicians and ask how many use Facebook, the response is substantially less than among consumers—and I'm not even talking about using it for communication with patients. And when I ask about Twitter, I get the sense they see me as an alien from another planet.

But that may be changing. The Mayo Clinic announced a new Center for Social Media and pointed out that they have over 60,000 followers on Twitter and 20,000 connections on Facebook. The CEO of Mayo declared, "Through this center we intend to lead the health care community in applying these revolutionary tools to spread knowledge and encourage collaboration among providers, improving health care quality everywhere."[60] When asked by the *Wall Street Journal* to identify the real goal, Center director Lee Aase said, "To help patients. Sometimes that means providing information directly to them, and sometimes it means disseminating information more rapidly to the medical community."[61]

Like the policy for electronic communication, the American Medical Association issued a policy statement on the use of social media emphasizing privacy and professionalism: "Physicians must recognize that actions online and content posted may negatively affect their reputations among patients and colleagues, may have consequences for their medical careers, and can undermine public trust in the medical profession."[62]

Physicians have various opinions about connecting with patients on Facebook. Dr. Katherine Chretien, in *USA Today*, wrote, "But, please, don't ask me to be your friend. That is, your *Facebook* friend." She describes one type of physicians: "the mere mention of the F-word sends shivers down their spines: it is too personal, too much potential risk, a frivolous time suck." She would not want to be entangled with patients on Facebook and asserts, "We need professional boundaries to do our job well."[63] In the *New York Times*, Dr. Daniel Lamas wrote about making friends on Facebook with a young man who was dying in an intensive care unit. When Lamas eventually went to contact the man again, he had died, and the physician lamented not having responded to an earlier message the patient had sent. Dr. Sean Khozin, an internist with Hello Health (see below) quoted in a *New York Times* column, "Medicine in the Age of Twitter," said, "We can use social media to coordinate care with patients and with different specialists, all using the same platform (his practice uses a secure website). I can monitor, and they can also use these tools to become empowered through a better understanding of their own disease and active engagement."[64] But, on the prospect of Facebook, one doctor said, "Is it appropriate for a patient that I'm caring for to see comments that my high school friends are talking about from thirty years ago?"[65]

A case in Rhode Island attracted national media attention when a forty-eight-year-old physician was fired from a hospital and reprimanded by the state medical board for posting information about a trauma patient. Although there was no name provided by the blog, there was enough information for the community to easily identify her. Commenting on this and related cases in 2011, Mostaghimi and Crotty suggested in the *Annals of Internal Medicine* that "dual online citizenship" should be considered, whereby physicians create and maintain a professional (public) and personal (private) social networking profile. The need for separation was highlighted by the 17 percent of physician blogs that contained information that could potentially identify patients. They described social networking as the elevator of the new millennium, in which you have little control over who overhears what you are saying.[66]

One social network has been specifically designed for physicians. Conceived by a surgeon, Daniel Palestrant, in 2006, Sermo was set up to simulate a doctor's lounge for conversations and sharing of information and has evolved to be a way for physicians to get fairly rapid, multidisciplinary "curbside" consultation. As Dr. Palestrant said, "Doctors can log in with a clinical picture of a question and then within a few hours have dozens of physicians weighing in on what they think is the right answer."[67] It currently has 115,000 doctors registered, or more than 15 percent of U.S. physicians. Of course, in practice it might be a bit different, but the site seemed to overly emphasize aspects of the business of clinical practice and how physicians can receive honoraria for completing industry-sponsored surveys.

Social media networking does not appear to have relevance for direct patient interactions—certainly not on nonsecure, public platforms, and perhaps not at all. But it is important for physicians to recognize the enormous popularity of these networks with their patients.[68] We have already seen how big a role they play in how patients get health information and advice, and they may prove a particularly useful means of communication for educational initiatives, especially when such information needs to be disseminated rapidly and broadly, as in the event of a flu epidemic or a bad lot of a particular medication.

TELEMEDICINE

Even if many of the most common electronic modes of communication are not suitable for doctor-patient relationships, some seem very promising. One is bidirectional video link between the patient and the doctor. Direct eye contact helps a physician see the individual's expression and get a better sense of his or her well-being, and it certainly enhances the interaction in terms of letting the doctor see whether the patient is being attentive and whether the patient's facial expression suggests confusion or comprehension. Apple's introduction of Facetime on the iPhone in 2010 is one way to make such video connections possible, although it relies on both the doctor and the patient having an iPhone 4.0 or later iteration. No doubt, as smart phones continue to evolve, such capabilities will be more widely available on multiple platforms, and this form of communication may become quite popular for a means to reduce office visits.

Other services such as Skype allow computer-to-computer videoconferencing. Unlike the Kaiser Permanente secure email assessment,[69] there have not yet been any studies to document whether video communication reduces the need for outpatient or emergency room visits. In light of the high impact of emails, which typically have significant delays between messages and are limited to the written word and still images, it is likely to have a significant effect. Since at least 50 percent of office visits may not be necessary, video visits could prove to be important for increasing the efficiency and productivity of *both* patients and doctors. One of the stumbling blocks, as with email, is the problem with reimbursement for physicians. While this works in a closed health system like Kaiser Permanente, it might be considered an abject failure in the typical fee-for-service model. The fact that some insurers, such as Cigna and UnitedHealth, are starting to compensate for e-visits and video chats (or v-visits) should enhance their value. Wireless biosensors and devices like the Vscan in the hands of patients will only make v-visits more attractive.

While emails and social networking have not been popular with physicians to date, it is a different case with another form of telemedicine: the use of robots in surgery. The Da Vinci robot system (a product of Intuitive Surgical, Santa Barbara, California), frequently used for open-heart surgery, prostate resection, gynecological operations, and many other procedures to avoid large incisions, has had a remarkable adoption rate throughout the United States. Rather than making direct contact with the patient, the surgeon moves a joystick to direct a robot in performing fine motor activities with powerful magnification of the surgical field. Surgeons must get extensive training to become proficient in using these robotic systems. Even though there is little evidence that such expensive technology—the hardware alone costs from $1.5 to $2 million—improves outcomes for patients, the uptake rate by surgeons has been extraordinary. Theories to explain why this one form of digital medicine is so captivating include the competitiveness among surgeons and hospitals for patients (through advertising use of the robots to the public) and the video game–like feel (they have been compared to the game system Kinect).

NEW MODELS

In 2009, *Fast Company* published a feature article, "The Doctor of the Future," with an illustration of a big leather medical black bag hanging by a bunch of wires connected to USB ports and electrical outlets. The opening sentences were: "Cost, access, quality—the prognosis for American health care may look grim but innovation is the cure. The medicine of tomorrow is being born today."[70]

Enter thirty-three-year-old Dr. Jay Parkinson, with $240,000 of medical school debt, who started a virtual medicine practice in Brooklyn in 2007.[71] He had been disillusioned by the typical eight-minute appointment time with patients with high pressure and high volume. He wanted to combine modern digital tools—instant messaging, email, video chat—with the antiquated practice of house calls. To get rid of the administrative hassles, he did not take insurance payments, and patients paid out-of-pocket via PayPal. Within three months he had three hundred patients. As he recruited additional physicians, the practice became known as Hello Health, with the slogan "the freedom to simply practice medicine." The publicity that came from this new style of medicine led to a connection with a Canadian software company named Myca Health to ultimately power all of the electronics. There is a $35 subscription fee, and the practice charges $100 to $200 per hour for online or office visits; brief emails are free.[72] Appointments are set up online (originally through Parkinson's Google calendar), most follow-up visits are not in person, and there is no need for a receptionist. The video of the visits are archived and linked to lab results for the patient to review. With each visit there is a request for comments and feedback, as well as sharing information through the Hello Health patient network. For prescriptions, the patient gets a text message indicating the cheapest pharmacy for the drug in the patient's neighborhood. All of this is done electronically. While Parkinson's disease is well-known, it seems we'll soon be talking about Parkinson's health care, too.

Another new model is exemplified by One Medical Group, an organization with five offices in San Francisco and one in Manhattan, which promises to be "not your typical doctor's office."[73] It doesn't feature video chatting or instant messaging, but it does encourage email and online scheduling, follow-ups by email and phone, and a $200 annual fee.[74] In return for the money, appointments are guaranteed to start on time, and same-day appointments are always accommodated. Nevertheless, the highest number of patients per physician day is sixteen, 40 percent fewer than the national average of twenty-five. Collectively, several thousand patients are being seen, and there is a greater than 50 percent annual growth rate. In contrast, in the real world of U.S. outpatient medicine, the average wait time to see a physician is currently twenty-three minutes, and almost half of 3,200 patients polled believe they should receive a discount if they have to wait.[75] It is no wonder the new models are rapidly gaining traction.

While Hello Health and One Medical Group are regional offerings, the MDVIP concierge program is national and has grown from 146 physicians in 2005 to 756 in 2010. Instead of the typical 3,000 or more patients to care for, each MDVIP doctor accepts only 300 patients. There is a $1,800 yearly fee for this concierge service, of which a third goes to the MDVIP organization for administrative support of the program. That's not a small amount of money and has led to concerns about a two-tiered medical system, but participants have been happy with the service they get, which includes communication by cell phone, email, and texting. The national renewal rate is about 92 percent, which is unusually high.[76]

DIGITAL DOCTORS OF THE FUTURE?

These new models are just the beginning of where medicine can go when it is wireless, connected to an Internet on steroids, and capable of tracking comprehensive physiologic metrics and even some forms of digital imaging. Data never previously available to both patients and physicians are now streaming. With the individual's biologic and DNA sequence data in the EHR and the patient's PHR, indicating all pharmacogenomic interactions, and with this information shared with the patient's pharmacist, prescriptions will eventually have a whole new look of precision. The need for in-person office visits will be substantially reduced over time, along with those to emergency rooms. Those physicians who can emerge as the medical digerati will have a decided advantage—their accurate data on outcomes, quality, and cost will be posted on the Web and automatically updated on a frequent basis. House calls of yesteryear will be making an incredible comeback, but done through the Web. Like all aspects of medicine, these virtual interactions will only be successfully accomplished by individuals with superior people skills. It is interesting to note that some medical schools have recently incorporated testing for such skills as part of the application process.[77]

Throughout this chapter I have highlighted the significant obstacles that need to be confronted, and it remains unclear whether there is adequate plasticity of a plurality of physicians to embrace the digital world and acknowledge that the era of paternalism is passé. My sense is that young physicians who are digital natives will be likely to assimilate but that it will be quite difficult for the vast majority who are in practice and inculcated with an older idea of how medical care should be rendered. Eventually there will be enough digital native physicians to take charge, but that will take decades to be accomplished. In the meantime, consumers are fully capable of leading the movement and contributing to medicine's creative destruction. And so they must.

REBOOTING THE LIFE SCIENCE INDUSTRY

The business model clearly worked—and up until 2001,
ironically at about the time of the human-genome breakthroughs,
most would have expected this trend to continue. It has not.
So now we are having to reinvent our industry.
—Andrew Wittey, CEO, Glaxo[1]

THE PHARMACEUTICAL INDUSTRY is the biggest component of the
life science industry, which includes biotechnology, medical devices, and
diagnostics. If there was ever an industry in peril, this is it. It faces a triple
whammy—research and development costs have increased from $15 billion
in 1995 to $85 billion in 2010; the number of new prescription medications
(known as new molecular entities) approved per year by the Food and Drug
Administration (FDA) has fallen from fifty-six in 1996 to about twenty in
each of the past few years (including twenty-one in 2010); and the "patent
cliff" of lost revenue as a result of branded drugs going generic is $267 billion
through 2016, with $52 billion in 2011 alone.[2] In 2011, a *New York Times*

article on the patent woes summed it up: "this year alone, the drug industry will lose control over more than ten megamedicines whose combined annual sales have neared $50 billion."[3] As John Lechleiter, the CEO of Eli Lilly, wrote in a recent *Wall Street Journal* op-ed, "The evidence is certainly mounting that we are facing today nothing short of an innovation crisis in America's life sciences."[4] The director of the study of drug development at Tufts University said, "This is panic time, this is truly panic time for the industry."[5]

The pharmaceutical industry, once considered the ultimate blue chip and extraordinarily profitable, has gone from a blockbuster to a busted model. There was change of an order of magnitude in the amount of research dollars invested in each new molecular entity: in the fifteen-year period from 1995 to 2010, the approximate expenditure for a newly approved drug for the overall industry went from $250 million to over $4 billion, a sixteenfold increase. The hypo-innovative landscape set up the profound patent expiration problem, best exemplified by Pfizer, the world's largest pharmaceutical company; before its drug Lipitor lost its proprietary status in 2011, it had been generating more than $13 billion in sales per year.[6]

Rather than innovate, at least in the short term, the industry has been going into consolidation, much like the airlines. Pfizer bought Wyeth, Warner Lambert, and Searle; Merck bought Schering-Plough; Roche bought Genentech; and many companies were formed by coalescence: Glaxo plus Smith Kline, Bristol Myers plus Squibb, Astra plus Zeneca, Sanofi plus Aventis (which itself had previously formed by Hoechst plus Rhone Poulenc), and Novartis, comprised of the former Ciba-Geigy and Sandoz. Furthermore, the big pharmaceutical companies have been buying up large biotechnology companies that make highly profitable injectable biologic drugs: Takeda bought Millennium for $8.8 billion, Astra Zeneca bought Medimmune for $15.6 billion, Sanofi-Aventis acquired Genzyme for $20.1 billion, and Roche's purchase of Genentech cost $46.8 billion.[7] This "land grab" has been predicated on the high prices and sales of biologics, which are typically monoclonal antibodies or other bioengineered protein injectable drugs, as summarized in Tables 10.1 and 10.2.[8]

These companies have also been buying up generic drug manufacturers, once their dreaded competitors: Pfizer acquired King Pharmaceuticals, Novartis acquired Hexal and Eon, Daiichi acquired Ranbaxy, and Sanofi acquired Medley and Zentiva. Besides actual purchasing, other large pharma firms have gone the joint venture marketing route, with large India-based generic drug firms such as Glaxo combining efforts with Dr. Reddy and AstraZeneca with Torrent.[9]

Drug Name	Indication	Annual Cost $
Etanercept (Enbrel, Amgen and Wyeth)	Rheumatoid arthritis	26,247
Trastuzumab (Herceptin, Genentech)	Breast cancer	37,180
Interferon beta-1a (Rebif, EMD Serono and Pfizer)	Multiple sclerosis	39,505
Adalimumab (Humira, Abbott)	Rheumatoid arthritis and Crohn's disease	50,933
Imatinib (Gleevec, Novartis)	Leukemia and gastro-intestinal stromal tumor	56,424
Epoetin alfa (Epogen, Amgen)	Anemia of chronic renal disease	84,467
Imiglucerase (Cerezyme, Genzyme)	Gaucher's disease	200,000

TABLE 10.1: Cost of Some Common Biologic Drugs.
Source: A. B. Engelberg, "Balancing Innovation, Access, and Profits: Market Exclusivity for Biologics," *New England Journal of Medicine* 361 (2009):1917–19.

Product	Sales value ($ billions)	Company
Enbrel (etanercept)	6.58	Amgen, Wyeth, Takeda Pharmaceuticals
Remicade (infliximab)	5.93	Centocor (Johnson & Johnson), Schering-Plough, Mitsubishi Tanabe Pharma
Avastin (bevacizumab)	5.77	Genentech, Roche, Chugai
Rituxan/MabThera (rituximab)	5.65	Genentech, Biogen-IDEC, Roche
Humira (adalimumab)	5.48	Abbott, Eisai
Epogen/Procrit/Eprex/ESPO (epoetin alfa)	5.03	Amgen, Ortho, Janssen-Cilag, Kyowa Hakko Kirin
Herceptin (trastuzumab)	4.89	Genentech, Chugai, Roche
Lantus (insulin glargine)	4.18	Sanofi-aventis
Neulasta (pegfilgrastim)	3.35	Amgen
Aranesp/Nespo (darbepoetin alfa)	2.65	Amgen, Kyowa Hakko Kirin

TABLE 10.2: Top-Selling Biologic Drugs in 2009.
Source: G. Walsh, "Biopharmaceutical Benchmarks 2010," *Nature* 28 (2010): 917–21.

Maybe all of this constitutes innovation in business deals, but even that is questionable. Where is the innovation to develop exciting new drugs and confront the real challenges of public health?

One of the most frequent explanations from big pharma for the lack of innovation in terms of deliverables—new, unique molecular entities approved by the FDA—is the high threshold set by the regulatory process. That the success of the life science industry is inextricably linked to the

FDA and the European Medicines Agency (EMA) is indisputable. Fulfilling the regulatory requirement for testing a drug in large randomized trials with "hard" clinical outpoints—such as preventing death or heart attacks, along with assurance of its safety—is remarkably expensive. Even as the industry moves toward its digital future, severe regulatory challenges will be present throughout.

A second reason for the lack of success in bringing new drugs to market is how difficult it is to find therapies that really work. In a wonderful story about a small biotech company's failed attempt of a cancer drug development, Malcolm Gladwell wrote in the *New Yorker* about why pharmaceutical firms have such a hard time developing drugs to fight cancer. A biotech scientist told Gladwell that "drug development is still so hard and so expensive because the human body is such a black box. We are totally shooting in the dark. You have to have good science, sure. But once you shoot the drug in humans you go home and pray."[10] This sentiment was echoed by Severin Schwan, the CEO of Roche, who expressed his view that half of all diseases can be considered untreatable and for the other half the drugs only work half the time and with major side effects. He said, "Imagine a car that starts only half the time, and whose brakes often don't work."[11]

The pessimism in the pharmaceutical industry in recent years has been further exacerbated by scandals of marketed drugs with serious side effects or aggressive promotion of medications for indications that had never been approved or studied. Perhaps the most famous case that changed the reputation of the industry was related to Vioxx, a drug and story that gave me exquisite familiarity with the potential downside of the life science industry. Having worked as a physician-researcher collaboratively with most of the major drug and device company firms since 1985, I was familiar with the ethical conduct of the people in the life science industry. By virtue of leading many of the pivotal clinical trials in heart disease over this extended period of time, I had the fortune of working with several of the largest companies in the sector—Pfizer, Lilly, Merck, Bristol Myers Squibb, Genentech, Roche, SmithKline, AstraZeneca, Novartis, Medtronic, Sanofi-Aventis, Johnson & Johnson, Schering Plough, and many others. In fact, with Merck, the manufacturer of Vioxx, I was leading a large trial of over 5,000 patients with heart disease to test a new medicine called Aggrastat (subsequently published in the *New England Journal of Medicine* in 2001),[12] when my colleagues and I came across worrisome data on Vioxx that had been presented to the FDA but had not sounded off alarm bells.

In 2001, we published the first paper registering significant heart attack and stroke concerns for both Vioxx and Celebrex, another drug in the same class of so-called Cox-2 inhibitors of inflammation.[13] An accompanying feature article in the *Wall Street Journal* on the day of that publication quoted me as saying, "we're staring at a major public health issue."[14] But that warning did not really play out until three years later, on September 30, 2004, when Merck suddenly withdrew Vioxx from the market because of the risks of heart attack and strokes.[15] By then, Vioxx was used by more than twenty million people and had attained sales of more than $2.5 billion in the previous year; it was by far the largest prescription drug withdrawal in history.[16]

The day of the withdrawal I watched a video of the CEO of Merck, Raymond Gilmartin, declaring that Merck had done everything right and that this was the first time they had seen any evidence that Vioxx was inducing heart attacks. Merck also published full-page ads in the major newspapers, claiming for the firm a "consistent and rigorous adherence to scientific investigation, transparency and integrity" and that "we promptly disclosed the clinical data on Vioxx."[17] Having followed the data and evidence for more than three years, I knew that the truth was not being told. So I penned an op-ed and sent it to the *New York Times*. I called it "Vioxx Vanquished" and concluded it with the claim that "our two most common deadly diseases should not be caused by a drug." It was promptly accepted, although the editor told me that the newspaper would write the headline. On Saturday morning, October 2, 2004, I picked up the newspaper from our driveway and found my article had been retitled to "Good Riddance to a Bad Drug."[18] That should have been a lesson to quit while I was ahead, but unfortunately this story had a life of its own.

Standing up for the lack of public health concern that had been breached by both Merck and the FDA, which had approved the drug in 1999, I was invited to prepare a commentary for the *New England Journal of Medicine* shortly after my op-ed appeared. The essay was published online a few days later.[19] Within weeks I was interviewed by Ed Bradley for a segment on *60 Minutes*; I learned that he had suffered a stroke on Vioxx, which he never publicly disclosed (only to me before the on-camera interview). Over time, extensive evidence via internal Merck emails emerged, such as Dr. Alice Reicin's 1997 message that read, "the possibility of increased C.V. [cardiovascular] events is of great concern," which clearly meant that Merck had been well aware for several years that Vioxx promoted blood clots. Mark Lanier, one of the attorneys who sued Merck on behalf of a large group of plaintiffs, nailed the reason for the cover-up. In a courtroom in Atlantic City, New Jersey, he enacted a short drama called *Desperate Executives*, derived from the television show *Desperate Housewives*, to portray why

Merck's management had gone off the tracks with Vioxx. In fact, I do believe that it was desperation that led to their conduct, which was qualitatively different from any that Merck had ever previously manifested. I had known Roy Vagelos, a physician-scientist and the prior CEO of Merck, and still have the highest regard for him today. He later said, "They have dropped enormously, and it's very sad to watch from the outside. You wonder how these companies survive when they don't know what to do. . . . It couldn't have happened when I was there."[20]

At a Congressional hearing the following month, in November 2004, Dr. David Graham of the FDA said, "We are faced with what may be the single greatest drug safety catastrophe in the history of this country or the history of the world."[21] This statement reverberated throughout the world, especially since an FDA official, who was classified as a whistle-blower and protected by a Senate committee, had uttered it. The hearing set off a chain of events: the NIH canceled all clinical trials with Cox-2 inhibitors, and the FDA placed a "black box" warning on the label of all commonly used nonsteroidal anti-inflammatory drugs (a group that includes Advil, Motrin, and Aleve).[22]

The Kafkaesque period had extraordinary moment-to-moment media coverage, ranging from the cover of *Fortune*, which featured "Merck's $27 Billion Heart Attack," to a *Wall Street Journal* editorial called "The Painkiller Panic," which claimed the response to the case was overblown: "The likes of Drs. Marcia Angell, Eric Topol, and David Graham have been shown up for the Luddites they are, willing to make grand pronouncements about the public health with nothing more than their anti-industry reflexes to support them." On the cover of *The Lancet* the editor, Richard Horton, wrote, "with Vioxx, Merck and the FDA acted out of ruthless, short-sighted, and irresponsible self-interest." Later the FDA admitted lack of adequate oversight, and one of its senior directors, Dr. Janet Woodcock, said, "This system has obviously broken down to some extent, as far as the fully informed provider and the fully informed patient."[23]

Lessons for me from this storm were plentiful. Along the way, I was wrongfully and outrageously accused of having financially profited from the Vioxx withdrawal; my family and I received death-threat phone calls at our home from people who demanded that I stop my public criticism of Vioxx and Merck; and although I refused to testify in any of the subsequent litigation, subpoenas from Merck lawyers were repeatedly nailed to our front door. Ultimately, despite my objections, I was compelled to give a full day of videotape testimony in the first federal Vioxx class action suit. Just a few days following the release of that testimony in early December

2005, my employer of fourteen years took away my title as provost of the medical school I had founded along with the chief academic officer role I had played for the institution for the prior five years.[24]

I don't bring up Vioxx to remind anyone of the thousands of people who unknowingly suffered heart attacks from the drug, nor to relive the emotional torment I went through. This case remains emblematic, even seven and more years after the drug was withdrawn, of much that is wrong with the drug discovery process—suppression of data from publication or public release, manipulation of the data in key clinical trials, use of ghost-writers to publish favorable articles in top-tier medical journals, exerting undue pressure on investigators and opinion leaders, and all along employing hyperaggressive sales and marketing tactics.[25] As a result of these actions, all of which were present in the Vioxx case, there has been a loss of public trust in the life science industry, the FDA regulatory body has become exceptionally risk-averse and more concerned about assuring complete safety (which is impossible) than efficacy,[26] and a new culture of whistleblowers within the FDA, academia, and industry has been spawned.[27]

Future model programs of drug discovery and development can learn a great deal from the Vioxx debacle. It occurred precisely at a time when a major pharmaceutical had lost the race to be first (Pfizer introduced Celebrex several months before Merck's Vioxx), many blockbuster drugs were coming off patent protection (such as Zocor), and there was little in the innovation pipeline. Ironically, Vioxx was a highly effective drug with potent anti-inflammatory action; it could have been saved if pharmacogenomic research was done to understand which patients were predisposed to blood clots or even if there had been a proactive warning of its cardiovascular risks by the company. Instead, with hypo-innovation came desperation, and a high price was paid.

NEW TOOLS DRIVE A NEW MODEL

A decade later, the science for discovering new drugs has never been so sophisticated and promising. This is especially true in the case of monoclonal antibodies (the chemical, but not trade, names of which end in the suffix "-mabs") that target a specific molecule. The drug Herceptin (see Table 10.1) is the one most frequently cited—it is used specifically for women with breast cancer that overproduces a protein called HER2; targeting HER2 leads to improved survival. The second targeted biologic for cancer to come along was Gleevec, while not an antibody or injectable. Jointly developed by scientists at Oregon Health and Science University and Novartis, it is a small molecule pill (suffix "-ib") that targets the fusion gene product found in patients with chronic myelogenous leukemia.[28] Also on the list are the antibody drugs etanercept, infliximab, and adalimumab, which target tumor necrosis factor, a molecule found in high quantities in autoimmune diseases such as rheumatoid arthritis, psoriatic arthritis, and Crohn's disease.

We discussed the BRAF directed drug for malignant melanoma in Chapter 6, which targets a frequently found mutation in this cancer and has been more than 80 percent effective in patients whose tumor carries the mutation.[29] At the 2010 and 2011 meetings of the American Society of Clinical Oncology (ASCO) many similar success stories were presented and simultaneously published in the *New England Journal of Medicine*.[30] One was on Pfizer's drug Crizotinib, which targets the gene ALK in patients with a type of lung cancer (non–small-cell) and lymphoma (anaplastic large-cell).[31] The other drug was Ipilimumab, developed by Bristol Myers Squibb, which targets a protein on a specific type of white blood cell of the immune system—cytotoxic T lymphocyte-associated antigen (CTLA-4)—and was shown to extend the lives of patients with malignant melanoma.[32] One patient who was treated with Crizotinib for lung cancer was quoted in the *New York Times* saying, "For someone who's been on chemo before, this is like a miracle drug. You feel yourself. You look yourself."[33]

The serial successes of drug discovery that are indexed to finding the right target, and then demonstrated to favorably change the natural history of patients with serious diseases like cancer and autoimmune disorders, are clear. While for years there were only Herceptin and Gleevec, the list is now getting long, and the optimism for more of the same to come is growing. Such targeting or drug discovery can be achieved in one of two ways.

One way is to identify the root cause of, or critical aberrant protein that is associated with, a particular disease, and then use that template to design an antibody or small molecule to block it. This fits what is known as the rational design model, driven by the lessons of patients with the disease and capable of employing elegant, structure-based drug discovery when the 3-D crystalline structure of the target is available.[34]

The other way to go is to use molecular screening libraries, which most large drug companies and several academic centers have invested in. These robotic, ultra–high-throughput screening (HTS) systems can screen more than a hundred million reactions (e.g., how a molecule binds to a protein) in ten hours at a millionth of the cost of traditional ways of screening potential drug compounds.[35] In some sense this mass-screening approach is the "backward" model of drug design, whereby a molecule is selected and then the disease it could be useful for is hunted down.[36]

On top of rational drug design and molecular screening libraries, hundreds of new genes have cropped up from gene peeks, the genome-wide association studies across more than a hundred complex, polygenic diseases, as reviewed in Chapter 5. All of this work, involving hundreds of thousands of patients worldwide, was performed without bias or a hypothesis, and as a result, the genes and pathways found to be linked to diseases were, for the most part, previously unknown or unsuspected. The avalanche of discovery of the underlying gene variants (the SNPs) has been denigrated by some scientists and journalists both for identifying only modest effects and for finding only statistical noise and so not clinically meaningful findings. The gene peeks, of course, are only the first step: once these zip codes of the genome get fully sequenced, comprehensive investigation has generated many examples of finding powerful signals—for example, gene variants that carry a very high rate of autoimmune disease risk, diabetes, or heart disease—which are the potential basis for new drug classes. Furthermore, the extensive work in cancer sequencing is identifying key mutations for targeting drugs and coinciding with the accelerated success recently demonstrated for such drugs in development. Thus many aspects of drug discovery can be viewed as innovative, and the ability to do systematic analysis of DNA, proteins, and metabolites of patients with a disease can accelerate rational drug design.

A major problem, however, is the difference between drug discovery, per se, and drug development—taking a putative drug from showing promise to having been proven to help patients. This requires a new approach that leverages all the assets of digital medicine. In my view this will involve three major components: (1) Wikimedicine, or the enablement of a collaborative brain trust and networking; (2) the guaranteed-to-succeed model of clinical development; and (3) innovative digital marketing and tracking of new products.

WIKIMEDICINE

Wikimedicine is not simply a website anyone can contribute to. Rather, it means making what has been previously considered extraordinary—collaboration, interaction, and networking among the life science industry, academia, government regulatory bodies, and the public—into something ordinary. One helpful factor in this process is using the Internet to support the fundamental goals of transparency and integrity. In the past, trials with negative data were suppressed, and many trials were still not published five years after FDA approval (as in the case of Vioxx). Often the data in the published trials were discrepant with what had been submitted and reviewed by the FDA. In 2008, this led a group at the University of California, San Francisco, after systematically reviewing the studies, to conclude that information "readily available in the scientific literature to health care professionals is incomplete and potentially biased."[37]

The website Clinicaltrials.gov was mandated in 1997 by the FDA Modernization Act, but it was not until 2005 that most clinical trials were getting registered on the site.[38] The number jumped from fewer than 2,000 in 2004 to nearly 13,000 in 2005, when the top-tier biomedical research journals began refusing to publish papers if the clinical trials had not been properly registered. To that point, industry had been reluctant to register trials because they considered them and their findings to be "trade secrets." Nevertheless, by 2008 over 60,000 trials were registered.[39] Not only is the global life science industry responsible for getting the data entered, but also all the academic centers must register the large number of clinical trials that they originate. Although a bit difficult to navigate, this information is available to anyone who seeks it, and any patient or family member can easily search for a condition of interest and find out which trials are ongoing, which are complete, the bottom-line results when available, and the people and institutions responsible for the work. That is progress—a clear sign that Wikimedicine can work.

Another use of the Internet to support transparency relates to the sunshine law—the Physician Payment Sunshine Act, originally introduced in 2007—that requires life science industry companies to disclose any financial payments to doctors or academic medical centers.[40] Although some companies and states have adopted this voluntarily, it has not yet been mandated throughout the international life science industry, or even within the United States. As part of American health care reform legislation, and in response to the Institute of Medicine recommendations, beginning in September 2013 any gifts, meals, or financial compensation from pharmaceutical or device manufacturers to physicians must be reported.[41] This information will be readily available to the public online and can be used to evaluate whether a physician's opinion or recommendations might be influenced by such a relationship.

There are many other signs of progress to establish open, collaborative networking.[42] In the past, it was almost impossible for two different life science companies to work together on any initiative. The concerns over intellectual property, competitiveness, and trade secrets and the insular nature of industry made this essentially taboo. But things have radically changed. Pharmaceutical companies are partnering with academia and with small start-up companies. In 2011, a five-year, $85-million strategic alliance between the University of California, San Francisco, and Pfizer to accelerate discovery of new medicines and a second one between the university and Sanofi were announced. Gilead Sciences partnered with Yale University for a molecular basis of cancer research program for $40 million over four years.[43] At Scripps, we initiated a partnership with Sanofi-Aventis in 2010 that previously would have been unthinkable. On specific joint projects, such as the pharmacogenomics of diabetes, we share intellectual property, and in every respect it has the feel of an ideal collaboration—complementary expertise, bilateral exposure and access to technology and resources that otherwise would not be available to either partner, and intellectually charged discussions on the science with the goal of meaningful discovery to improve patient care. Besides its relationship with Scripps, Sanofi set up parallel partnerships with the Salk Institute, Caltech, MIT, and others. For any large pharma company to do this, no less one from France operating in the United States, this reflects a real change. One could interpret this as either a sign of desperation or a sign of real progress—I believe it represents the latter— and there have been many others like it, some very extensive.

In 2011 leaders from the University of California, San Francisco, penned an article in *Nature Medicine* that pled for enhanced academic-industry ties and pointed out that "with rare exception, the public benefits of discoveries made in academia are realized only when they have been translated into use through industry." A 2011 *Nature* review of this startling increase in big pharma ties with academia suggested this was a means of "outsourcing the earliest phase of drug discovery" and included one researcher's perspective: "All the drug companies are looking for a new model."[44]

The intercompany partnerships and joint initiatives are similarly new. Pfizer and GSK (Glaxo), the two largest companies, announced ViiV Healthcare, an initiative to pool their HIV products into a specialized company to treat and prevent AIDS.[45] In 2007, the biomedical research community was stunned when Novartis, which had invested millions of dollars over three years into genomics research in Type 2 diabetes, put all of their raw data on the Internet for the rest of the life science industry and academic centers to use. The head of the Novartis research programs, Dr. Mark Fishman, said, "To translate this study's provocative identification of diabetes-related genes into the invention of new medicines will require a global effort."[46]

In the field of Alzheimer's disease, a group of pharmaceutical companies have agreed to share their data from eleven failed clinical trials. In 2010, Johnson & Johnson, Glaxo, Abbott Laboratories, and Sanofi put data from 4,000 patients in these trials on the Internet, with more to come, accessible to academic researchers and the whole life science industry. As Chapter 6 discussed, a large number of drug trials have attacked the problem of Alzheimer's, and to date they have been overwhelmingly disappointing. But there are more than a hundred new drugs in development throughout the industry, and the unprecedented collaboration to pool all the data and learn lessons from postmortem examinations of failed clinical trials is encouraging. Bristol-Myers Squibb embraced a collaborative approach to develop their new melanoma drug Yervoy by working with academic scientists at the University of California, Berkeley, and a biotech company, Medarex, and the company now depends on "a bigger universe of innovation."[47] Recently, large pharmaceutical companies like Pfizer have made their long list of drugs that either failed in the clinic or were not fully developed available to the academic community for "repurposing." These are just a few examples of a new model of sharing, enabling, and engendering mass collaboration throughout the biomedical research community for the first time.

But with these signs of progress, there are some formidable obstacles to confront. Collaboration between academic centers has greatly improved with the need for "big science," bringing together multidisciplinary expertise to accomplish ambitious projects. Witness the human genome project, the International Haplotype Map, ENCODE, and all of the international genome, proteome, metabolome, and microbiome programs reviewed in Chapter 5. A 2011 *Wall Street Journal* article, "Sunset of the Solo Scientist," tells the story: the prospects for mass collaboration among academic researchers has never been greater than with platforms such as ResearchGATE, representing a community of 700,000 scientists, and the Nature Innovation Pavilion, fostering increased data sharing and crowdsourcing. Team science has gone to a new level, and the data supporting team science are certainly impressive. The most highly cited publications—more than 1,000 and referred to as "home run papers"—are six times more likely to come from a team as from an individual. The steady trend of team takeover was evident on nearly 20 million peer-reviewed publications and over two million patents.[48]

While the sum is infinitely greater than the parts, the concern is still about the parts. Many young investigators work extremely hard and do not get recognized for such efforts. Their academic careers are dependent on having publications in which they are prominent authors and competing favorably for peer-reviewed research grants. The team science approach markedly diminishes their individual chances to succeed. And the rivalry between academic institutions to get recognition for important discoveries can be fierce, often reducing the likelihood that important relationships will be established. To foster collaborations in the rebooted world of big science, it will be important for academic institutions to fully recognize the individuals for participation and to encourage the sharing and pooling of efforts to promote synergic discovery opportunities.[49]

The other major challenge relates to the "academic-industrial complex," a largely pejorative term used to express concern about the closeness of working relationships between faculty researchers and the life science industry. This issue is especially polarizing, with some who believe that the system is corrupt and that there should be strict limitation of engagements, and others who find close and extensive collaboration essential for making substantive advances in medicine. While the debate has been intense and long-standing, the field has unfortunately been moving in the former direction in recent years.

The topic reminds me of an incident when I was a junior faculty member at the University of Michigan in 1987. I had been invited to give a lecture at Harvard Medical School's Brigham and Women's Hospital—an honor to receive just a couple of years into my first job. During my visit, I had the extraordinary opportunity to meet with Dr. Eugene Braunwald, the chair of Medicine at the time, in his large, wood-paneled office. Braunwald is the father of modern cardiology and one of the most respected physician-investigators in the world. In the microcosm of cardiology, it was the equivalent to sitting down with the pope. While we were talking, one of the senior Harvard faculty members came in to say good-bye. It was Dr. Arthur Sasahara, who was leaving Harvard to head up research efforts at Abbott Laboratories in Chicago. Gene Braunwald got up to shake his hand and reiterated that it would be great to collaborate. Then he joked, "Remember, Art, we can't be bought, but we can be rented." And we all laughed.

Decades later this concern has not gone away. The "sunshine" regulations will undoubtedly discourage financial links between physicians and the life science industry, for these data will be Internet-ready for media use. Indeed, such information has already been used to damage the reputations of several physician-researchers.[50] Nevertheless it is remarkably difficult to find an expert physician-researcher in any field who is not in some way connected to the life science industry. Of course, that connection may not be the same from one researcher to the next: in some cases, it might have come about through a research or educational grant, while in others through serving as an advisor or consultant, receiving royalties, serving on a company's board of directors or its scientific advisory board, or giving speeches or being part of a drug company speaker's bureau. The last category is the one most potentially problematic, as for many years physicians were used as a conduit or tool to put out drug or device information using the company's slides—a practice that should be banned. In contrast, research and educational grants are critical to academic medical centers for execution of research projects and conducting educational programs not related to any company product. A recent report from fifty universities with medical schools and more than 3,000 life science faculty showed the average faculty member received $33,417 per year in industry funding, reflecting the extensive research support that the industry currently provides.[51] Nevertheless, having the data publicly provided by the companies and the individual physicians themselves should serve a useful purpose as long as it does not inhibit worthy collaborations from occurring. Most of the greatest discoveries in medicine have been a result of fruitful collaboration between academia and the life science industry.

Government regulation is another big topic in its own right. The FDA relies heavily on the life science industry for its funding, through "user" fees assessed to the biopharmaceutical and device companies to review their applications. The user fee for pharma has increased from $100,000 in 1993 to $1,542,000 in 2010 for a new drug application with clinical trial data.[52] The fees now represent $1.25 billion, or 46 percent of the whole FDA drug program budget in 2010, as compared to 30 percent of the budget in the first few years of the century.[53] Despite the trend in fees, the time it takes for FDA review has been steadily increasing, which naturally is a source of profound frustration from the industry side, representing lost opportunity in the commercialization of a new product. In a 2011 *Wall Street Journal* op-ed, "America's Innovation Agency: The FDA," the FDA commissioner boasted that there had already been approval for twenty-one new medicines for the calendar year and that the approval process timing compared favorably with the European Medicines Agency.[54] However, the attributes of efficiency and "innovation agency" are not generally perceived by the life science industry, medical community, and even the public as related to the FDA. As an integral part of Wikimedicine, the FDA needs to considerably step up its efforts to support innovation, speed up review times, foster enhanced communication and true collaboration with industry, and override previous accusations that have been leveled by members of Congress and the media of being "too cozy" with corporations, especially common in the midst of the Vioxx affair. Ideally, the FDA would be supportive of a new way to conduct definitive clinical research in the era of digital medicine.

THE GUARANTEED-TO-SUCCEED MODEL

The current cost of prescription medications in the United States is over $300 billion per year and steadily rising.[55] Out of the population as a whole, 48 percent take at least one drug, and 31 percent take at least two every day. For those age sixty and older, the proportion taking at least two prescription drugs each day rises to 76 percent, and more than 37 percent take at least five different drugs per day.[56] While many of these prescriptions are helping the people who are taking them, most are not. Our response to every drug is controlled, at least in part, by our genome, and this explains why there is such marked variability in the effect of medications, both in terms of efficacy of the drug and also with respect to major side effects. As reviewed in Chapter 2 in the discussion of population medicine, most of the drugs that have been commercialized were developed with the blockbuster model: that is, with the aim of treating as many people as possible even if the beneficial effect was modest. Very large randomized trials, typically enrolling more than 10,000 people, were often required to demonstrate the small benefit. And virtually no work was done to define which of these patients derived a particularly strong therapeutic response or, on the other hand, a severe adverse drug reaction—or why.

In the Vioxx example, in which about 1 in 200 people (0.5 percent) were at risk for having a heart attack or stroke, genome-wide scanning and sequencing would likely have determined specific gene variants that largely explained the risk, and could have been used to keep those at risk from ever having taken the drug in the first place. While this would not fully eradicate the risk of heart attack or stroke, and the patients would still need to be advised of the trade-offs between benefit and risk, such screenings would nonetheless be expected to markedly diminish the odds.

Another perfect example for avoidance of side effects plus achieving the desired efficacy involves the biologic drug PEG-interferon, which is used to treat hepatitis C. This drug is administered for forty-eight weeks at a cost of $50,000. All patients who receive it feel as if they have a severe flu-like illness. But the drug only works in half the people who get it, and that can be quickly determined by a simple genotype of the IL28B gene variant.[57] Similarly, patients with cancer who are potentially going to receive Erbitux or Vectibix need to have their tumor screened for mutations in the gene Kras. If particular mutations in Kras are found, these drugs won't work. They cost approximately $10,000 a month or nearly $80,000 for a full course of therapy.[58]

By contrast, t-PA for heart attacks, which cost $2,200 in 1987 (see Chapter 2) and still costs $2,200 in 2011, can actually save lives: the average person who survives a heart attack lives an additional eleven years. When t-PA first hit the market, the $2,200 per dose was on the front page of all major newspapers and stirred a national controversy, but it represents a paltry sum compared with today's cost of biologics for cancer. As a *New York Times* reporter explained in an in-depth article on the high cost of cancer drugs, "cancer is a uniquely frightening disease, and people will pay almost any price for treatment."[59]

The overall group of biologic drugs has considerably longer patent-protection than traditional medications, such as Herceptin, which was introduced in 1998 and is protected until 2019, and Avastin, with fifteen years of protection (until 2019).[60] The big-ticket items with long patent protection and billions of dollars of sales per year per drug, along with difficulty for generic "biosimilars" to get competitive, all make this a highly alluring area for biopharmaceutical companies to pursue.

This raises the question of the costs of biologic drugs. They have quickly ratcheted up to account for an ever-increasing proportion of the $300 billion per year cost of prescription medications. In 2008, they represented $46 billion; in 2012 they will account for over $75 billion of sales.[61] Many of these drugs are directed to types of cancer, and the data for survival benefit from pivotal clinical trials is only a matter of months (typically one to five) or even weeks, with no sign of cures except in rare individuals. The number one and two selling biologic drugs are Enbrel and Remicade (see Table 10.2, p. 198), typically used for the treatment of rheumatoid arthritis, but they only work in half of patients.[62]

As the cost of biologics has become high-profile, a telling precedent occurred that has not received nearly the notice it deserves. The drug Velcade, which costs about $35,000 for a year for treating multiple myeloma, was initially rejected by Great Britain's National Institute for Health and Clinical excellence (NICE), because of high cost relative to its marginal benefit. The manufacturer, Johnson & Johnson, agreed not to charge any patient who did not derive benefit from the drug.[63] This led to approval of the drug in the United Kingdom and the first guarantee program of a biologic in history. If you paid $35,000 to $100,000 for a treatment, wouldn't you want a guarantee?

The exciting future for drug and device therapies lies in the ability to digitize humans and movement toward guaranteeing their success. Rather than the modest benefit that takes large populations of patients to demonstrate, we can now focus on specific individuals who have the particular characteristics that could lead to a large impact. This approach has already been amply validated in a few disciplines of medicine such as oncology, with a BRAF mutation–directed drug for malignant melanoma in patients who carry the specific V600E mutation, and the treatment of rare, "orphan" diseases such as the use of Cerezyme (see Table 10.1) for Gaucher's disease. But now this method can be increasingly applied to common diseases that are being unraveled.

This represents a radically different approach from the current template for drug development, not only with respect to time required to acquire definitive evidence but also the remarkable costs. The large clinical trials of population-based medicine have a strong influence on the total price tag, frequently cited as at least $1.2 billion, to develop a drug.[64] Overall, the average, pivotal, so-called phase-3 clinical trial enrolls more than 5,000 patients, and in heart disease the trial size typically is twice that. These definitive trials for FDA registration cost between $300 million and $600 million to execute; the average is $400 million—tens of thousands of dollars per patient in the trial.[65] As one researcher, Bart Denys, put it, "We're wasting too much time and money on trials that are poorly designed and difficult to execute. They take too long. They produce trivial information and not enough important treatments. They don't ask relevant scientific questions. Clinical trials are broken, just broken."[66] The exorbitant costs of doing these trials has led much of the life science industry to move the clinical development program overseas, to China, India, and Eastern Europe in particular.[67] And some trials adding to the cost of developing a drug have little or nothing to do with getting a drug approved but are primarily directed to promoting a drug once it is on the market. Over six hundred physicians performed a "seeding" trial with Vioxx in over 5,500 patients with arthritis, which was "intended to lure leading physicians into the habit of prescribing Vioxx in the month leading up to the U.S. FDA's approval of the drug—although the purpose of the trial was not made clear to patients or doctors."[68] Such seeding trials are also not ethical because the patients enrolled are led to believe that the results will benefit others. Bioethicist Carl Elliott wrote about the ultimate outcome: "When a company deceives them into volunteering for a useless study, it cynically exploits their good will, undermining the cause of legitimate research everywhere."[69]

The primary way to reboot clinical trials is to leverage the ability to define individuals in ways that were previously not possible. Let's start with one of the most common diseases—hypertension, which affects more than seventy million people in the United States. Currently, at least six different drug classes and more than a hundred different drugs, including various combination pills, are used to treat blood pressure. We know that among those being treated, only about half are receiving well-managed care. In part that's because the root cause of hypertension is different from one person to the next. The common gene variants associated with high blood pressure are known; one involves the gene adducin, present in about 20 percent of hypertensive patients. A recent study investigating specific treatments for patients with the adducin gene variants showed a striking reduction of blood pressure—an average of 14 mm of mercury—whereas conventional drugs like a water pill (diuretics) or a commonly used angiotensin receptor blocker, losartan, had no effect in such patients. The effectiveness of the drug for the patients carrying the gene variant could easily be monitored by continuous wireless sensor over an extended period of time (a week or a month) so as to be certain that the drug had a therapeutic effect and also to quantify the effect very precisely.[70]

Treatment of Type 2 diabetes, and the care of the more than three hundred million people with this disease, could similarly benefit from such approaches. We now know from extensive genomic studies that these patients can be divided into two major groups: those who have problems with making or secreting insulin and those who have problems with the action of insulin in the body's tissues—so-called insulin resistance.[71] Some patients have both of these processes in play. But in treating diabetes there are eleven different classes of drugs and no rationale for picking one of the many drugs in each class. The eleven different drug classes have to represent the all-time record for options for a particular condition; most patients receive at least two or three different drugs for diabetes, so the number of potential permutations is almost four million. In the United States alone, there is a $29-billion market for the sea of anti-diabetic medications.[72] We know the most commonly used drug, metformin, doesn't even work in about 25 percent of patients. The map of which common gene risk variants are present in an individual can more precisely guide the right drug selection, such as a sulfonylurea for individuals with the TCF7L2 common gene variant.

Performing a genomic and molecular biologic analysis of an individual with diabetes would provide insight as to what is precisely accounting for the alteration of glucose handling in that patient. Not only could this lead to far better and intelligent treatment, such as using insulin in patients who cannot make it as opposed to using it in patients who are resistant to it, but the effect, such as changes to blood pressure, can be continuously monitored. The wireless sensor could be used to define a special type of diabetes that is only manifest at night, for example, and this would be part of the entry criteria rather than just the monitoring or surveillance application of sensors.

Glucose levels or blood pressure measurements are considered surrogate end points, as discussed in Chapter 2, because the important outcomes are clinical events like heart attacks and strokes. In recent years, the FDA has moved away from accepting data on surrogate end points because there have been many instances when these do not track with "hard" clinical events. But there is considerable room for a comeback of sophisticated real-time, continuous-metric surrogate end points; the category should not be dismissed. For glucose, the laboratory end point that currently commands the highest regard is glycosalated hemoglobin (HbA1C). However, this only provides an overall view or gestalt of the blood glucose elevation over several weeks or months and is thus insensitive to either low blood glucoses or day-to-day patterns of abnormal glucose. By continuous tracking with wireless sensors, a far better surrogate end point may emerge that does correlate quite well with major clinical outcomes. Moreover, as with hypertension or diabetes, for each individual there are considerably more data ready to capture than we currently measure to guide therapies more intelligently.

A third example is cancer, which will undoubtedly be prototypic in the future. Besides checking the BRAF V600E driver mutation for a BRAF-directed drug in malignant melanoma (and other cancers for which it can be present), the ALK-gene (anaplastic lymphoma kinase) mutation in a type of lung cancer can be screened before giving the Crizotinib drug directed to this driver mutation.[73] Of note, Pfizer developed the drug, but it was Abbott Laboratories that developed the screening for the ALK gene mutation, and they are collaborating—a good example of Wikimedicine.[74]

Another good example is the collaboration of Takeda with Zinfandel, to use the latter's gene test for TOMM40 in the former's Alzheimer's drug development.[75] Indeed, tackling Alzheimer's disease—for which no treatment has worked to date—would be a crowning success for digitized medicine. As discussed in relation to brain imaging, it is possible to detect amyloid deposits ten to twenty years before even mild cognitive impairment, the earliest stage at which Alzheimer's starts to manifest. The use of brain imaging or other biomarkers could be the basis for selecting patients for a new drug. An example would be to select patients with one or two copies of the apoε4 allele, use PET scanning with Pittsburgh Compound B to find beta-amyloid plaques in the characteristic parts of the brain, and perform cognitive testing to detect very early memory loss. Combining genomics and digital imaging, along with sensors that detect cognitive ability on a frequent or continuous basis—the sort of thing one could program to run on a smart phone—could collectively be used to identify a drug intervention with particular promise and precision for preventing or markedly delaying the onset of Alzheimer's.

These examples all fall under the heading of "theranostics," or the integrated used of a diagnostic with a therapy. The gene variants in hypertension and diabetes and driver gene mutations in cancer are forms of diagnostic biomarkers that, when coupled with a therapy, can enable the right treatment for the right patients. The third limb of this digitized approach is confirmation or titration of the desired effect with the use of wireless sensors. We don't even have a word for that yet, but the triad package of some type of biomarker, a therapy, and wireless sensor would be an exceptionally powerful means for catapulting medicine into the future.

Once this digital package enters the clinical trial arena, the massive trials of population-based medicine would no longer be needed; instead we could design trials of a few hundred patients at most. The right patient group is defined at the front end. For example, in patients with Type 2 diabetes, a genomic panel that includes variants in the key genes like TCF7L2, which is associated with abnormal function of pancreatic beta, or insulin-producing, cells, would indicate whom to include. TCF7L2, which is the most common gene variant accompanying Type 2 diabetes, is found in more than 20 percent of these patients.[76] Only the patients with the risk variants of this gene would get the new drug, and the effect would be measured continuously via a glucose sensor. If in one hundred patients there is a dramatic effect of normalizing glucose throughout the day, at night, after meals, and at virtually all times assessed, and there have not been any side effects, then the drug looks like a winner. And of course, the new drug would have already been tested in many different animal models ("preclinical") to be sure there were no worrisome toxic effects of the drug. Now the question is: is it ready for FDA approval?

In our new guaranteed-to-succeed model, the answer is yes—a conditional yes. The drug can go forward under the heading of conditional approval, but as you saw in one of the first examples in this book, one would still need large patient exposure to be able to know whether important side effects will emerge. There is really no accurate way to know that without going to the real world—recent prescription drug withdrawals besides Vioxx include ceruvastatin (Baychol), fen-phen, rezulin, lumiracoxib, ximelagatran, and rimonabant (the latter three had been approved by EMA but not the FDA). The artificial construct of clinical trials demands careful screening and a long list of entry criteria, besides the use of a biomarker as stipulated in this new design. When the incidence of a serious side effect is less than 1 percent, it may take, by play of chance, more than 1,000 patients to see it. Similarly, when the side effect actually occurs at a rate of less than one per 500 patients, which has been demonstrated for many drugs, it could

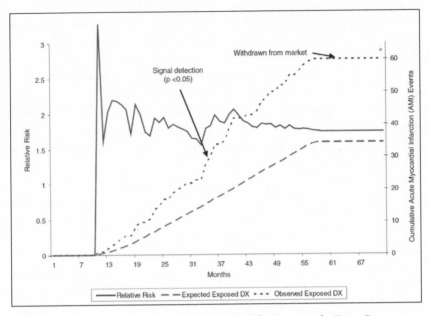

FIGURE 10.1: Detection of the signal of heart attack risk for Vioxx via the Kaiser Permanente health information system compared with the withdrawal of the drug by the manufacturer. Source: Institute of Medicine, *Challenges for the FDA: The Future of Drug Safety, Workshop Summary* (Washington, DC: National Academies Press, 2007).

take tens of thousands of patients to see it. Currently, the FDA wants to see safety assured in as many patients as possible and therefore encourages trials of 10,000 patients or more. In such trials, with a placebo being randomly administered to half the patients, there is a basis for comparison to say if the drug has induced a side effect and to quantify it. Nevertheless, many drugs that have passed muster by the FDA with this template have been shown to induce serious, fatal adverse events in the postapproval phase.

In the digital world, this should not happen. Beyond the potential of wireless sensors to track each individual in the assessment of effectiveness of a new drug, there should be exquisite means of detecting side effects in the population. Let's go back to Vioxx. The drug was approved in 1999 and mass-marketed. As shown in Figure 10.1, it took more than sixty months before it was withdrawn for a heart attack risk that was greater than twofold over baseline. But a database of seven million people demonstrated the effect at thirty-four months, or twice as quickly. If we had a hundred million people in a database, or one-third of the population in the United States, the heart attack risk would have been detected within two months. If the whole U.S. population was being tracked, it might have been detected in a week or two.[77]

Accordingly, we need mass postmarketing—"conditional approval" surveillance—before a new drug is given the green light of full approval. The tricky part of this detection is when the unanticipated side effect is something that occurs commonly in the population not taking the drug, such as heart attack in the case of Vioxx, which can mask the ill effect of the drug. It is much easier when the problem is liver toxicity or liver failure, which is exceptionally rare in the absence of hepatitis. But with large numbers of patients taking different medications, even heart attack can be correctly ascribed to a drug, as shown with Vioxx. We already have precedents for mass-population digital detection of disease in the H1N1 "swine" flu epidemic[78] and with food outbreaks with salmonella. Using crude tools such as Google's "Flu Trends"—or, in the case of the peanut better outbreak, Internet searches for food poisoning, peanut butter, diarrhea, or recall—provided much more rapid detection and geographic definition of the outbreaks than the public announcements that followed four to five weeks later.[79] I use the term "crude" tools here because search engines like Google and Bing, while exceptionally powerful, were not even designed for this purpose, and the methods used are quite indirect; nevertheless, they do perform well beyond expectations. Leveraging the Internet to capture the adverse event data in a comprehensive, targeted, and meticulous manner is an integral part of the guaranteed-to-succeed initiative.

Another tactic that could optionally be integrated with the digital package of new drugs is the use of wirelessly tagged pills to track whether, when, and where a drug was taken. It's yet another potent means of tracking a new drug in a conditional approval phase—one that takes into account for the first time not just the real world of prescription medicines but the reality of whether and to what extent the patient has been compliant with the medication.

Obviously this conditional approval would require many restrictions: such a drug could not be mass-marketed, and direct-to-consumer advertising would be prohibited. Patients who took the new drug in the phase of conditional approval would have to be advised that it is still being researched and there is a risk of serious side effects. Presumably only patients who had failed preexisting therapies, whether for lack of effectiveness or intolerable side effects, would be the ones largely taking the new drug in this phase. When an adequate sample of thousands of patients had been exposed to the drug for a long enough period of time and the side effect profile in the real world was defined, then the drug could be given final, full-fledged approval.

This program leverages many aspects of digital medicine—the Internet, high-resolution imaging, wireless sensors, biomarkers, and tagged pills. But there is one more essential feature that takes us back to DNA, and that is pharmacogenomics. Every new drug needs to have a pharmacogenomic study embedded into its development to identify genomic variation in individuals who derive particular benefit and those who have an unanticipated major side effect. While the guaranteed-to-succeed premise is that patients are only entered into a clinical trial after they fulfill individualized entry criteria, and scientific rationale must be given for coupling these criteria with the drug, there is also much to learn from actually giving the drug and monitoring the effects. Other types of individuals may derive benefit: testing the drug in one form of autoimmune disease like rheumatoid arthritis may result in finding it works in preventing Type 1 diabetes—the common thread being particular DNA sequence variants. Importantly, if any serious side effects occurred in the conditional approval phase, patients with DNA would be available for proper genetic studies that would define regions in the genome associated with side effects.

Ultimately, with deep sequencing of these regions, or downstream biologic markers like an alternatively spliced or sick, misfolded or aggregated protein, or an altered metabolite level, one could screen for such biomarkers and greatly improve the safety of the new drug. In the cases of Vioxx, all of the prescription drugs that had to be withdrawn from the market, and many that failed to get approval after late, large phase-3 trials (such as omapatrilat for blood pressure or torcetrapib for heart disease), it is conceivable or even likely that each of them could be commercially available today and successful if pharmacogenomic programs had been incorporated. Each of these drugs worked well in most patients but had rare, serious side effects that may well have been detected by DNA analysis. If they had been defined, the benefit-to-risk ratio could have been greatly improved.

The guaranteed-to-succeed model has potential impact at many levels, depending on the perspective. It could save immense costs on the part of the life science industry by only having to demonstrate efficacy via small clinical trials. At the national level, in terms of ensuring health care for all citizens, paying for performance of the therapy, like the Velcade model in Great Britain, is an attractive model. For patients, they can be confident that the drug they are taking is going to work and they are indeed going to feel better. But inching toward such a guarantee requires the full adoption and embrace of the power of pharmacogenomics. The biologic drugs, which have been emphasized in this chapter because their approach typically exploits the knowledge of the underlying biologic process, along with their marked popularity and expense, have, for the most part, not even touched on pharmacogenomics. Herceptin for breast cancer is a classic exception to that unfortunate trend.

The drugs for rheumatoid arthritis, on the other hand, are another unhappy example of pharmacogenomic ignorance. These cost over $14 billion per year—some $30,000 to $50,000 per patient—for use of the tumor necrosis factor blockers Enbrel, Remicade, and Humira. But only half the patients with rheumatoid arthritis have benefit! This means we are wasting more than $7 billion per year just to treat a single condition.[80] Shouldn't defining who benefits from these drugs be obligatory from a national and global perspective, let alone for the sake of the individuals who suffer unnecessarily because they are nonresponders?

Everything in this new model applies not only to pharmaceuticals and biologic drugs but equally well to devices and vaccines. Each year 250,000 defibrillators—or about $6 billion worth of these devices—are implanted, but only about 10 percent ever activate for the life of the patient. Using biomarkers to guide the use of these implants would potentially promote enormous savings, but also the information may prove useful to guide individuals who aren't even currently being considered for such devices—accounting for 300,000 to 400,000 sudden deaths in the United States per year. Similarly, a vaccine intended to benefit patients with a particular diagnosis could be better employed, such as Provenge from the biotechnology company Dendreon, commercially approved in 2010 for treatment of prostate cancer. The vaccine, which revs up the immune function of men with prostate cancer, costs $93,000 and was associated with an overall four-month improvement in survival.[81] To succeed in medicine, we need to know which patients are the ones who actually derive benefit. That has to be part of all new therapies of the future. It's eminently attainable but just not getting done.

Conditional approval requires reconditioning of the FDA approval process, which currently does not allow for such capability. The FDA is far more risk-averse than the EMA, its European counterpart. In the past few years, there were eighty-two drugs (new molecular entities) considered for approval by both the FDA and EMA; eleven were approved in Europe only.[82] Recognition of the powerful digital platform that extends from the ability to define any individual at the granular, molecular, and pixel level to the ability to track a drug at the macro, mass-population level needs to be the basis for a new regulatory process. In the United States, harnessing the full potential of digital medicine requires, for the first time, extensive cooperation between two governmental regulatory agencies—the FDA and the Federal Communication Commission (FCC). The FCC needs to be involved because of the oversight of body area networks—wireless sensors, the Internet, and cell phones. From past experience of two different government agencies each wanting to exert control, such as the FBI and CIA in national security, this may be a recipe for "guarantee to fail." Harmonization of federal agency efforts to support the many exciting innovative solutions will be essential. As we move forward, we will be asymptotically approaching the ideal of matching up a drug or device with the right individual and guaranteeing its success. This must be what we strive for; we are being endowed with remarkable digital tools that have yet to be exploited for reshaping the future of medicine.

DIGITAL MARKETING, TRACKING, AND SALES

Until recently there were over 100,000 drug sales representatives for the 700,000 doctors in the United States. That doesn't even take into account the sales reps for the other components of the life science industry, particularly device companies. Do we need a sales rep for every five to seven doctors to come by on a frequent basis and distribute materials or even, as is currently in vogue, show slides on an iPad? Or how about the near $5 billion per year spent on direct-to-consumer television advertising to the entire American population?[83] Is a thirty- or sixty-second slot the best way to convey the virtues and long list of side effects of a medication or device? Isn't there a more efficient way for these companies to get their word out in the era of personalized advertising and social media networking?

The pharmaceutical industry spent only 4 percent of its direct-to-consumer marketing budget on the Internet in 2008, but undoubtedly that will be changing. There have been some false starts: in 2009 the FDA sent warning letters to fourteen pharmaceutical manufacturers that had bought drug ads on Google and other search engines but did not include the list of side effects.[84] The agency also warned Novartis for using the Facebook "share" icon on its company's website to promote a leukemia drug, Tasigna, without indicating the risks of the drug. A Facebook user provided the following testimonial: "This saved my mom's life with no side effects." Here is the FDA's warning on that incident: "The shared content is misleading because it makes representations about the efficacy of Tasigna but fails to communicate any risk information associated with the use of this drug. . . . It's the first time the FDA has issued an enforcement letter over a Facebook widget."[85]

Similarly, there have been violations involving sponsored tweets by drug companies. But these represent the earliest and most rudimentary attempts by the life science industry to target its promotions. Eventually, it will figure out that the key to individualized medicine is not only to develop medicine for a specific patient subgroup, defined by a variety of digital signatures, but also to market to these individuals in a highly targeted way. Here the opportunity for the life science industry is to customize a message for a particular person at a particular moment. For example, it is easy to access patient online communities like PatientsLikeMe and CureTogether for specific conditions or medications or both. Companies have already launched their own sponsored social networking sites, as Sanofi did for patients with diabetes, and others will likely follow. Unlike television direct-to-consumer marketing, the Internet, in particular online communities like Facebook, provides instant feedback and quantification.

Now that we can know each other's location, advertising on mobile platforms specifically to patients, hospitals, doctors, medical centers, and drugstores would be easy. There could even be Groupon deals and Foursquare connections for people looking for discounts on their prescription drugs while en route or browsing in the drugstore.

The power of social networking is likely to have a profound impact on the life science industry. The companies have previously heavily relied on key opinion leaders to help transmit their story, tapping high-volume prescribers who were easily identifiable through prescription-tracking databases, and even at times resorting to celebrities to engage the public. But now case examples of the intersection between the life science industry and social networking are providing new insights. For example, recent studies have indicated that social networks were critical to the 40 percent decline in the prescriptions in Lipitor. But perhaps most illuminating is the use of Merck's Januvia drug for diabetes.[86] One of the first lessons on the power of social networking in drug prescriptions came via 610 doctors in the Raleigh-Durham Research Triangle in North Carolina, home of both the Duke University and the University of North Carolina medical centers. Prescribers of Januvia were highly influenced by adopters with one degree of separation in their network neighborhood, and the influence on prescribing Januvia extended even to three degrees of separation in the social network (see Figure 10.2). Similar effects were found in other social networks of doctors. This has led to the formation of a new company, MedNetworks, to foster the study and implementation of social networking for promoting new pharmaceuticals and devices.[87]

Over time, the efficiency of the life science industry will clearly be enhanced with the use of digital marketing and sales. The Epocrates app, available for mobile phones and tablets, reaches nearly half of the physicians in

FIGURE 10.2: The large dots represent doctors prescribing Januvia, and the size of those dots correspond to the number of prescriptions; the small dots represent the doctors who did not prescribe Januvia. Source: "Case Study 1: Adoption of Januvia," MedNetworks Inc., n.d., www.mednetworks.com/case-studies.html.

the United States; doctors use it to check medication doses and side effects in the midst of office visits or on hospital rounds. The app even helps determine what pill a patient is taking if only the color and shape are known. But now physicians are complaining about all the "DocAlert" advertisements that appear as they use the app. Epocrates claims that for every dollar spent on these ads, the drug manufacturer gets three dollars of increased sales. As one of the company's executives said, "You have a drug industry that spends $14 billion a year to influence people who prescribe drugs. There are only 600,000 people who are allowed to prescribe drugs, so there is $14 billion spent against 600,000 people. If you have a channel to reach these physicians, it is a gold mine." The chief medical officer of Pfizer explained the singular advantage of this particular advertising medium: "The beauty of the work we do with Epocrates is that we literally put ourselves in the palm of their hand."[88]

While this will markedly reduce the need for sales representatives, it also demonstrates the shifting of priority from reaching doctors to targeting patients. It is just one component of reinventing the future of the life science industry. It won't take long for these companies to assemble teams of social-media gurus, media planners, creative, and user-experience experts to efficiently get the word out.[89] For it to work, the marketing of the future cannot look like the marketing programs of today. Moreover, the same digital social networking analysis can be leveraged to track side effect and efficacy information, so the new opportunity of social networking to study how clinicians are adopting a drug can also provide an attractive means of filling in this major gap about how the drug is really working among its consumers. If consumers can have their iPhones and Droids track their location wherever they go, can't we design a simple methodology to track the effects of a new drug as it rolls out to masses of people?

THE NEW MODEL TAKES HOLD

In 2011, two drug development programs in my view validated the rebooting of the life science industry.[90] Francis Collins discovered the CFTR (for cystic fibrosis transmembrane conductance regulator) gene mutation for cystic fibrosis in 1989. In affected individuals, who number roughly 80,000 worldwide, chloride ions cannot normally pass through the cell membrane. For twenty-two years hope for an effective gene-based therapy went unfulfilled. But that changed in 2011, when Vertex Pharmaceuticals announced the results of a randomized trial of 161 patients with cystic fibrosis. This wasn't a trial of all comers with cystic fibrosis. Over 1,800 mutations in the CTFR gene are associated with the disease, and the various mutations have different molecular defects in the chloride transport pathway. Instead, only a subgroup of about 3 to 4 percent of individuals with this disease were selected on the basis of a specific mutation called G551D. The results were striking, with marked improvement in lung function, breathing, and normalization of the sweat chloride level, a sensitive indicator of the physiology of chloride transport. Getting the chloride channel to operate normally had never been seen before in a clinical trial for cystic fibrosis. This has resulted in an exceptionally rapid approval process, from announcement of results to commercial availability in a matter of months.

The second example I have touched on previously; it not only reinforces individualized targeting according to the molecular biologic defect but also brings up key concepts about the ethics of drug development. The diagnosis of malignant melanoma, which is made in about 68,000 individuals each year in the United States, can be likened to a death sentence within one year for most people, since the median survival is only eight months. Then came a revolutionary new approach. The BRAF inhibitor drug known as

PLX4032 from the biotech company Plexxicon, given in a pill form and directed at the driver BRAF gene mutation, which occurs in about 50–60 percent of patients with malignant melanoma, was remarkably successful: 81 percent had a marked, rapid tumor shrinkage response in a Phase-1 trial of 38 patients. And the benefit was confined to those patients carrying the BRAF mutation; those who did not have the mutation actually fared worse with the drug. The comparator drug, dacarbazine, had a very minimal 15 percent response rate with considerable toxicity. Was this convincing? It has been referred to by many physicians as a "Lazarus effect," a term I had not heard ever ascribed to a drug in my career. A leading cancer specialist, Dr. Keith Flaherty of the Massachusetts General Hospital in Boston, said, "I know all that I need to know based on the results we already have. My use of this drug is not going to be informed by testing it against a drug we all hate and would rather never give a dose of again in our lives."[91] But more trials were to be done, including a Phase-3 trial of 676 patients, with half assigned to receiving PLX4032 and the other half to get dacarbazine. Why was this trial undertaken? Because the FDA wants to have data on the survival advantage of the new drug. More crudely, the FDA insists on a body count to be able to quantify how much and how long the new drug improves survival.

The Pulitzer Prize–winning journalist Amy Harmon wrote a series of articles in 2010 and 2011 on this drug and whether new rules should be adopted for a highly targeted gene-based drug.[92] Perhaps the story from this series that captured the most attention was of two cousins, Thomas McLaughlin, age twenty-four, and Brandon Ryan, age twenty-two, each with a new diagnosis of malignant melanoma. Thomas was randomly assigned to the BRAF inhibitor, but his cousin was assigned to the control arm. And you guessed it: Thomas did well; Brandon died.

Dr. Charles Sawyers of Memorial Sloan-Kettering offered perspective. "With chemotherapy, you're subjecting patients to a toxic treatment, and the response rates are much lower, so it's important to answer 'are you really helping the patient?' But with these drugs that have minimal side effects and dramatic response rates, where we understand the biology, I wonder, why do we have to be so rigorous? This could be one of those defining cases that says, 'look, our system has to change.'"[93]

Yes, our system has to change. It has to reboot. We need creative destruction of the old rules. For we are talking about not only elegant science driving the therapeutic in these examples of cystic fibrosis and malignant melanoma but also the ethics of continuing to demand placebo trials. Recall the evidence base for current cancer treatments: most of the placebo-controlled trials have only shown minimal gains—just one to three months of extended survival. One could certainly make a strong case for very accelerated approval after an early trial and use subsequent data that are accrued to gauge benefit compared to historical controls.[94] For example, in malignant melanoma we know that the response rate for the standard chemotherapy, dacarbazine, is 15 percent and a two-month average survival. When we can combine a highly targeted drug with an extensive body of historical data, that ought to seal the deal. Make a new rule. Move the whole development process forward to progressive, and avoid assigning patients to toxic, ineffective therapies. The old ways do nothing for our knowledge but lay bare some pathetic moral reasoning.

Here I have tried to preview how a new model for the life science industry is taking hold. It moves the concept of match.com to a new level. The individual's differentiating features, such as a particular gene mutation, are matched with a highly specific therapy. This match has to be at the granular level, since if we just target a gene like cystic fibrosis CFTR, it wouldn't be adequate. It has to get to the precise point mutation of the individual— G551D for the Vertex drug or V600E mutation in BRAF for the BRAF inhibitor. Then we're onto overwhelming efficacy and rapid regulatory approval. A rebooted life science industry would leverage the science of individuality, getting the relevant digital readout from a person to fashion a therapy, instead of a mass-population–directed strategy. We now have the tools to do this on a broad basis throughout medicine and for the first time promote a level of prescription precision we have never seen before.

HOMO DIGITUS AND THE INDIVIDUAL

*By putting our physical bodies inside our extended nervous systems,
by means of electric media, we set up a dynamic by which all previous
technologies that are mere extensions of . . . —our bodies, including cities—
will be translated into information systems.*
—Marshall McLuhan, *Understanding Media*, 1964[1]

*As with previous revolutions driven by technology—whether it is the rise
of literate and scientific culture with the spread of the printing press or the
economic and social globalization that followed the invention of the
telegraph—what matters now is not the new capabilities we have, but how
we turn those capabilities, both technical and social, into opportunities.*
—Clay Shirky, *Cognitive Surplus*, 2010[2]

IN ORDER TO develop the notion of a super-convergence, whereby the
digital world finally infiltrates the medical cocoon, we have gone through
a series of lesser convergences in the course of this journey.

Up until now the medical community has been the privileged, nearly exclusive source, purveyor, and reservoir of all health and medical information. The Internet and the unprecedented growth of online, health-oriented peer-to-peer networking, however, have forced a rapidly approaching parity of knowledge between the public and the medical profession. As more individuals become privy to their relevant DNA data and can view principal physiologic metrics in real time on their cell phones, the advent of parity will only be accelerated. The group of tools not only each provide new ways to digitize a human being, but as we have seen, these tools in combination yield even greater power and flexibility (see Figure 11.1). Collectively, for each person we have a way to obtain data on his or her anatomy, physiology, and biology in a way never possible before. When we put all of these capabilities together, we have created a virtual human being that, although not real, replicates many of an individual's essential characteristics.

Now we are ready to discuss the implications of this series of convergences and perhaps the greatest convergence in our history: the one that finally coalesces the rapidly maturing digital, nonmedical world of mobile devices, cloud computing, and social networking with the emerging digital medical world of genomics, biosensors, and advancing imaging.

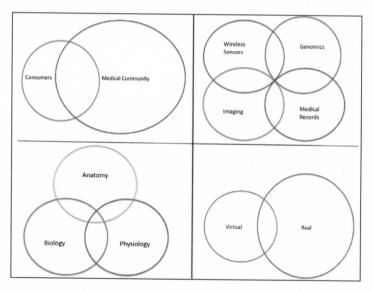

FIGURE 11.1: A series of convergences toward digitizing humans. Each diagram is meant to represent the trend toward fusion of the formerly discrete entities.

THE SCIENCE OF INDIVIDUALITY

The intent of this book has *not* been to provide a "techno-tour" but rather to describe both how the creative destruction of medicine can and will be achieved and how we will arrive at a knowledge of individuals so fine-grained that we can speak of a science of individuality. Just the DNA sequence and genomic profiling of an individual provide our ability to identify an individual's unique identity and bar code, for even identical twins have important differences in their epigenomic markings. Superimposed on these molecular biological matchless features are the other dimensions of what makes us tick: a window into every organ system and the integrated function of how we respond to our environment. Many are captured in the new "-omics" fields: the proteome, for proteins; the transcriptome, for genetic material transcribed into RNA; the metabolome, for molecules, such as hormones, synthesized by our bodies; the glycome, for all our sugars; the lipome, for lipids; the interactome, for how proteins relate to one another; and the exposome, denoting our environment. With the remarkable convergence of the digital medical tools we now have the "individualome." We are on the threshold of determining how each person in our universe is indeed distinct.

We are also, thanks to the science of individuality, on the threshold of eliminating much fundamental ignorance from medicine. The medical term "idiopathic" is quite commonly used in medicine for not knowing the diagnosis or cause of a condition. The word is derived from the Greek *idios*, which means "personal," "separate," or "distinct," combined with *pathic*, which refers to suffering from a condition. Ironically, with progress of the "individualome," we are moving back to the original connotation of *idios*: we are learning about the uniqueness of each person and will be much more likely to unravel the diagnosis or root cause of the individual's condition. Moreover, "essential" hypertension denotes our lack of understanding about why seventy million Americans have high blood pressure, and the term "cryptogenic" (literally "of obscure or unknown origin") suggests there is some mystery about a medical condition we haven't figured out. Even for those conditions for which a diagnosis is assigned, such as diabetes or inflammatory bowel disease, the molecular basis will vary considerably from one person to the next. The era of individualized medicine ultimately promises to do away with terms like "cryptogenic" and "essential," fully recognizing that each human being is *idios* and needs to be seen and treated with utter respect for his or her individuality. It will not be long until digitizing a person unlocks the root cause for what is wrong, creating valuable knowledge that can save a life or markedly improve the quality of a life. This is a major outgrowth of the science of individuality—but not the only one.

The entire classification system of medical conditions and diagnoses is about to be rewritten. Instead of our current reductionist model for which individuals are unwisely assigned to such categories as one of two types of diabetes or cancer of a particular organ, the science of individuality will promote a new molecular taxonomy that invokes the principal biologic basis, in terms of genes or pathways, along with physiologic phenotypic factors, such as glucose dysregulation that occurs only at night. So we will be hearing about Type 5b diabetes characterized by sick transport of the insulin protein (because of binding problems with zinc), as compared with Type 8 diabetes, which is attributable to a melatonin receptor malfunction and possibly sensitive to blue light. Or we would identify BRAF V600E cancer as what manifests in the skin, thyroid, or a number of other organs. Or interleukin-17 receptor immune disease will be known to show up as multiple sclerosis, Crohn's disease, asthma, or lupus.

Virtually all of the tools to digitize human beings involve networks, whether they are mobile sensor networks, the World Wide Web, gene regulatory networks, neural networks, or social networks. The nodes of these networks differ from people in a social network to DNA loci in a cell, but the key concepts of driver nodes and hubs are shared. Another common feature is how data are generated to understand a network. Whether it is sequencing human genomes or processing the data from a wireless biosensor, it involves massively parallel computation using concurrent platforms. The enormity of data, and the potential to upgrade it to information, relies on multicore processing and increasingly the use of cloud computing. What is fascinating here is the paradox that our contemporary capacity to understand an individual relies on network science—the more data that can be captured and processed, the sharper the definition of a particular individual.

THE *N* OF 1 TO THE *N* OF BILLIONS

In clinical research we typically refer to *n* as the number of patients enrolled in a given project. The "*n* of 1" represents the smallest sample of human beings—the individual. It would ordinarily be unthinkable in clinical trials to consider a single subject as adequate for meaningfully evaluating any therapy, although in testing educational or behavioral interventions this approach has been used. In the medical arena, n-of-1 trials have been used to assess different medications for pain management, whereby different drugs and doses are tested in an individual on a serial basis to find which regimen is effective. More recently, the n-of-1 approach has been used to test a new medication for Parkinson's disease. Some researchers have used n-of-1 self-experimentation on a formal basis to understand their patterns of sleep, mood, and weight.[3] To some extent, we all do that, especially when there are abundant data to help provide guidance.

A combined series of n-of-1 trials can be particularly informative, but that has not yet led to the routine use of such a study design. But the ability to digitize humans changes this situation with a considerable new body of data that can be derived from any individual, both at baseline and after an intervention. Such a data set is referred to as P, for the number of variables being investigated. The shift to "large P, small n" in medical science has extraordinary promise for the future of research and discovery.[4] Whether it is in the form of a series of n-of-1 trials that are combined, or just much smaller clinical trials to test innovative treatments for a chronic disease, there is a new large P, small n path of research going forward. This opportunity leverages the immense molecular biological, physiologic, and anatomic data that can be determined for any individual, and reinforces that the ultimate goal of an intervention is to have a markedly favorable impact on each n-of-1, rather than the current model, which emphasizes population medicine with the relatively small chance that any individual may derive benefit.

But what happens when we aggregate all such data from individuals? An n of billions is at least theoretically possible in the years ahead, when massive data sets are amalgamated from health information technology systems throughout and across different countries. With our principal interest in the individual, it might seem that such a herculean effort would be misdirected. However, the ability to extract exceptionally useful information from the ultimate clinical sampling of people is exponentially increased.

For example, the only reliable way we will be able to determine whether the millions of variations of the human genome are functionally important will be to study the effects of a particular base change, a copy number variation, or an insertion or deletion in the largest possible sample (thousands or tens of thousands) of people who have the condition of interest. We are also keenly aware of marked interancestry differences of the human genome, such that the three major ancestries—African, Asian, and European—have specific patterns of variation. Access to enormous "cosmopolitan" clinical HIT systems that have millions of individuals from each major ancestry represented would provide an invaluable resource to understand such vital interancestry relationships.

The same concept holds true for interpretation of physiological monitoring and rare anatomical or molecular variations that are picked up by digital imaging. Since we have such little experience with remote, continuous monitoring of most physiologic metrics, completely new patterns will be observed, and the question of what is "normal" can be best settled with very large populations undergoing the same assessment.

The n-of-1/n-of-billions idea will be vital to the program of Wikimedicine. Massive pooling of the granular but "pixelated" data from individuals creates a positive feedback loop, such that the overabundant granular data becomes more valuable and defined—transforming the extensive data to real information and knowledge that can ultimately be used to improve the health of individuals. The enhancement of health in a large number of individuals is the precursor to an upgrade in population health. This is how to connect the dots from the science of individuality to improving health on a large scale—a bottom-up approach that could not have been previously contemplated.

THE DELIVERABLES

Promoting Prevention and Precision

Our health care approach is reactive, and, as a result, we have a world of chronic diseases, most of which are poorly managed, such as congestive heart failure, high blood pressure, and diabetes, or not managed at all, as in the case of Alzheimer's. More than half of Americans age sixty-five and up have five or more chronic diseases, and more than 75 percent of the health care budget is apportioned to this massive burden.[5] These chronic diseases can largely be viewed as an end-stage phenomenon, since once they are manifest there is often irrevocable damage to vital organs and tissues in the body—such as the heart muscle, the pancreatic beta-islet cells that make insulin, the lungs, the kidneys, or the brain.

Now comes a new wave of technology to not only improve the outlook for the chronic diseases of today but shift the capability, for the first time, to true prevention. It has been rare in medicine that we have been able to prevent diseases from occurring. The fundamental reason is that we have not had the root cause defined. Once it was determined that most cervical cancer in women was due to the human papillomavirus (HPV), vaccines were produced that have afforded remarkable protection from developing this form of cancer. Vaccines tied to the essential knowledge of the pathogen are the prototypes for prevention of many diseases, such as polio, influenza, meningococcal meningitis, diphtheria, rubella, mumps, hepatitis A, and smallpox. In each case some biological material related to the bacteria or virus is administered, either from dead or inactivated organisms or purified protein derivatives, and the immune system is revved up specifically to handle the pathogen if infected. There have even been attempts of vaccines for common diseases that are not tied to bacterial or viral infection, such as Alzheimer's or high blood pressure, but these efforts have failed.

The lack of attention and priority for vaccine development is tied to our relative void of knowledge of the true underpinnings of the disease and a mind-set geared toward treatment rather than prevention. The problem is compounded, as we have discussed, by the fact that each common disease is a mosaic of varying perturbations at the molecular level, involving different genes and pathways in any given patient. For example, the ability to develop a vaccine to prevent autoimmune-based (Type 1) diabetes is waiting to happen, now that we know some of the principal genes and biologic pathways responsible in some individuals. However, it is unlikely that the same vaccine would work for all children at risk for autoimmune diabetes.

Once the root cause of susceptibility for a condition is known at the individual level, many approaches besides vaccines may be used to prevent the disease from ever occurring. One could argue that real prevention starts before conception. We know of 1,139 recessive genetic disorders, many of which lead to serious malformations and account for 20 percent of infant deaths. Current advances in genome sequencing enable screening for more than 570 of these genetic disorders, not just by checking an infant's genotype for a known mutation but by sequencing the genomic regions of interest in both prospective parents.[6] This can reveal whether both carry important recessive mutations and help reduce the toll of these serious diseases, much as screening has led to marked decline in the incidences of Tay-Sachs disease and cystic fibrosis.

After conception, the next step toward true prevention is to sequence the whole genome of the unborn baby at the earliest possible time, which is now possible even in the early weeks of pregnancy by capturing fetal DNA fragments in a maternal blood sample. Heretofore, we have had no way to anticipate or preempt most of the risk of perinatal and infant mortality. The current standard today is to perform a heel stick of the baby for a blood sample to screen for rare, monogenic, predominantly metabolic diseases such as phenylketonuria; the method misses some of infants' most important vulnerabilities, such as sudden infant death syndrome (SIDS), respiratory distress, and neonatal hypoglycemia. The risks for such conditions could be, at least in part, ascertained through genomic sequencing. Once it is known a baby has a high risk of SIDS, wireless vital sign monitoring could be used to prevent a fatal event. Digitizing the fetus in such a way provides an opportunity to significantly reduce the toll of infant mortality and serious morbidity.

The risk for various autoimmune conditions that typically affect younger individuals, such as inflammatory bowel disease, multiple sclerosis, lupus, juvenile rheumatoid arthritis, and Type 1 diabetes, will also be gauged using comprehensive "omic" profiling. Once the conditions are identified, methods to modulate the immune system can be mounted long before there has been destruction of vital tissue.

Most human diseases, including heart disease, cancer, and neurodegenerative conditions, are late onset. That essentially means that we have a forty- or fifty-year head start in preventing them, which should be an extraordinary advantage—one that has never been leveraged before. The information from digitizing a person early in life, and the appropriate surveillance with biosensors and imaging, lay the groundwork for true prevention. We already saw in Chapter 8 how heart attacks and many different cancers could be prevented. Nothing would be more precise, or more hyperindividualized, than creating the person's disease in a dish. While this may not ultimately become a routine means of determining the root biologic cause of an individual's disease susceptibility, or a method to screen various drugs that would be effective, the technology of induced pluripotent stem cells (iPS) can be used selectively for people with serious conditions or those who are refractory to treatments. The potential of moving from n of 1 to n of thousands of individuals with iPS studies will undoubtedly make our prevention and treatments more precise in the future.

The biggest immediate impact for resetting our approaches to chronic diseases relates to the routine use of pharmacogenomics. Recent acceleration of one-to-one medical knowledge for gene-specific effectiveness or side effects (or both) has been truly remarkable and now provides insights for precision prescribing of Plavix, Tegretol, PEG-interferon, statins, flucloxacillin, warfarin, and several other commonly used drugs. This list will greatly expand in coming years, and eventually the lag in uptake by physicians will be addressed. The wide-scale use of inexpensive pharmacogenomic panel data for each individual, which are stored and made available as needed by his or her pharmacy supplier, will undoubtedly catalyze this process. It is part of a consumer empowerment model and will redefine medication errors with a new category—a drug or dose of a drug that was not properly matched to the individual's genome (not that we needed any more categories of prescription errors!).

The Steady Demise of Hospitals and Clinics

Radical transformation involves an overhaul of the infrastructure of conventional medicine. The emblematic places to start with are the icon facilities of medicine—hospitals and doctors' offices. I am not advocating DIY medicine; there will always be a critical need for the doctor-patient relationship. Nevertheless, its context will change.

The need for hospitals in the future will be substantially reduced and restricted to the care of the most acutely ill patients who require intensive care and monitoring. There are multiple reasons why hospitals should and will be avoided. Their expense is extraordinary, for one. They present real risks, with some 80,000 hospital-acquired, dangerous infections, as well as 150,000 unnecessary procedures and medical errors per year, leading to more than 25,000 deaths.[7] Those numbers have remained stubbornly high despite a ten-year effort to lower them.[8] In 1946, George Orwell characterized a hospital as the "antechamber to the tomb."[9] In some ways that is not so far off today.

The most frequent cause of hospitalizations, such as congestive heart failure, asthma, and chronic obstructive lung disease, are all eminently amenable to digital medical strategies that forgo inpatient facilities. Obviously, not all the infrastructure is yet in place to enable widespread home monitoring, but I expect it to become very common in the next five years. One immediate means of hospital avoidance is the elimination of sleep laboratories, used to diagnose and treat sleep apnea and other related disorders. Anything that would be done in hospital-based sleep labs can be done in a home.

In the years ahead I expect some 50 to 70 percent of office visits to become redundant, replaced by remote monitoring, digital health records, and virtual house calls. Already in 2011, virtual doctors are ready to take hold with such technology as Cisco's HealthPresence, which not only includes videoconferencing but also integrates a high-resolution magnifying video camera, transmission of medical images, a telephonic stethoscope, an ear-nose-throat scope (the otolaryngoscope), and tracking of oxygen concentration, blood pressure, respiratory rate, and heart rhythm.[10] Today it is being used for reaching individuals who are far from medical facilities; tomorrow it will be used routinely to make interactions between doctors and patients far more efficient and convenient.

There's something ironic about the pace of this creative destruction. Why has the retail industry been "Amazonized"—with bricks overtaken by clicks—and medicine not? After all, many people find shopping in a book-store or browsing a retail store pleasurable. I have not met anyone who de-rives such pleasure from being in the hospital or visiting the doctor. In some sense, the answer is simple: multiple studies have shown that it has taken about seventeen years for a medical discovery or new, validated clinical knowledge to become a fixture of daily clinical practice.[11] Fortunately the means for accelerating that process are at hand.

Mind Control

This idea is a bit "way out," and certainly not imminent the way prevention, precision, or changes in medical infrastructure are. We've seen that biopsied skin can be transformed to pluripotent stem cells, reprogrammed to neural cells, and genomically edited. We've seen the exquisite imaging of the brain via functional MRI and PET and how activation mapping can lead to suc-cessful treatments, including the awakening of an individual who was min-imally conscious for more than six years. The treatment in that case, deep brain stimulation, is also successfully being applied to cases of obsessive-compulsive disorder and depression to alter behavior.

Despite the magnitude of those advances, this effort is still in a nascent stage. The book *The Soul Hypothesis* warns that neuroscientists "will pinpoint the exact three neurons whose firing accompanies the thought of our de-ciding to make a phone call or, if you prefer, deciding to get up and get a beer from the refrigerator."[12] The Human Connectome Project is aimed at precisely defining how human memories, personality traits, and skills are stored and processed.[13] The "connectome," a term intentionally derived from and related to the genome and all of the "omics" movement, refers to a granular mapping of the brain to understand its hundred billion neurons and its hundred trillion to a quadrillion synapses, or connections.[14]

Taking mind science to this level, especially while being aware of the potent impact a brain pacemaker can have, should at least engender concern. As opposed to the "Listening to Prozac" era of the 1990s, we could be lis-tening to brain chips that were afforded wireless activation through our cell phones on demand to uplift our mood, control our temper, or suddenly be-come romantic. Imagine you are struggling to remember a word or an event, and you simply go to the "mybrain" app on your phone and tap "activate memory function." The hippocampus gets a tiny tweak of wireless-mediated electrical impulse, and faster than you can currently search online for the word, it is in the front of your mind and spoken. Or you decide to delete a specific memory, and you inform your network of friends that it no longer exists—no, sorry, please forget that idea. It may take a while for these apps to appear on your phone, but the early phase of mind control is already here. Its expansion, as a function of digitizing our brains, is inescapable.

Democratized Data and Socialized Medicine

The prevailing theme of this book has been to explain what digitizing humans is all about—why it is desperately needed, how it is being accomplished, and its broad impact on radically changing the future of medicine. But vital to the actualization of this new medicine is your involvement.

With the personal montage of your DNA, your cell phone, your social network—aggregated with your lifelong health information and physiological and anatomic data—you are positioned to reboot the future of medicine. Who could possibly be more interested and more vested in your data? For the first time, the medical world is getting democratized. Think of the priests before the Gutenberg printing press. Now, nearly six hundred years later, think of physicians and the creative destruction of medicine.

It will not be simply a solo process, however. The pejorative term "socialized medicine" got firmly rooted in 1947, when the American Medical Association was taking on President Truman's initiative of public-funded, governmentally controlled health care. While the term has as many definitions and versions as there are countries, a new one is about to take hold. The exponential and unanticipated uptake of social-media networking has set the stage for a new approach to chronic diseases. That members of a social network trust their peers and friends more than their doctors is an important revelation. The concept of "birds of a feather flock together"—the affinity of humans to segregate in a bubble or colony, known as homophily—is much bigger than anyone anticipated.[15] And the trust level is profound. A survey of 25,000 consumers from over fifty countries found social network members trusted their friends, family, and peers for product recommendations and brands 90 percent of the time.[16] Being able to share critical health and medical data with one's circle of friends represents a new opportunity.

One of the earliest signs of this trend has been manifest with expectant parents who post their unborn baby's ultrasound pictures, or even the ultrasound movie loop, online and create the baby's personal Facebook page well before he is born. With the Internet, the possibilities are unlimited. Whether it is sleep brain wave data or around-the-clock glucose or blood pressure measurements, these data will be widely circulated. With Facebook and even more simply with Google+ one can specify sharing with a "preferred" subgroup of friends; this makes such sharing all the more likely. Competitions for the best sleeper, the best glucose or blood pressure management, and the like will be spawned.

The potential of leveraging online communities to improve health has already been suggested by early studies, and this work will certainly continue. Some clinical research programs have been based on these online communities, such as on the lack of efficacy of lithium for amyotrophic lateral sclerosis (ALS), predicting the response to Imitrex for migraines, or the optimal dose of Baclofen in patients with multiple sclerosis.[17] With tens of thousands of individuals affected by a particular disorder, and their willingness to participate in research programs, the ability to rapidly do crowdsourcing and answer unsettled research questions is both novel and unprecedented. On the flip side, there looms the concern that online health communities will be exploited or controlled by entrepreneurial interest or the life science industry, which would likely detract from individuals' willingness to freely share their experiences about a particular condition.

The extent of engagement in social networking certainly varies from person to person, and some of that is likely to be under genetic control. Of note, a part of the brain in the amygdala hypertrophies as a function of increased social networking activity.[18] Without question, the popularity and power of social networking have far exceeded all expectations. The revolutions in 2011 that were seeded in Tunisia, Egypt, Yemen, and other North African and Middle Eastern countries among young people hyperconnected via dense social networks provide proof of principle. With a digital social network platform, the power of impassioned people to stage a revolution is now unprecedented. A new path in history has been charted. Let's take the digital world's profound ability to influence and apply it to our health.

THE NEED FOR ACTIVISM

I hope that I have convinced you by now of the transformative and exciting features of digitizing human beings and that you are ready to participate in the democratization of medicine. Although it is inevitable that the digital revolution will stick in medicine, I worry about how long it might be before it succeeds. Take the stethoscope; ubiquitous for two centuries, it was still two decades after its invention before it became a standard tool. We cannot afford to wait that long—we'd be passing up too much and letting too many people die too soon. To penetrate the conservative medical cocoon, to make individualized medicine real, will require a mass movement. We have already seen how the American Medical Association and governmental agencies of the United States want to prohibit people from having direct access to their DNA data and require a physician's prescription and involvement.

Concerned about inaccuracy and interference with mobile telecom, the FDA is unwilling to allow continuous physiologic monitoring, such as blood glucose, to be displayed on your cell phone. We know the medical community is profoundly resistant to change. But a *radical* change is necessary to take medicine and health care where it needs to go, where it can go. As Michael J. Fox has said, "There is no Department or Secretary of Cures. It's us."[19]

The same techniques in democratization that worked in the Middle East can be harnessed to bring in this new socialized medicine. Instead of access to health care, this is about access to innovative technologies that make for precise medicine, the avoidance of vast waste, the reduction of medical and medication errors, and a fresh individual-centric approach. Surely this movement can be facilitated by the hundreds of millions of people on Facebook, on Twitter, in patient advocacy groups, and participating in online peer-to-peer medical communities. It is time for the public outcry "Show me the data!" And to be more precise: "Show me my data!"

This book cannot possibly engender such a movement on its own, but it can help lay out the preexisting infrastructure and tools to promote self-organizing of informed, collaborative individuals to transcend "dumbed-down medicine." To make a difference this can't be a passive slacktivism, a pure social-media play, as profiled by Malcolm Gladwell[20] or what Jonathan Franzen has spoken about in a recent college commencement address: "Liking [as in Facebook's "Like"] is for cowards, go for what hurts."[21] Become an interconnected information seeker, with instrumentation as needed. I won't try to provide a long list of action items here, as they will greatly expand in coming years, but a few examples may be useful.

The next time you are set up for a population-based screening test (such as mammography or PSA), find out if it is necessary for you. If it's an ultrasound test in a medical facility, why can't this be performed in the course of the office visit's physical examination with a pocket device? If it's a test involving nuclear imaging or CT, determine your radiation exposure in MSv units and whether there are suitable alternatives to avoid radiation. If you have a new prescription or are taking medications for a particular condition that have clear genetic interactions (which you can check with your search engine), think about having a pharmacogenomic panel ordered through the Web (via 23andMe, Pathway Genomics, or Navigenics—undoubtedly more to come).

If you or a loved one is diagnosed with cancer, consider having paired exome or whole-genome sequencing to determine the driver mutations to target therapy precisely. Similarly, if you or a family member has a serious but undiagnosed condition, whole-genome sequencing may demystify it and potentially provide an effective treatment. For monitoring many conditions like heart arrhythmias, hypertension, and sleep disorders, mobile apps and new sensors are or soon will be available. If you are one of the 346 million people with diabetes or one of the 70 million Americans with pre-diabetes, you might benefit from wireless continuous glucose monitoring to learn precisely what food, activity, and lifestyle help your blood sugar regulation.[22]

Keep an eye on the websites that cover these developments, like mobihealthnews.com, wired.com, fastcompany.com, and gizmodo.com, and I'll be tweeting this information along with other significant developments. By all means, tap into online health communities, especially those not sponsored by industry, to crowdsource useful information. For prevention of various conditions, it may be worthwhile to have whole-genome sequencing at some point in the future. We're not there yet, as the ability to sequence is far out in front of the yield of actionable information, but in the next few years this step will become increasingly worthwhile.

DIGITAL DYSTOPIA

Of course, a big part of making this program of revolution work will be a full reckoning with the potential downsides of digitizing humans. With the ability to digitize so many essential aspects of a person comes the paradox of depersonalization. Physicians will be tempted to treat the scan or the DNA data or the biosensor output, and not the patient. And they will not only treat the scan and the digital data but order it to make the practice of medicine more efficient, for it takes time to listen to, examine, and interact with a patient. Remote monitoring and diminished face-to-face visits between a doctor and patient can contribute to a significant degradation of intimacy, the loss of the literal healing touch. At a time when any human being can be essentially reduced to six characters (even if there are quadrillions of 0s, 1s, A's, Cs, Ts, and Gs) the dehumanizing concerns must be registered front and center.

The ubiquity of information, wireless networks, and biosensors are pushing the "virtualization of the real world."[23] As conveyed by David Gelernter in *Mirror Worlds* back in 1991, "You will look into a computer screen and see reality."[24] The MIT Media Lab is currently working on a project that "will make it possible to monitor humans' vital signs at a digital glance." As one of the lab's graduate students said, "This can happen, and in the future it will be in mirrors."[25] The theme of mirrors extends to mirror neurons, which are arguably the "specialized circuitry for social cognition" that differentiates humans from apes,[26] and extends to our genes, billed as "mirrors of life experiences."[27]

How can we tell what is real and virtual? When nonbiological circuits such as prosthetic limbs, cochlear implants,[28] artificial vision systems, and wearable sensors are integrated with the body, the brain, and our environment, the lines get blurred. In order to understand human beings at a far more fundamental and in-depth level, we will be cultivating cyborgs, a fusion of real humans and biosensors. Undoubtedly, the technology becoming available is formidable and impressive, but at the same time it creates a part-synthetic, virtual human, a reflection of an individual. Will we be readily able to differentiate the digitized human, *Homo digitus*, from the real human in the future? Just as Nicholas Carr wrote about the "slow erosion of our humanness and our humanity,"[29] referring to the impact of the digital world on behavior, the shakeup in the world of medicine may contribute to this pattern of attrition.

It is not possible to fully allay this concern, but I will at least attempt to calibrate it. Whatever digital tools are used to simulate and understand a human being, we will never fully replicate the individual, the person. The virtual human is not a real human. The digitized human being is an extension of the real person in particular ways, a convergence and approximation that could not even be conceived or strived for in the past. No matter how comprehensively, deeply, and finely humans can be digitized, the human factor[30] and each person's complexities cannot ever be fully captured; there is an intrinsic and critical chasm with reality. Ray Kruzweil's concept of singularity, that the dawning of a new civilization in which the distinction between human and machine, real reality and virtual reality, is eradicated,[31] is simply off base. *Homo digitus* will never equate with the individual. The Turing test as it applies to health and medicine is a nonstarter. Supercomputers and artificial intelligence are in the mix of how digital medicine moves forward, but maintaining the distinction between an individual and one's avatar is not only possible but essential.

Next on our list of concerns, and many would put them front and center above all, are the privacy and security of the data of human beings. Ironically, there are now thirty-two closed-circuit cameras within two hundred yards of the London flat where George Orwell wrote *1984*, the master forecast for a technological dystopia.[32] We know there will always be those who seek to violate our privacy in the name of laudable goals and evil ones. The forecast for the future of digital medicine is darkened when one considers breaches of leaked or hacked individualized information, exposing all genomic liabilities, stigmatized health conditions such as a psychiatric illness, or highly confidential biosensor and scan data. Indeed, some see the medical records of a lifetime, stored in the cloud, as the harbinger of a potential "horror world."[33] We have only begun to confront this vital issue at the genetic level, with passage of legislation to protect against using such data for discrimination by employers and health insurers. But that effort is incomplete, not covering life insurance and long-term disability insurance, and it ignores all other forms of digital medical data. On the privacy side, there has been aggressive protection of the rights of individuals through the HIPAA (Health Insurance Portability and Accountability Act) set of laws passed in 1996. While it cannot be guaranteed that there would never be an infringement of secure data, every possible attempt needs to be made to mitigate that risk and its serious consequences.

Perhaps an overriding concern is whether all the newly available granular information—every byte of physiologic metrics, every pixel of digital imaging, every base of a genome—will be a help or a hindrance. Will it preserve and palpably improve health or cultivate a culture of cyberchondriacs, of people in profound fear of a disease susceptibility that showed up in their genome sequence, or who monitor themselves the way some check their email every couple of minutes.

It's likely both will happen, much like someone with a family history of cancer or Huntington's might worry excessively about such things today, with just the Web and their fears to enable them. But that doesn't mean the information shouldn't be there to access—after all, most of us are not obsessed with self-diagnosis of imagined ills via WebMD, just as most of us had no compunction to do so with medical textbooks or psychiatry's *Diagnostic and Statistical Manual* (the DSM). The medical profession has never given consumers adequate credit for their remarkable ability to assimilate and exploit new knowledge. It is vital that the Luddite argument—really, the ignorance argument—which assumes no ordinary person can handle the truth, not be allowed to win out.

The ethical dilemmas and controversies that lie ahead in the era of digitized humans are undoubtedly going to be polarizing. When a couple can readily screen for thousands of rare mutations they may carry before conceiving a baby, drawing the line as to what constitutes eugenics or appropriate "planned parenthood" will inevitably lead to a divisive debate. When the elderly are attached to a boatload of sensors and continuously monitored by their caregivers, will the sense of "Big Brother" prevail and engender serious depression rather than the intended assurance of safety and preservation of relative autonomy of aging?

With digital medicine dependent on the Internet and broadband access, there are significant problems with lack of access across the world—not just in the developing world. In California, one in five adults do not use the Internet, and about a third do not have home access to broadband. This is a global limitation that flies in the face of a flattened world where virtually all of the technology and tools discussed in this book could be evenly applied without geographic constraints. Digital medicine should be considered an evolving humanitarian story that both parallels and transcends communication, the core platform upon which much of it is built; it will one day become a global standard means of prevention and care. We already have the exciting capability to convert a mobile phone to a diagnostic laboratory, to rapidly and accurately diagnosis HIV and other communicable diseases, or through biosensors and pocket imaging to provide state-of-the-art diagnosis and management for such conditions as heart disease or diabetes. As the dominant burden of chronic disease in the developing world continues its shift to noncommunicable causes, especially heart disease, diabetes, and cancer, the case for sensors and imaging devices will become increasingly urgent.[34] While transitioning from cell phones to smart phones will help improve global access, an all-out effort needs to be mounted to diminish the digital divide. Although each of the over seven billion individuals on the planet are biologically and physiologically unique, every one of us stands to benefit from being digitized.

Other questions follow this logic train and loom large. Will the access to and up-front costs associated with individualized medicine exacerbate the disturbing disparities that already exist in health care? Will this approach become a new substrate for malpractice litigation by shifting the standard of care?[35] Will the tsunami of data create havoc among health professionals as devices constantly beep and demand a response when a group of patients are being actively monitored and hospitals are no longer the place for patients unless they are critically ill? The signal processing of real-time, continuous physiologic data from remote monitoring will need to be efficient and automate the appropriate responses accurately. Otherwise this technological movement may be considered a blatant failure. And what about the rapid wireless transmission of digital ultrasound movie loops and the ever-increasing need for expanded bandwidth of the Internet? Will YouTube and Netflix video traffic be handled in the same way as bona fide health-related Internet transmissions in the brave new medicine world? These questions and many more will arise as the digital medicine era takes hold. Radical change is intrinsically controversial—and all the more so when applied to the world of medicine.

HOMO DIGITUS AND MEDICINE SHUMPETERED

In spite of all of these concerns, I hope that you have been inspired by the prospects for digitizing human beings, for defining individuals at a more granular and molecular level than ever imaginable. Digitized, virtual humans, or the human "in the mirror," is the prerequisite for medicine to be turned upside down. Whether it is mapping the mind to awaken an individual who has been minimally conscious for several years, or mapping the genome of a person to diagnose an idiopathic, life-threatening condition or prevent an otherwise inevitable, premature death from cancer or heart attack, the technological capabilities are with us now—and emerging at a breakneck, unprecedented pace, eventually leading to the ability to print organs and even to control aspects of the mind. Humans digitizing humans is the ultimate life changer. This is much bigger than a change; this is the essence of creative destruction as conceptualized by Schumpeter. Not a single aspect of health and medicine today will ultimately be spared or unaffected in some way. Doctors, hospitals, the life science industry, government, and its regulatory bodies: all are subject to radical transformation.

Digital high definition of humans will shape the great inflection of medicine, producing a reflection of human beings through the unparalleled super-convergence of DNA sequencing, mobile smart phones and digital devices, wearable and embedded wireless nanosensors, the Internet, cloud computing, information systems, and social networking. Collectively the billions of bytes, bases, and pixels define each human being in four dimensions, a composite picture that transcends what any of us previously considered to be personal uniqueness. Without these remarkable tools, hyperpersonalization of health care and fulfilling the dream of true prevention of diseases would not be consummated. A new individualistic ideology for how medicine goes forward will not happen soon enough without the people who have the greatest stake—the individuals—to be fully participatory. There will be titanic changes ahead—medicine can and will be rebooted and reinvented one individual at a time.

AFTERWORD

IN THE SUMMER of 2008, I had the great fortune to meet Gary and Mary West, the couple who had built West Communications from the ground up, to one of the leading telecommunication companies in the United States, offering large-scale telemarketing and teleconferencing and handling the majority of 911 calls in the country. Although they had visited San Diego each year for nearly a decade, when they had sold their majority stake of the company in 2006, they became multibillionaires and moved their residence from Omaha, Nebraska, to San Diego.

Gary, whom I met first through a mutual friend, was raised in Iowa, is in his early sixties, and is about as down-to-earth a person as I have ever met. With a full head of graying hair, brown eyes, and a muscular frame of medium stature, he looks like he could be a professional football coach, and his hard-driving aura reflects the extreme work ethic he has exhibited throughout his life. He wanted to do something with the newfound wealth they had acquired, and he was well aware of the out-of-control costs of health care, having run a large company that had faced ever-escalating coverage for its employees. He and Mary are dog lovers, big time San Diego Chargers football fans, and great humanitarians. Mary is a petite, remarkably compassionate, and thoughtful lady who has deep brown eyes and shoulder length brown hair with a flip, exercises religiously on awakening early each morning, and, like Gary, had unlimited capacity to work and grow their company. In 2010 in downtown San Diego, they built a major center for hundreds of indigent seniors, who are fed meals and taught by local students to use computers and the Internet. The Wests' foundation is dedicated to seniors and to promoting aging in place, avoiding nursing homes or assisted-living situations.

When we first got together to explore a transformative project in health care, we discussed a new medical school that would have departments in genomics and wireless and have an unprecedented commitment to fostering individualized medicine. But institutional governance issues eventually put this initiative on hold.

While our further discussions were incubating, the relationship I had developed with Qualcomm, which had begun in 2007, was going in to high gear. Back in late 2007 the Scripps Translational Science Institute and Qualcomm submitted a major grant to the National Institutes of Health. It called for innovative research to change the future of medicine. That was the year I started at Scripps, and although I knew wireless was the number one industry in San Diego, I hadn't foreseen convergence with our genomics efforts.

Qualcomm is not only the largest company in San Diego but also the world's largest producer of chips for the wireless industry. In its twenty-five-year history, it has led to a remarkable proliferation of wireless companies in the region—over six hundred by 2010 and more than a hundred of these working on health-related products. The director of Qualcomm's health and medical division is Don Jones, who had been driving this new business area for over five years when we first met.

Don is a particularly affable fellow, bald with wireless glasses, the look of a professor, and he is one of the best-known figures in the worldwide mobile health field. He travels the world incessantly, proselytizing to both the tech and the medical worlds on the impact of wireless on the future of health care. He loves to use the phrase "Every Body on the Net" to capture the sense of the Internet of medical things. A few months before submitting our NIH grant in 2007, Don and I got together to see whether we could fold wireless medicine into our application. Acknowledging that it was fairly loose and undeveloped, we nevertheless incorporated a few sections on how we would jointly train physicians to be active in wireless medicine research. When the grant was reviewed in the spring of 2008, the review team was enthusiastic about the unique prospects of leveraging wireless technology for innovative medical applications. That turned out to be an important external peer validation of what we were thinking and a green light to pursue this further.

With the idea of a new medical school put aside, and the nascent but exciting future of wireless medicine spurring us on, the concept of the first dedicated wireless health institute was spawned. I approached Gary and Mary West with this possibility, and they were immediately enthusiastic. The Institute could be a major catalyst in advancing wireless health by performing clinical validation, dealing with regulatory and reimbursement issues, developing sensors and system solutions for unmet medical needs, and confronting the difficulties of adopting new technology into medical practice.

The Wests decided to develop a nonprofit medical research institute to drive this initiative, which was ultimately announced in April 2009 at the International Wireless meeting (CTIA) in Las Vegas.[1] At the plenary session, it was a special delight for me to be able to recognize them; they were given thunderous applause by the 5,000 attendees. The Wests had made a philanthropic investment of $100 million in this cause and opened a beautiful three-story building in January 2010, overlooking the Torrey Pines Golf Course and the Pacific, just a few hundred yards away from our genomics institute.

The overwhelming excitement in the new field of wireless medicine, the San Diego location as the country's hub of wireless technology, and a new dedicated institute with outstanding resources made it a magnet for attracting top talent. The charismatic Don Casey, a twenty-five-year veteran from Johnson & Johnson with considerable executive experience in pharma, medical devices, e-health, and consumer products, was recruited as the CEO. Also from Johnson & Johnson, and previously from Guidant, a medical device company, and academic cardiology, Dr. Joseph Smith joined as the chief medical and science officer. Joe is a brilliant physician-engineer who trained at Harvard medical school and MIT. Soon thereafter, they were joined by Dr. Mohit Kaushal, a physician who had been leading the mHealth unit at the Federal Communications Commission. The list of all stars included an exceptional group of engineers from academia and industry, along with Shelley Valentine as executive vice president and Nicole Boramanand from Medtronic.

This multidisciplinary team, with expertise in such diverse areas as regulatory science, reimbursement, clinical trials, and economics, coalesced their efforts to get key projects initiated. In the first year of operation, the institute developed a sensor for remote monitoring of high-risk pregnancy, capturing real-time data on uterine contractions and fetal heart rate. The stage has been set for catalyzing wireless innovative solutions to transform the future of medicine.

ACKNOWLEDGMENTS

More than three decades ago, when I graduated from medical school, there were no cell phones, no personal computers, and certainly no such thing as the Internet. No things digital—the term "digital" referred exclusively to the rectal examination.

Thirty years later, medicine continues to resist the digital revolution. In too many ways medicine is stuck with the original digital context, with its head in the wrong place. But the convergence of the digital world and medicine is inevitable, setting the stage for a radical disruption we desperately need. This has extraordinary potential for hyperinnovative means of precision in medical care and for preventive strategies we have never seen before.

Over these past three decades, I have edited over thirty medical textbooks, all for the medical community microcosm. I never thought I would be the author of a book broadly directed to the public. That changed with my realization that the convergence of the digital and medical worlds has to go forward and that the participation and primacy of consumers—before they ever become patients—would be essential for this new era of medicine to unfold.

For me, as a "rookie" author, the project took a remarkably serendipitous path. With a long-standing interest in genetics (my major in college), I came to Scripps at the end of 2006 to start a genomics institute. But residing in San Diego, the world hub of wireless technology, led to unforeseen opportunities, eventually coupling the major digital tools of genomics and wireless in two nearly adjacent, interconnected research institutes—a unique digital medical cluster.

A number of individuals set the foundation, enabling this whole project and perspective. I am deeply grateful to my closest friend, Dr. Paul Teirstein, who is the director of the Prebys Cardiovascular Institute and was instrumental in recruiting me to Scripps. The president and chief executive officer of Scripps Health, Chris Van Gorder, deserves exceptional credit for believing in what we could accomplish at Scripps and giving me carte blanche to build the programs in digital medicine. The Scripps chief medical officer, Dr. Brent Eastman, has been the consummate champion and ultimate support to help make it happen.

At our Scripps Translational Science Institute, with its emphasis on human genomics and being fully integrated with the Scripps Research Institute, I want to express my sincere gratitude to my colleagues for helping me edit the manuscript—Drs. Nicholas Schork, Samuel Levy, and Sarah Murray, and for all the animated interactions that we've had over the years to bring genomics into day-to-day medical practice. Katrina Schreiber, my executive assistant, provided immense help in physically getting the book together and also has been my right-hand person in an exceptionally high-throughput academic environment. Our research is supported by a flagship grant by the National Institutes of Health called the Clinical and Translational Science Award (CTSA), and without that grant many of the projects described in this book would not even exist. Our genomics efforts have been catapulted by the presence of the world's two largest genomic science companies in San Diego—Life Technologies and Illumina. I am especially grateful to Greg Lucier, CEO of Life Technologies, for his unwavering support for our work in genomics.

On the wireless side, I am genuinely indebted to Gary and Mary West, who not only are extraordinary individuals and friends but have philanthropically invested in an effort to build the first dedicated wireless health institute and have supported my work in an incomparable fashion, including endowing a chair of innovative medicine and providing research backing for our ambitious efforts to transform the future of medicine. Many of the individuals we brought on at West Wireless Health Institute have been extremely helpful to me in editing the manuscript and supporting the project, including Dr. Joseph Smith, chief medical officer; Shelley Valentine, executive vice president; and Erin Bateman, my assistant. The Wests deserve utmost recognition for their profound commitment to better the future of health care.

The team at Qualcomm, the leading worldwide force in wireless technology, has been especially supportive, right from the top—Dr. Paul Jacobs, the chair and CEO; its founder, Dr. Irwin Jacobs; its president, Steve Altman; the vice president for Wireless Health, Don Jones; and all of their colleagues. They have funded training slots in our program of wireless medicine at the Scripps Translational Science Institute and connected me with the global innovations that are occurring in the digital health arena. Although I had been involved with a pioneering wireless cardiac rhythm device many years ago (CardioNet), it was the Qualcomm team that stirred my excitement about the transformative impact of wireless medicine.

All these folks helped build my knowledge base and gave me a panoramic view of the opportunities in digital medicine. But actually writing and publishing a book was a big challenge—-the hardest I have ever worked on a project in my career. Major kudos go to Thomas Kelleher—TJ—who is the executive editor at Basic Books and provided unparalleled editing and idea crystallization for the book. He deftly sharpened the message and made this a much better product. I am particularly thankful to Katinka Matson, president at Brockman, Inc., who served as my publishing agent and was instrumental in helping me develop the proposal, requiring more drafts than I'm willing to admit, along with the link to TJ. Kay Mariea and Beth Wright were vital to the editing process. My author friends—Thomas Goetz, the executive editor at *Wired*, and James Fowler, author of *Connected*—were key to referring me to Katinka Matson and Brockman. Other authors who helped prod and inspire me along the way include Malcolm Gladwell, Atul Gawande, Michael Specter, and Dean Ornish.

I want to also acknowledge any perceptual conflicts of interest. I am on the board of directors of two wireless medical device companies, DexCom, which makes sensors for diabetes, and Sotera, which makes vital sign sensors. I also serve as an advisor to Zeo, which makes a sensor to detect brain waves and promote better sleep. The experience in working with these companies has greatly enriched my understanding of the challenges in wireless medicine, along with the untapped potential. I have no interest in promoting these companies or their products in this book and have tried to be as objective as possible in describing their efforts. But certainly these relationships are important for readers to know about.

As I learned throughout this project, which took a couple of years, it takes a lot of isolation and solitude to do the research and think, no less to write. My family—my wife, Susan, and our grown-up children, Sarah and Evan—were exceptionally patient with me and supported the whole initiative. I'm so thankful to them and especially to all the folks named here to make this endeavor a reality.

Finally, let me express my appreciation to all of you who read this book and actively contribute to the creative destruction of medicine, as we know it today, by promoting our capability of digitizing human beings. My modus operandi: Think big and act bigger!

NOTES

Introduction

1. "The Leaky Corporation," *Economist*, February 26, 2011, 75–77.
2. R. G. Baraniuk, "More Is Less: Signal Processing and the Data Deluge," *Science* 331 (2011): 717–19.
3. John Markoff, "Computer Wins on 'Jeopardy!': Trivial It's Not," *New York Times*, February 10, 2011.
4. Ibid.
5. David Gelernter, "Coming Next: A Supercomputer Saves Your Life," *Wall Street Journal*, February 5, 2011, C2.
6. S. J. Gould, "The Median Isn't the Message," *Discover* 6 (1985): 40–42.

Chapter 1

1. Marshall McLuhan, *Understanding Media: The Extensions of Man* (New York: McGraw-Hill, 1964), 57.
2. Gary Wolf, "The World According to Woz," *Wired* (June 2009), www.wired.com/wired/archive/6.09/woz_pr.html.
3. Chris Anderson and Michael Wolff, "The Web Is Dead: Long Live the Internet," *Wired* (September 2010), www.wired.com/magazine/2010/08/ff_webrip/all/1.
4. Clay Shirky, *Cognitive Surplus: Creativity and Generosity in a Connected Age* (New York: Penguin Press, 2010); "Building with Big Data," *Economist*, May 29, 2001, 74.
5. Nicholas Carr, *The Shallows: What the Internet Is Doing to Our Brains* (New York: Norton, 2010); Bill Keller, "The Twitter Trap," *New York Times Magazine*, May 22, 2011, 12.
6. Shirky, *Cognitive Surplus*.
7. Anderson and Wolff, "The Web Is Dead."
8. Tim O'Reilly and John Battelle, "Web Squared: Web 2.0 Five Years On," Web 2.0 Summit, 2009, www.web2summit.com/web2009/public/schedule/detail/10194.
9. David Pogue, "Kinect Pushes Users into a Sweaty New Dimension," *New York Times*, November 4, 2010, www.nytimes.com/2010/11/04/technology/personaltech/04pogue.html; Ashlee Vance, "Microsoft's Push into Gesture Technology," *New York Times*, October 29, 2010, www.nytimes.com/2010/10/30/technology/30chip.html; Susan Orlean, "Connected," *New Yorker* (December 2010).
10. "Person of the Year 2010: Mark Zuckerberg," *Time* (December 2010); Richard Stengel, "Only Connect," *Time* (December 2010); Claire Cain Miller, "Another Try by Google to Take on Facebook," *New York Times*, June 28, 2011; William Powers, *Hamlet's BlackBerry: A Practical Philosophy for Building a Good Life in the Digital Age* (New York: Harper, 2010), 32.
11. Nicholas Negroponte, *Being Digital* (New York: Vintage, 1996), 221.
12. "The Book of Jobs: Hope, Hype, and Apple's iPad," *Economist*, January 30–February 5, 2010.
13. David Pogue, "Apps We Wish We Had," *New York Times*, July 14, 2010.
14. Carr, *The Shallows*, 116.
15. Aaron Smith, *Mobile Access 2010*, Pew Internet & American Life Project Report, July 2010, www.pewinternet.org/Reports/2010/Mobile-Access-2010.aspx.
16. R. Kwok, "Personal Technology: Phoning in Data," *Nature* 458 (2009): 959–61; Anand Giridharadas, "Where a Cellphone Is Still Cutting Edge," *New York Times*, April 10, 2010, www.nytimes.com/2010/04/11/weekinreview/11giridharadas.html.

253

17. "Mobile Marvels: A Special Report on Telecoms in Emerging Markets," *Economist*, September 24, 2009, www.economist.com/node/14483896.

18. Kwok, "Personal Technology."

19. "Apple Scrambles to Secure iPad Deal," *Wall Street Journal*, March 18, 2010, online.wsj.com/article/SB10001424052748703532204575129862264704190.html.

20. Michael Malone and Tom Hayes, "Bye-Bye, PCs and Laptops: Smart Phones and Tablets Will Soon Handle the Majority of Our Personal Computing Needs," *Wall Street Journal*, January 7, 2011, online.wsj.com/article/SB10001424052970204527804576043803826627110.html.

21. Phyllis Korkki, "Internet, Mobile Phones Named Most Important Inventions," *New York Times*, March 9, 2009, www.nytimes.com/2009/03/08/business/08count.html.

22. D. Cantwell, "Attention Deficit Disorder: A Review of the Past 10 Years," *Journal of American Academy of Child & Adolescent Psychiatry* 35 (1996): 978–87.

23. Matt Richtel, "Attached to Technology and Paying a Price," *New York Times*, June 6, 2010, www.nytimes.com/2010/06/07/technology/07brain.html; Matt Richtel, "Growing Up Digital, Wired for Distraction," *New York Times*, November 21, 2010, www.nytimes.com/2010/11/21/technology/21brain.html; Elizabeth Bernstein, "Your BlackBerry or Your Wife: When the Whole Family Is Staring at Screens, Time to Try a Tech Detox," *Wall Street Journal*, January 11, 2011, online.wsj.com/article/SB10001424052748703779704576073801833991620.html.

24. Nick Bilton, *I Live in the Future & Here's How It Works: Why Your World, Work, and Brain Are Being Creatively Disrupted* (New York: Crown Business, 2010), Kindle edition, Introduction.

25. Powers, *Hamlet's BlackBerry*, 159.

26. Richtel, "Attached to Technology."

27. Bill Keller, "The Twitter Trap," *New York Times*, May 18, 2011; B. Sparrow et al., "Google Effects on Memory: Cognitive Consequences of Having Information at Our Fingertips," *Science*, July 14, 2011.

28. Richtel, "Attached to Technology"; Richtel, "Growing Up Digital"; Bernstein, "Your Black-Berry"; Powers, *Hamlet's BlackBerry*.

29. Bilton, *I Live in the Future*.

30. Don Tapscott and Anthony D. Williams, *Macrowikinomics: Rebooting Business and the World* (New York: Portfolio/Penguin, 2010).

31. Anne Eisenberg, "When a Camcorder Becomes a Life Partner," *New York Times*, November 6, 2010, www.nytimes.com/2010/11/07/business/07novel.html.

32. Bilton, *I Live in the Future*; Powers, *Hamlet's BlackBerry*; "How Long Will Google's Magic Last?" *Economist*, December 2, 2010, www.economist.com/node/17633138; "A Sea of Sensors: Everything Will Become a Sensor—and Humans May Be the Best of All," *Economist*, November 4, 2010, www.economist.com/node/17388356; #numbers, *Twitter Blog*, March 14, 2011, blog.twitter.com/2011/03/numbers.html; Jose Antonio Vargas, "The Face of Facebook: Mark Zuckerberg Opens Up," *New Yorker*, September 20, 2010; "Groupon Anxiety: The Online-Coupon Firm Will Have to Move Fast to Retain Its Impressive Lead," *Economist*, March 17, 2011, www.economist.com/node/18388904; Austin Carr, "Why Kevin Systrom Turned Down Zuckerberg, Left Twitter to Start Instagram," *Fast Company*, June 13, 2001; Virginia Heffernan, "When Shilling on the Web, Think Small," *New York Times*, July 31, 2011,opinionator.blogs.nytimes.com/2011/07/29/when-shilling-on-the-web-think-small.

33. Nicholas Wade, "A Decade Later, Gene Map Yields Few New Cures," *New York Times*, June 13, 2010, www.nytimes.com/2010/06/13/health/research/13genome.html.

34. Teddy Wayne, "Age Gap Narrows on Social Networks," *New York Times*, December 26, 2010, www.nytimes.com/2010/12/27/business/media/27drill.html; David Kirkpatrick, *The Facebook Effect: The Inside Story of the Company That Is Connecting the World* (New York: Simon & Schuster, 2010); Stengel, "Only Connect"; Miller, "Another Try by Google"; Geoffrey A. Fowler and Ian Sherr, "Google Missed 'Friend Thing,'" *Wall Street Journal*, June 1, 2011.

35. Tapscott and Williams, *Macrowikinomics*, 69.

36. Steven Johnson, *Where Good Ideas Come From: The Natural History of Innovation* (New York: Riverhead Books, 2010).

37. Bilton, *I Live in the Future*.

38. Ibid.; "Profiting from Friendship: Social Networks Have a Better Chance of Making Money Than Their Critics Think," *Economist*, January 28, 2010, www.economist.com/node/15351026.

39. Bilton, *I Live in the Future*, Chapter 4.

40. Chris Anderson, *The Long Tail: Why the Future of Business is Selling Less of More* (New York: Hyperion, 2006).

41. Holman Jenkins Jr., "Google and the Search for the Future," *Wall Street Journal*, August 15, 2010, online.wsj.com/article/SB10001424052748704901104575423294099527212.html.

42. "Arthur Sulzberger: We Will Stop Printing the *New York Times* Sometime in the Future," *Huffington Post*, January 9, 2011, www.huffingtonpost.com/2010/09/09/arthur-sulzberger-we-will _n_710251.html.

43. "Internet TV Looks Set to Take Control in the New Media Age," WorldTVPC, September 27, 2010, http://www.worldtvpc.com/blog/internet-tv-set-control-era/.

44. "Arthur Sulzberger."

45. "Internet TV Looks Set."

46. Ray Kurzweil, "Room for Debate: A Running Commentary on the News," *New York Times*, December 27, 2010, www.nytimes.com/roomfordebate.

47. "Biology 2.0: A Special Report on the Genome," *Economist*, June 17, 2010, www.economist .com/node/16349358/comments.

48. Tapscott and Williams, *Macrowikinomics*, 368.

49. "It's A Smart World: A Special Report on Smart Systems," *Economist*, November 4, 2010, www.economist.com/node/17388368.

50. Andy Kessler, "How Videogamers Are Changing the Economy," *Wall Street Journal*, January 3, 2011, online.wsj.com/article/SB10001424052970203418804576040103609214400.html.

51. "Genomes by the Thousand," *Nature* 467 (2010): 1026–27.

52. "A Sea of Sensors."

53. Bilton, *I Live in the Future*.

54. J-B. Michel et al., "Quantitative Analysis of Culture Using Millions of Digitized Books," *Science*, December 16, 2010; Patricia Cohen, "In 500 Billion Words, New Window on Culture," *New York Times*, December 16, 2010, www.nytimes.com/2010/12/17/books/17words.html.

55. Marshall McLuhan, *The Gutenberg Galaxy: The Making of Typographic Man* (Toronto: University of Toronto Press, 1962), 21.

56. David Gelernter, *Mirror Worlds, or, The Day Software Puts the Universe in a Shoebox: How It Will Happen and What It Will Mean* (New York: Oxford University Press, 1991), 27.

Chapter 2

1. Matthew Herper, "Pfizer Wins Longer Life for Lipitor," *Forbes*, June 18, 2008, www.forbes .com/2008/06/18/pfizer-ranbaxy-lipitor-biz-healthcare-cx_mh_0618bizpfizer.html.

2. "Lipitor's Pitchman Gets the Boot," *New York Times*, February 27, 2008, A26, www.nytimes .com/2008/02/27/opinion/27wed3.htm.

3. John Carey, "Do Cholesterol Drugs Do Any Good?" *Business Week*, January 28, 2008, www .businessweek.com/magazine/content/08_04/b4068052092994.htm.

4. E. J. Topol, "Pharmacy Benefit Managers, Pharmacies, and Pharmacogenetic Testing: Prescription for Progress?" *Science Translational Medicine* 2, no. 44 (2010): 44.

5. P. M. Ridker, "Rosuvastatin to Prevent Vascular Events in Men and Women with Elevated C-Reactive Protein," *New England Journal of Medicine* 359 (2008): 2195–207; James Le Fanu, "'Statins for All'—and Billions for Drug Firms," *Telegraph* (London), January 19, 2011, www .telegraph.co.uk/health/8270156/Statins-for-all-and-billions-for-drug-firms.html; Katherine Hobson, "Doctors Launch 'TheNNT.com' to Give Treatment Info," *Wall Street Journal*, October 5, 2010, blogs.wsj.com/health/2010/10/05/doctors-launch-thenntcom-to-give-treatment-info; *The NNT: Quick Summaries of Evidence-Based Medicine*, www.theNNT.com; F. Taylor, K. Ward, et al., "Statins for the Primary Prevention of Cardiovascular Disease," *Cochrane Collaboration* (2011), www2.cochrane.org/reviews/en/ab004816.html.

6. Le Fanu, "'Statins for All.'"

7. Peter Orszag, "Malpractice Methodology," *New York Times*, October 20, 2010, www.nytimes.com/2010/10/21/opinion/21orszag.html.

8. P. R. Orszag and E. J. Emanuel, "Health Care Reform and Cost Control," *Health Policy and Reform: Remaking Health Care* (*New England Journal of Medicine*), June 16, 2010, healthpolicy andreform.nejm.org/?p=3564.

9. The CAPRIE Steering Committee, "A Randomised, Blinded, Trial of Clopidogrel Versus Aspirin in Patients at Risk of Ischemic Events (CAPRIE)," *Lancet* 348, no. 9038 (1996): 1329–39.

10. "FDA Announces New Boxed Warning on Plavix: Alerts Patients, Health Care Professionals to Potential for Reduced Effectiveness," press release, March 12, 2010, www.fda.gov/NewsEvents/Newsroom/PressAnnouncements/ucm204253.htm.

11. S. B. Damani and E. J. Topol, "The Case for Routine Genotyping in Dual-Anti-Platelet Therapy," *Journal of the American College of Cardiology* 56 (2010): 109–11.

12. J-S. Hulot et al., "Cytochrome P450 2C19 Loss-of-Function Polymorphism Is a Major Determinant of Clopidogrel Responsiveness in Healthy Subject," *Blood* 108, no. 7 (2006): 2244–47.

13. Eric J. Topol, "Shotgun Medicine," *Los Angeles Times*, March 28, 2007, articles.latimes.com/2007/mar/28/opinion/oe-topol28.

14. M. Prewitt, "Two Hospitals Study New Drug for Heart Attack Patients," *Baltimore Sun*, February 22, 1984, F16.

15. Gruppo Italiano per lo Studio della Streptochinasi nell'Infarto Miocardico (GISSI), "Effectiveness of Intravenous Thrombolytic Treatment of Acute Myocardial Infarction," *Lancet* 327, no. 8478 (1986): 397–402.

16. "The Flat Earth Committee," *Wall Street Journal*, July 13, 1987, 22.

17. The GUSTO Investigators, "An International Randomized Trial Comparing Four Thrombolytic Strategies for Acute Myocardial Infarction," *New England Journal of Medicine* 329 (1993): 673–82.

18. Ralph H. Blum and Mark Scholz, *Invasion of the Prostate Snatchers: No More Unnecessary Biopsies, Radical Treatment or Loss of Sexual Potency* (New York: Other Press, 2010), Kindle edition, Chapter 1.

19. Richard Ablin, "The Great Prostate Mistake," *New York Times*, March 9, 2010, www.nytimes.com/2010/03/10/opinion/10Ablin.html.

20. Ibid.

21. Roni Caryn Rabin, "Patterns: Rethinking Prostate Cancer Treatment," *New York Times*, August 2, 2010, www.nytimes.com/2010/08/03/health/03patt.html.

22. S. Woloshin and L. M. Schwartz, "The Benefits and Harms of Mammography Screening: Understanding the Trade-offs," *Journal of the American Medical Association* 303, no. 2 (2010): 164–365; E. Marshall, "Brawling Over Mammography," *Science* 327 (2010): 936–38.

23. N. J. Wald and M. R. Law, "A Strategy to Reduce Cardiovascular Disease by More than 80%," *British Medical Journal* 326 (2003): 1419–24.

24. Timothy W. Martin, "Government Advises Less Fluoride in Water," *Wall Street Journal*, January 8, 2011, A3, online.wsj.com/article/SB10001424052748704739504576068162146159004.html.

25. E. Lonn et al., "The Polypill in the Prevention of Cardiovascular Disease: Key Concepts, Current Status, Challenges, and Future Directions," *Circulation* 122 (2010): 2078–88.

26. Jeff Donn, Martha Mendoza, and Justin Pritchard, "Prescription Drugs Found in Drinking Water Across U.S.," *USA Today*, March 10, 2008, www.usatoday.com/news/nation/2008-03-10-drugs-tap-water_N.htm; "Drug Traces Common in Tap Water," Associated Press, March 10, 2008; Marianne English, "Antidepressants Kill Bacteria in Great Lakes," *DiscoveryNews*, May 27, 2011, news.discovery.com/human/antidepressants-kill-bacteria-in-great-lakes-110527.html. It is notable that a link has already been established between the water supply and prescription drugs. In 2008, studies of the water sources conducted in twenty-four major metropolitan areas in the United States found that a wide array of pharmaceuticals were nearly omnipresent. Drug such as antidepressants, antibiotics, and hormone replacements were found in twenty-eight of sixty-two drinking water sources, which supplied water for over forty million Americans. While these medications were detected at very low levels, this reflects the outgrowth of population medicine today—the cumulative effect of wide-scale use of prescription drugs, often administered to the wrong individuals or at incorrect doses.

27. M. J. Stampfer et al., "Vitamin E Consumption and the Risk of Coronary Disease in Women," *New England Journal of Medicine* 328 (1993): 1444–49; P. Knekt et al., "Antioxidant Vitamin Intake and Coronary Mortality in a Longitudinal Population Study," *American Journal of Epidemiology* 139, no. 12 (1994): 1180–89; D. P. Vivekananthan et al., "Use of Antioxidant Vitamins for the Prevention of Cardiovascular Disease: Meta-analysis of Randomized Trials," *Lancet* 361 (2003): 2017–23; E. Lonn et al., "Effects of Long-term Vitamin E Supplementation on Cardiovascular Events and Cancer," *Journal of the American Medical Association* 293, no. 11 (2005): 1338–47.

28. The Writing Group for the WHI Investigators, "Risks and Benefits of Estrogen Plus Progestin in Healthy Post-Menopausal Women: Principal Results of the Women's Health Initiative Randomized Controlled Trial," *Journal of the American Medical Association* 288, no. 3 (2002): 321–33; Tara Parker-Pope, "The Women's Health Initiative and the Body Politic," *New York Times*, April 8, 2011, www.nytimes.com/2011/04/10/weekinreview/10estrogen.html; A. J. Fugh-Berman, "The Haunting of Medical Journals: How Ghostwriting Sold 'HRT,'" *PLoS Medicine* 7, no. 9 (2010): 1–11; Gail Collins, "Medicine on the Move," *New York Times*, April 7, 2011, A23.

29. J. Ioannidis, "Why Most Published Research Findings Are False," *PLoS Medicine* 2, no. 8 (2005): 696; David H. Freedman, "Lies, Damned Lies, and Medical Science," *Atlantic*, November 2010, www.theatlantic.com/magazine/archive/2010/11/lies-damned-lies-and-medical-science/8269; Sharon Begley, "Why Almost Everything You Hear About Medicine Is Wrong," *Newsweek*, January 24, 2011.

30. Jonah Lehrer, "The Truth Wears Off," *New Yorker*, December 13, 2010, www.newyorker.com/reporting/2010/12/13/101213fa_fact_lehrer; Gina Kolata, "Trial in a Vacuum: Study of Studies Show Few Citations," *New York Times*, January 17, 2011; K. A. Robinson and S. N. Goodman, "A Systematic Examination of Citation of Prior Research Reports of Randomized, Controlled Trials," *Annals of Internal Medicine* 154, no. 1 (2011): 50–56; J. Loscalzo, "Can Scientific Quality Be Quantified?" *Circulation* 123 (2011): 947–50.

31. Kolata, "Trial in a Vacuum"; Robinson and Goodman, "A Systematic Examination"; Loscalzo, "Scientific Quality."

32. U. Landmesser et al., "Simvastatin Versus Ezetimibe: Pleiotropic and Lipid-Lowering Effects on Endothelial Function in Humans," *Circulation* 111 (2005): 2356–63.

33. J. Kastelein, "Simvastatin With or Without Ezetimibe in Familial Hypercholesterolemia," *New England Journal of Medicine* 358 (2008): 1431–43; A. Rossebo, "Intensive Lipid Lowering with Simvastatin and Ezetimibe in Aortic Stenosis," *New England Journal of Medicine* 359 (2008): 1343–56; R. Peto, "Analyses of Cancer Data from Three Ezetimibe Trials," *New England Journal of Medicine* 359 (2008): 1357–66.

Chapter 3

1. P. Hartzband et al., "Untangling the Web: Patients, Doctors, and the Internet," *New England Journal of Medicine* 363 (2010): 1063–66.

2. Atul Gawande, "The Velluvial Matrix," *New Yorker*, June 16, 2010, www.newyorker.com/online/blogs/newsdesk/2010/06/gawande-stanford-speech.html.

3. R. N. Khouzam et al., "A Heart with 67 Stents," *Journal of American College of Cardiology* 56 (2010): 1605–7.

4. Atul Gawande, "The Cost Conundrum," *New Yorker*, June 1, 2009, www.newyorker.com/reporting/2009/06/01/090601fa_fact_gawande.

5. Hannah Fairfield, "Health Spending Versus Results," *New York Times*, June 6, 2010, www.nytimes.com/interactive/2010/06/06/business/metrics-health-care-outlier.html.

6. J. M. Donohue et al., "A Decade of Direct-to-Consumer Advertising of Prescription Drugs," *New England Journal of Medicine* 357 (2007): 673–81; K. M. Lovett et al., "Direct-to-Consumer Cardiac Screening and Suspect Risk Evaluation," *Journal of the American Medical Association* 305, no. 24 (2011): 2567–69.

7. J. Greene, "Pharmaceutical Marketing and the New Social Media," *New England Journal of Medicine* 363 (2010): 2087–89.

8. "Twelfth Annual Survey on Consumer Reaction to DTC Advertising of Prescription Drugs Reveals: Nearly 50% of Online Consumers Report Health Videos a Top Resource," press release,

Rodale, July 7, 2009, www.rodaleinc.com/newsroom/12th-annual-survey-iconsumer-reaction
-dtc-advertising-prescription-drugsi-reveals.

9. *Labeling and Advertising for Prescription Drugs,* S. 502-505, 109th Cong., 1st Session (September 8, 2005), www.theorator.com/bills109/hr3696.html.

10. "Royal Jelly," Wikipedia, n.d., en.wikipedia.org/wiki/Royal_jelly.

11. "Dangerous Supplements," *Consumer Reports,* August 3, 2010, www.consumerreports.org/health/natural-health/dietary-supplements/overview/index.htm; "The Growing Case Against Herbs: More Research Questions Safety, Effectiveness," *Wall Street Journal,* August 29, 2002; "There Is No Alternative," *Economist,* May 21, 2011, 16; "Alternative Medicine: Think Yourself Better," *Economist,* May 21, 2011.

12. "There Is No Alternative."

13. J. Y. Reginster et al., "Long-Term Effects of Glucosamine Sulphate on Osteoarthritis Progression: A Randomized, Placebo-Controlled Clinical Trial," *Lancet* 357 (2001): 251–56.

14. Maggie Fox, "Many Dietary Supplements Are Contaminated," MSNBC, August 3, 2010, www.msnbc.msn.com/id/38542031/ns/health-alternative_medicine.

15. D. Vivekananthan, "Use of Antioxidant Vitamins for the Prevention of Cardiovascular Disease: Meta-Analysis of Randomized Trials," *Lancet* 361 (2003): 2017–23.

16. Ibid.; J. M. Armitage et al., "Effects of Homocysteine-Lowering with Folic Acid Plus Vitamin B_{12} Versus Placebo on Mortality and Major Morbidity in Myocardial Infarction Survivors," *Journal of the American Medical Association* 303, no. 24 (2010): 2486–94.

17. K. M. Sanders et al., "Single Annual High-Dose Oral Vitamin D and Falls and Fractures in Older Women," *Journal of the American Medical Association* 303, no. 18 (2010): 1815–22.

18. Alice Park, "The Vitamin-D Debate: How Much Is OK?" *Time,* August 30, 2010, 66; A. Slomski, "IOM Endorses Vitamin D, Calcium Only for Bone Health, Dispels Deficiency Claims," *Journal of the American Medical Association* 305, no. 5 (2011): 453–56; Institute of Medicine Committee to Review Dietary Reference Intakes for Vitamin D and Calcium, *Dietary Reference Intakes for Calcium and Vitamin D,* ed. A. Catharine Ross et al. (Washington, DC: National Academies Press, 2011); S. Shapses and J. Manson, "Vitamin D and Prevention of Cardiovascular Disease and Diabetes: Why the Evidence Falls Short," *Journal of the American Medical Association* 305, no. 24 (2011): 2565–66; A. Maxmen, "The Vitamin D-lemma," *Nature* 475 (2011): 23–26.

19. D. Kronhout et al., "N-3 Fatty Acids and Cardiovascular Events After Myocardial Infarction," *New England Journal of Medicine* 363 (2010): 2015–26; GISSI-Prevenzione Investigators, "Dietary Supplementation with N-3 Polyunsaturated Fatty Acids and Vitamin E After myocardial Infarction: Results of the GISSI-Prevenzione Trial," *Lancet* 354 (1999): 447–55; R. De Caterina, "N-3 Fatty Acids in Cardiovascular Disease," *New England Journal of Medicine* 364 (2011): 2439–50.

20. S. Basaria et al., "Adverse Events Associated with Testosterone Administration," *New England Journal of Medicine* 363 (2010):109–22.

21. Catherine Saint Louis, "UVA Reform: It's Not PDQ," *New York Times,* June 24, 2010, www.nytimes.com/2010/06/24/fashion/24Skin.html.

22. Ibid.; K. Schweitzer, "Can Your Sunscreen Cause Skin Cancer?" MSNBC, July 7, 2007; Gardiner Harris, "FDA Unveils New Rules About Sunscreen Claims," *New York Times,* June 14, 2011, www.nytimes.com/2011/06/15/science/15sun.html.

23. Saint Louis, "UVA Reform."

24. L. Nainggolan, "Cardiologists Don't Ask About Nutraceutical/OTC Meds," theheart.org, July 5, 2010; J. Glisson and L. A. Walker, "How Physicians Should Evaluate Dietary Supplements," *American Journal of Medicine* 123, no. 7 (2010): 577–82.

25. *Labeling and Advertising for Prescription Drugs.*

26. David H. Freedman, "The Triumph of New-Age Medicine," *Atlantic* (July/August 2011); Robert Capps, "Author Simon Singh Puts Up a Fight in the War on Science," *Wired* (September 2010): 112–15.

27. Freedman, "The Triumph of New-Age Medicine."

28. Capps, "Author Simon Singh."

29. P. Goode et al., "Behavioral Therapy With or Without Biofeedback and Pelvic Floor Electrical Stimulation for Persistent Postprostatectomy Incontinence: A Randomized Controlled Trial," *Journal*

of the American Medical Association 305, no. 2 (2011): 151–59; Lesley Alderman, "Using Hypnosis to Gain More Control Over Your Illness," *New York Times*, April 16, 2011, B6.

30. Ibid.

31. "NEHI Research Shows Patient Medication Nonadherence Costs Health Care System $290 Billion Annually," press release, www.nehi.net/news/press_releases/110/nehi_research_shows _patient_medication_nonadherence_costs_health_care_system_290_billion_annually.

32. E. J. Topol, "Pharmacy Benefit Managers, Pharmacies, and Pharmacogenomic Testing: Prescription for Progress?" *Science Translational Medicine* 2, no. 44 (2010): 22.

33. "Twelfth Annual Survey on Consumer Reaction."

34. R. Chafe, "The Rise of People Power," *Nature* 472 (2011): 410–11; Denise Grady, "Vein-Opening Procedure Attracts Adherents, Through Theory Is Unproved," *New York Times*, June 29, 2010, 1, 4; K. Moisse, "The YouTube Cure," *Scientific American* (February 2011): 34–37; Thomas M. Burton, "Studies Raise Doubts on MS Theory," *Wall Street Journal*, August 2, 2010, online.wsj.com/article/SB10001424052748703787904575403160155710380.html; Thomas M. Burton, "MS Study Debunks Blocked-Vein Theory," *Wall Street Journal*, April 13, 2011, online .wsj.com/article/SB10001424052748703551304576261130285337192.html.

35. Chafe, "The Rise of People Power."

36. Ibid.

37. Ibid.

38. Moisse, "The YouTube Cure"; Burton, "Studies Raise Doubts"; Burton, "MS Study Debunks."

39. David Gelber, "Twenty-first Century Snake Oil," CBS News, April 18, 2010, www.cbsnews .com/stories/2010/04/16/60minutes/main6402854.shtml.

40. R. Miller, "Finally Final? New STITCH Analysis Finds No Subset That Benefit from Ventricular Reconstruction," theheart.org, March 16, 2010.

41. R. Kerber, "Automatic External Defibrillators for Public Access Defibrillation: Recommendations for Specifying and Reporting Arrhythmia Analysis Algorithm Performance, Incorporating New Waveforms, and Enhancing Safety," *Circulation* 95 (1997): 1677–82.

42. Ibid.

43. D. J. Elias and E. J. Topol, "A Big Step Forward for Individualized Medicine: Enlightened Dosing of Warfarin," *European Journal of Human Genetics* 16, no. 5 (2008): 532–34.

44. D. Matchar, "Effect of Home Testing of International Normalized Ratio on Clinical Events," *New England Journal of Medicine* 363 (2010):1608–20.

45. N. R. Kleinfield, "Flood of Health Kits Widens Home Tests for Early Symptoms," *New York Times*, October 1, 1984, www.nytimes.com/1984/10/01/business/flood-of-health-kits-widens -home-tests-for-early-symptoms.html; Anna Wilde Matthews, "Worried About Cholesterol? Order Your Own Tests," *Wall Street Journal*, January 11, 2011.

46. "WebMD," *Wikipedia*, n.d., en.wikipedia.org/wiki/WebMD; "QuickStats: Percentage of Adults Aged ≥18 Years Who Looked Up Health Information on the Internet, by Age Group and Sex— National Health Interview Survey, United States, January–September 2009," *Morbidity and Mortality Weekly Report* 59, no. 15 (April 23, 2010): 461.

47. "Evaluating Health Information on the Internet," National Cancer Institute, March 6, 2009, www.cancer.gov/cancertopics/factsheet/Information/internet; Virginia Heffernan, "Online Medical Advice Can Be a Prescription for Fear," *New York Times*, February 4, 2011, www.nytimes.com/2011/ 02/06/magazine/06FOB-Medium-t.html.

48. Claire Cain Miller, "Social Networks a Lifeline for the Chronically Ill," *New York Times*, March 24, 2010, www.nytimes.com/2010/03/25/technology/25disable.html; J. R. Ingelfinger and J. M. Drazen, "Patient Organizations and Research on Rare Diseases," *New England Journal of Medicine* 364 (2011): 1670–71.

49. Deborah Copaken Kogan, "How Facebook Saved My Son's Life," *Slate*, July 13, 2011, www.slate.com/id/2297933; Riva Greenberg, "Are Doctors Losing Their Relevance Due to Social Media Health Sites?" *Huffington Post*, June 8, 2010, www.huffingtonpost.com/riva-greenberg/are -doctors-losing-their_b_596060.html; "PatientsLikeMe Poll Reveals Patients Share Health Data Online, Prefer to Keep Quiet with Doctors, Employers," PatientsLikeMe.com, April 13, 2011, www.patientslikeme.com/press/20110413/26-patientslikeme%C2%AE-poll-reveals-patients

-share-health-data-online-prefer-to-keep-quiet-with-doctors-employerspipatients-unveil-top
-reasons-not-to-share-health-information-i-; Don Tapscott and Anthony D. Williams, *Macrowiki-nomics: Rebooting Business and the World* (New York: Portfolio/Penguin, 2010); Natasha Singer, "When Patients Meet Online, Are There Side Effects?" *New York Times*, May 30, 2010, 3.

50. Brian Donnelly, "Patients Turn to Social Websites to Discuss Ailments, Worrying Some Doctors," Fox News, June 28, 2010, www.foxnews.com/health/2010/06/28/patients-turn-social-web sites-discuss-ailments-worrying-doctors.

51. Alliance Health Networks, "Diabetic Connect, the Largest Online Community for Diabetics, Rolls Out New Social Networking Features," press release, October 5, 2010.

52. Singer, "When Patients Meet Online"; Stuart Elliott, "Web Site to Offer Health Advice, Some of It From Marketers," *New York Times*, October 6, 2010, B4.

53. "Best Hospitals," *U.S. News & World Report* (August 2010): 58.

54. O. Wang, "'America's Best Hospitals' in the Treatment of Acute Myocardial Infarction," *Internal Medicine* 167, no. 13 (2007): 1345–51.

55. Malcolm Gladwell, "The Order of Things," *New Yorker*, February 14, 2011, 68–75.

56. Wang, "'America's Best Hospitals'"; S. Williams, "Performance of Top-Ranked Heart Care Hospitals on Evidence-Based Process Measures," *Circulation* 114 (2006): 558–64; H. Krumholz, "Evaluation of a Consumer-Oriented Internet Health Care Report Card," *Journal of the American Medical Association* 287, no. 10 (2002): 1277–87.

57. "Scripps Health Selected Amongst Nation's Top 10 Health Systems," press release, June 21, 2010, www.scripps.org/news_items/3717-scripps-health-selected-among-nation%E2%80%99s -top-10-health-systems.

58. Roni Caryn Rabin, "You Can Find Dr. Right, with Some Effort," *New York Times*, September 30, 2008, www.nytimes.com/2008/09/30/health/30find.html; T. Lagu and P. K. Lindenauer, "Putting the Public Back in Public Reporting of Health Care Quality," *Journal of the American Medical Association* 304, no. 15 (2010): 1711–12.

59. Amy Dockser Marcus, "Hiring Your Own Scientist to Find a Cure: Families of Terminally Ill Set Up Research Foundations; Here's Where to Get Help," *Wall Street Journal*, April 25, 2002, D1.

60. Jonathan Weiner, *His Brother's Keeper: A Story from the Edge of Medicine* (New York: Harper-Collins, 2004).

61. Marcus, "Hiring Your Own Scientist."

62. The series featured the following articles, all by Amy Dockser Marcus: "Diagnosed with Rare Disease He Studied for Years, Dr. Olney Struggles to Find Doctors," *Wall Street Journal*, November 21, 2006, A1; "Medical Student Takes on a Rare Disease—His Own," *Wall Street Journal*, April 2, 2004, A1, online.wsj.com/public/resources/documents/SB108077194020270691.htm; "A Patient's Quest to Save New Drug Hits Market Reality," *Wall Street Journal*, November 16, 2004, A1, online.wsj.com/public/resources/documents/SB110056934631375066.htm; "Patients with Rare Diseases Work to Jump-Start Research," *Wall Street Journal*, July 11, 2006, A1, www.michaeljfox.org /newsEvents_parkinsonsInTheNews_article.cfm?ID=102.

63. Marcus, "Medical Student Takes on a Rare Disease."

64. Amy Dockser Marcus, "New Approach to Lung Cancer: Being Aggressive," *Wall Street Journal*, June 24, 2004, A1, online.wsj.com/public/resources/documents/SB108845334318249415.htm.

65. Tara Parker-Pope, "A Father's Quest to Cure His Children," *New York Times*, January 22, 2010; Geeta Anand, *The Cure: How a Father Raised $100 Million—and Bucked the Medical Establishment—in a Quest to Save His Children* (New York: Harper Paperbacks, 2009); *Extraordinary Measures*, directed by Tom Vaughan (CBS Films, 2010).

Chapter 4

1. Don Tapscott and Anthony D. Williams, *Macrowikinomics: Rebooting Business and the World* (New York: Portfolio/Penguin, 2010), 167.

2. "A Sea of Sensors: Everything Will Become a Sensor—and Humans May Be the Best of All," *Economist*, November 4, 2010, www.economist.com/node/17388356.

3. "Appendix F: The Internet of Things (Background)," *Disruptive Technologies: Global Trend 2025*, SRI Consulting Business Intelligence, www.dni.gov/nic/PDF_GIF_confreports/disruptivetech/appendix_F.pdf.

4. Thomas Goetz, "The Feedback Loop," *Wired*, July 2010.

5. Tas Anjarwalla, "Inventor of Cell Phone: We Knew Someday Everybody Would Have One," *CNN.com*, July 9, 2010, articles.cnn.com/2010-07-09/tech/cooper.cell.phone.inventor_1_car-phone-cell-phone-building-phones.

6. Steven Johnson, *Where Good Ideas Come From: The Natural History of Innovation* (New York: Penguin, 2010), Kindle edition, Introduction.

7. R. Kwok, "Personal Technology: Phoning in Data," *Nature* 458 (2009): 959–61.

8. M. Cooper (Dr. Unplugged), "The Continuous Physical Exam: Cell Phone Inventor Talks About the Coming Revolution in Healthcare," Medscape.com, November 4, 2010; Bill Saporito, "Leaving Cell Hell," *Time*, February 28, 2011, 71.

9. "Living by Numbers," *Wired* (July 2009); Kashmir Hill "Adventures in Self-Surveillance, aka The Quantified Self, aka Extreme Navel-Gazing," *Forbes*, April 7, 2011, blogs.forbes.com/kashmirhill/2011/04/07/adventures-in-self-surveillance-aka-the-quantified-self-aka-extreme-navel-gazing.

10. "Living by Numbers."

11. Williams M. Digifit, "Withings and Zeo, Three Leaders in Health and Fitness Monitoring, Join Forces to Deliver the 'Health Triad'—Fitness, Weight and Sleep," *Digifit News*, November 22, 2010.

12. *Chronic Diseases: The Power to Prevent, the Call to Control: At a Glance 2009*, Centers for Disease Control and Prevention, December 17, 2009, www.cdc.gov/chronicdisease/resources/publications/AAG/chronic.htm.

13. L. Saxon, "Long-Term Outcome After ICD and CRAT Implantation and Influence of Remote Device Follow-Up," *Circulation* (2010): 2353–55; D. Matlock, "Big Brother Is Watching You," *Circulation* 122 (2010): 319–22.

14. "The United States of Diabetes: New Report Shows Half the Country Could Have Diabetes or Prediabetes at a Cost of $3.35 Trillion by 2010," United Health Group, November 23, 2010, www.unitedhealthgroup.com/newsroom/news.aspx?id=36df663f-f24d-443f-9250-9dfdc97cedc5; G. Danaei et al., "National, Regional, and Global Trends in Fasting Plasma Glucose Prevalence Since 1980: Systematic Analysis of Health Examination Surveys and Epidemiological Studies with 370 Country-Years and 2.7 Million Participants," *Lancet* 378, no. 9785 (2011): 31–40; C. Quinn et al., "Cluster-Randomized Trial of a Mobile Phone Personalized Behavioral Intervention for Blood Glucose Control," *Diabetes Care* (September 2011), doi: 10.2337/dc11-0366; Jennifer Dooren, "Ring! Time for Blood Test," *Wall Street Journal*, August 2, 2011, online.wsj.com/article/SB10001424053111903341404576482383432907992.html.

15. Jude Garvey, "Color-Shifting Contact Lenses Alert Diabetics to Glucose Levels," *Gizmag*, December 29, 2009, www.gizmag.com/color-changing-contact-lenses-diabetic-glucose/13682; "Look into My Eyes," *Economist*, June 4, 2011; Kenrick Vezina, "Tattoo Tracks Sodium and Glucose via an iPhone," *Technology Review*, July 20, 2011, www.technologyreview.com/computing/38065; Alexander George, "Digital Tattoo Gets Under Your Skin to Monitor Blood," *Wired*, July 25, 2011, www.wired.com/gadgetlab/2011/07/blood-monitor-tattoo-iphone.

16. "Holter Monitor (24h)," *MedlinePlus*, n.d., www.nlm.nih.gov/medlineplus/ency/article/003877.htm.

17. S. Rothman, "The Diagnosis of Cardiac Arrhythmias: A Prospective Multi-Center Randomized Study Comparing Mobile Cardiac Outpatient Telemetry Versus Standard Loop Event Monitoring," *Journal of Cardiovascular Electrophysiology* 18 (2007): 1–7.

18. D. H. Freedman, "How to Fix the Obesity Crisis," *Scientific American* (February 2011): 40–47; Lesley Alderman, "Losing Weight the Smartphone Way, with a Nutritionist in Your Pocket," *New York Times*, July 16, 2010, www.nytimes.com/2010/07/17/health/17patient.html.

19. Institute of Medicine, *A Population-Based Policy and Systems Change Approach to Prevent and Control Hypertension* (Washington, DC: National Academies Press, 2010); R. McManus,

"Telemonitoring and Self-Management in the Control of Hypertension (TASHINH2): A Randomised Controlled Trial," *Lancet* 346 (2010): 163–72.

20. P. Rothwell, "Limitations of the Usual Blood-Pressure Hypothesis and Importance of Variability, Instability, and Episodic Hypertension," *Lancet* 375 (2010): 938–48.

21. Institute of Medicine, *To Err Is Human: Building a Safer Health System* (Washington: Penguin, 2000); B. Starfield, "Is U.S. Health Really the Best in the World?" *Journal of the American Medical Association* 284 (2000): 483–85.

22. E. J. Topol, "Transforming Medicine via Digital Innovation," *Science Translational Medicine* 2 (2010): 16; "Inhaling Information," *Economist*, April 9, 2011, 90.

23. Ibid.

24. "Ranking America's Mental Health: An Analysis of Depression Across the States," Mental Health America, n.d., www.nmha.org/go/state-ranking.

25. C. Brauer, "Incidence and Mortality of Hip Fractures in the United States," *Journal of the American Medical Association* 302 (2009): 1573–79.

26. "When Your Carpet Calls Your Doctor," *Economist*, April 8, 2010, www.economist.com/node/15868133.

27. "Telemedicine Comes Home," *Economist*, June 5, 2008, www.economist.com/node/11482580.

28. John Leland, "Sensors Help Keep the Elderly Safe, and at Home," *New York Times*, February 12, 2009, A1.

29. D. Estrin, "Open mHealth Architecture: An Engine for Health Care Innovation," *Science* 330 (2010): 759–60.

30. D. Cutler, "Thinking Outside the Pillbox-Medication Adherence as a Priority for Health Care Reform," *New England Journal of Medicine* 362 (2010): 1553–55; "More Than a Quarter of Prescription Takers Cut Corners to Save Money," *Wall Street Journal*, August 24, 2010, blogs.wsj.com/health/2010/08/24/more-than-a-quarter-of-prescription-takers-cut-corners-to-save -money; Jonathan D. Rockoff, "More Balk at Cost of Prescriptions," *Wall Street Journal*, October 12, 2010, online.wsj.com/article/SB10001424052748703927504575540510224649150.html.

31. S. Jencks, "Rehospitalizations Among Patients in the Medicare Fee-for-Service Program," *New England Journal of Medicine* 360 (2009): 1418–28.

32. Brian Dolan, "Study: GlowCaps Up Adherence to 98 Percent," *Mobi Health News*, June 23, 2010, mobihealthnews.com/8069/study-glowcaps-up-adherence-to-98-percent.

33. "Potential Encapsulated," *Economist*, January 14, 2010, www.economist.com/node/15276730.

34. Anne Eisenberg "'Fantastic Voyage,' Revisited: The Pill That Navigates," *New York Times*, January 31, 2009. www.nytimes.com/2009/02/01/business/01novel.html; Avery Johnson, "The Do-It-Yourself House Call," *Wall Street Journal*, July 27, 2010.

35. B. Chi, "Mobile Phones to Improve HIV Treatment Adherence," *Lancet* 9755 (2010): 1807–8.

36. Denis Campbell, "Mobile Phone Kits to Diagnose STDs," *Guardian* (London), November 5, 2010, www.guardian.co.uk/uk/2010/nov/05/new-test-mobile-phones-diagnose-stds.

37. Ibid.

38. Campbell, "Mobile Phone Kits"; Anne Eisenberg, "Beyond the Breathalyzer: Seeking Telltale Signs of Disease," *New York Times*, July 3, 2011.

39. Chuck Salter, "The Doctor of the Future," *Fast Company* (May 2009).

40. Wouter Stomp, "Toyota to Integrate ECG Sensors into Steering Wheels," MedGadget, July 25, 2011, medgadget.com/2011/07/toyota-to-integrate-ecg-sensors-into-steering-wheels.html; Wouter Stump, "Ford Unveils Contactless ECG Sensing Driver Seat," MedGadget, May 30, 2011, medgadget.com/2011/05/ford-unveils-contactless-ecg-sensing-driver-seat.html.

Chapter 5

1. Kevin Davies, *The $1,000 Genome: The Revolution in DNA Sequencing and the New Era of Personalized Medicine* (New York: Free Press, 2010), Kindle edition, Chapter 8.

2. Mark Johnson, Kathleen Gallagher, "One in a Billion: A Boy's Life, a Medical Mystery," *Journal Sentinel* (Milwaukee, WI), December 27, 2010, www.jsonline.com/features/health/112518634.html; Matthew Herper, "Sequencing a Child's DNA—and Convincing an Insurance Company to Pay,"

Forbes, March 2, 2011, blogs.forbes.com/matthewherper/2011/03/02/sequencing-a-childs-dna
-and-convincing-an-insurance-company-to-pay; Mark Johnson and Kathleen Gallagher, "Hospitals,
Researchers Excited to Take DNA Sequencing to New Levels," *Journal Sentinel* (Milwaukee, WI),
July 19, 2011; E. Worthey, "Making a Definitive Diagnosis: Successful Clinical Application of Whole
Exome Sequencing in a Child with Intractable Inflammatory Bowel Disease," *Genetics in Medicine*
10 (2011): 1–8.

3. Nicholas Wade, "Genetic Code of Human Life Is Cracked by Scientists," *New York Times*, June
27, 2000, partners.nytimes.com/library/national/science/062700sci-genome.html; "Cracking the
Code," *Time*, July 3, 2000, 404.

4. "Biology's Big Bang," *Economist*, July 16–22, 2007.

5. T. Ley and R. Wilson, "DNA Sequencing of a Cytogenetically Normal Acute Myeloid Leukemia
Genome," *Nature* 456 (2008):66–72; E. Mardis, "Recurring Mutations Found by Sequencing an
Acute Myeloid Leukemia Genome," *New England Journal of Medicine* 361 (2009): 1058–66.

6. K. Frazer, "Human Genetic Variation and Its Contribution to Complex Traits," *Nature Review*
10 (2009): 241–51.

7. Ibid.; The International HapMap Consortium, "A Haplotype Map of the Human Genome," *Nature*
437 (2005): 1299–320; The HapMap Project, *Nature* (cover), October 27, 2005, 437; R. Strausberg
and S. Levy, "Human Genetics: Individual Genomes Diversify," *Nature* 456 (2008): 49–51.

8. Ibid.

9. Frazer, "Human Genetic Variation."

10. Ibid.; Strausberg and Levy, "Human Genetics."

11. Frazer, "Human Genetic Variation."

12. R. Klein, "Complement Factor H Polymorphism in Age-Related Macular Degeneration," *Science* 308 (2005): 385–89; A. Edwards, "Complement Factor H Polymorphism and Age-Related
Macular Degeneration," *Science* 308 (2005): 421–24; J. Haines, "Complement Factor H Polymorphism and Age-Related Macular Degeneration," *Science* 308 (2005): 419–21.

13. Klein, "Complement Factor H Polymorphism."

14. Ibid.; Edwards, "Complement Factor H Polymorphism"; Haines, "Complement Factor H
Polymorphism."

15. E. J. Topol, "The Genomics Gold Rush," *Journal of the American Medical Association* 298, no.
2 (2007): 218–21.

16. L. A. Hindorff et al., "A Catalog of Published Genome-Wide Association Studies," Office of
Population Genomics, National Human Genome Research Institute, National Institutes of Health,
n.d., www.genome.gov/gwastudies.

17. Klein, "Complement Factor H Polymorphism"; Edwards, "Complement Factor H Polymorphism"; Haines, "Complement Factor H Polymorphism"; Topol, "The Genomics Gold Rush"; Hindorff et al., "Genome-Wide Association Studies"; T. Manolio, "A HapMap Harvest of Insights into
the Genetics of Common Disease," *Journal of Clinical Investigation* 118 (2008): 1590–1605; M.
McCarthy, "Genome-Wide Association Studies for Complex Traits: Consensus, Uncertainty and
Challenges," *Nature Review* 9 (2008): 356–69; T. Manolio, "Genome-Wide Association Studies
and Assessment of the Risk of Disease," *New England Journal of Medicine* 363 (2010): 166–76.

18. D. Klionsky, "Crohn's Disease, Autophagy, and the Paneth Cell," *New England Journal of Medicine* 360 (2009): 1785–86.

19. A. Franke, "Genome-Wide Meta-Analysis Increases to 71 the Number of Confirmed Crohn's
Disease Susceptibility Loci," *Nature Genetics* 42 (2010): 1118–25.

20. M. McCarthy, "Genomics, Type 2 Diabetes, and Obesity," *New England Journal of Medicine*
363 (2010): 2339–50.

21. J. Bluestone, "Genetics, Pathogenesis and Clinical Interventions in Type 1 Diabetes," *Nature*
464 (2010): 1293–300.

22. J. Gudmundsson, "Two Variants on Chromosome 17 Confer Prostate Cancer Risk, and the
One in TCF2 Protects Against Type 2 Diabetes," *Nature Genetics* 39 (2007): 977–83.

23. Franke, "Genome-wide Meta-Analysis"; H. Allen, "Hundreds of Variants Clustered in Genomic
Loci and Biological Pathways Affect Human Height," *Nature* 467 (2010): 832–38; McCarthy, "Genomics, Type 2 Diabetes."

24. S. Lubitz, "Association Between Familial Atrial Fibrillation and Risk of New-Onset Atrial Fibrillation," *Journal of the American Medical Association* 304 (2010): 2263–69; N. P. Paynter, "Association Between a Literature-Based Genetic Risk Score and Cardiovascular Events in Women," *Journal of the American Medical Association* 303 (2010): 631–37.

25. Davies, *The $1,000 Genome*, Chapter 7.

26. T. A. Manolio, "Finding the Missing Heritability of Complex Diseases," *Nature* 461 (2009): 747–53.

27. Nicholas Wade, "A Decade Later, Genetic Map Yields Few New Cures," *New York Times*, June 12, 2010, www.nytimes.com/2010/06/13/health/research/13genome.html.

28. "The Genome, 10 Years Later," *New York Times*, June 21, 2010, A5, www.nytimes.com/2010/06/21/opinion/21mon2.html.

29. S. Hall, "Revolution Postponed," *Scientific American* (October 2010): 60–67.

30. David H. Freedman, "The Gene Bubble: Why We Still Aren't Disease-Free," *Fast Company* (November 2009): 116–22.

31. Victor K. McElheny, *Drawing the Map of Life: Inside the Human Genome Project* (New York: Basic Books, 2010).

32. Matt Ridley, "The Failed Promise of Genomics," *Wall Street Journal*, October 9, 2010, online.wsj.com/article/SB10001424052748703843800457553411197417550.html.

33. Steve Sternberg, "The Human Genome: Big Advances, Many Questions," *USA Today*, July 8, 2010, www.usatoday.com/news/health/2010-07-08-1Agenome08_CV_N.htm.

34. "Biology 2.0: A Special Report on the Human Genome," *Economist*, June 19, 2010, 1–14.

35. J. Couzin-Frankel, "Major Heart Disease Genes Prove Elusive," *Science* 328 (2010): 1220–21.

36. D. Ge, "Genetic Variation in IL28B Predicts Hepatitis C Treatment-Induced Viral Clearance," *Nature* 461 (2009): 399–401; V. Suppiah, "IL28B Is Associated with Response to Chronic Hepatitis C Interferon—and Ribavirin Therapy," *Nature Genetics* 41 (2009): 1100–104; Y. Tanaka, "Genome-Wide Association of IL28B with Response to Pegylated Interferon-a and Ribavirin Therapy for Chronic Hepatitis C," *Nature Genetics* 41 (2009):1105–9; S. Iadonato, "Hepatitis C Virus Gets Personal," *Nature* 461 (2009): 357–58; M. Enserink, "First Specific Drugs Raise Hopes for Hepatitis C," *Science* 332 (2011): 159–60.

37. Ge, "Genetic Variation in IL28B"; Suppiah, "IL28B Is Associated with Response"; Tanaka, "Genome-Wide Association of IL28B."

38. Ge, "Genetic Variation in IL28B."

39. Tanaka, "Genome-Wide Association of IL28B."

40. J. S. Hulot, "Cytochrome P450 2C19 Loss-of-Function Polymorphism Is a Major Determinant of Clopidogrel Responsiveness in Healthy Subjects," *Blood* 108, no. 7 (2006): 2244–47.

41. J. S. Hulot, "Cardiovascular Risk in Clopidogrel-Treated Patients to Cytochrome P450 2CP19*2 Loss-of-Function Allele or Proton Pump Inhibitor Co-administration: A Systematic Meta-Analysis," *Journal of American College of Cardiology* 56 (2010): 134–43; J. Mega, "Reduced-Function CYP2C19 Genotype and Risk of Adverse Clinical Outcomes Among Patients Treated with Clopidogrel Predominantly for PCI," *Journal of the American Medical Association* 304 (2010): 1821–30.

42. S. Damani and E. J. Topol, "The Case for Routine Genotyping in Dual-Antiplatelet Therapy," *Journal of American College of Cardiology* 56 (2010): 109–11.

43. E. J. Topol, "Pharmacy Benefit Managers, Pharmacies, and Pharmacogenomic Testing: Prescription for Progress?" *Science Translational Medicine* 2, no. 44 (2010): 22; E. J. Topol and N. Schork, "Catapulting Clopidogrel Pharmacogenomics Forward," *Nature Medicine* 17 (2011): 40–41.

44. A. Shuldiner, "Association of Cytochrome P450 2C19 Genotype with the Antiplatelet Effect and Clinical Efficacy of Clopidogrel Therapy," *Journal of the American Medical Association* 302 (2009): 849–58.

45. Ibid.

46. D. Elias, "Warfarin Pharmacogenomics: A Big Step Forward for Individualized Medicine: Enlightened Dosing of Warfarin," *European Journal of Human Genetics* 16, no. 5 (2008): 532–34.

47. G. Cooper, "A Genome-wide Scan for Common Genetic Variants with a Large Influence on Warfarin Maintenance Dose," *Blood* 112 (2008): 1022–27.

48. Davies, *The $1,000 Genome*.

49. J. Fellay, "ITPA Gene Variants Protect Against Anaemia in Patients Treated for Chronic Hepatitis C," *Nature* 464 (2010): 405–8.

50. The SEARCH Collaborative Group, "SLCO1B1 Variants and Statin-Induced Myopathy: A Genomewide Study," *New England Journal of Medicine* 359 (2008): 789–99.

51. A. Daly, "HLA-B*5701 Genotype Is a Major Determinant of Drug-Induced Liver Injury Due to Flucloxacillin," *Nature Genetics* 41 (2009): 816–19; M. McCormack, "HLA-A*3101 and Carbamazepine-Induced Hypersensitivity Reactions in Europeans," *New England Journal of Medicine* 364 (2011): 1134–43.

52. Ibid.; The SEARCH Collaborative Group, "SLCO1B1 Variants."

53. Daly, "HLA-B*5701 Genotype"; McCormack, "HLA-A*3101."

54. J. Singer, "A Genome-wide Study Identifies HLA Alleles Associated with Lumiracoxib-Related Liver Injury," *Nature Genetics* 42 (2010): 711–14.

55. F. W. Frueh, "Pharmacogenomic Biomarker Information in Drug Labels Approved by the United States Food and Drug Administration: Prevalence of Related Drug Use," *Pharmacotherapy* 28, no. 8 (2008): 992–98.

56. D. A. Flockhart, "Clinically Available Pharmacogenomics Tests," *Clinical Pharmacology & Therapeutics* 86, no. 1 (2009): 109–13; Andrew Pollack, "Patient's DNA May Be Signal to Tailor Medication," *New York Times*, December 29, 2008, www.nytimes.com/2008/12/30/business/30gene.html; A. Daly, "Genome-Wide Association Studies in Pharmacogenomics," *Nature Reviews* 11 (2010): 241–46.

57. K. Small, "Synergistic Polymorphisms of B1-a2c-Adrenergic Receptors and the Risk of Congestive Heart Failure," *New England Journal of Medicine* 347 (2002): 1135–42; S. Liggett, "A GRK5 Polymorphism that Inhibits B-adrenergic Receptor Signaling Is Protective in Heart Failure," *Nature Medicine* 14 (2008): 510–18; C. Ross, "Genetic Variants in *TPMT* and *COMT* Are Associated with Hearing Loss in Children Receiving Cisplatin Chemotherapy," *Nature Genetics* 41 (2009): 1345–49; The GoDARTS and UKPDS Diabetes Pharmacogenetics Study Group and the Wellcome Trust Consortium Case Control Consortium, "Common Variants near ATM Are Associated with Glycemic Response to Metformin in Type 2 Diabetes," *Nature Genetics* 43 (2010): 117–20; Flockhart, "Clinically Available Pharmacogenomics Tests"; Pollack, "Patient's DNA."

58. J. Veltman, "A De Novo Paradigm for Mental Retardation," *Nature Genetics* 4 (2010): 1109–12; K. Bilguvar, "Whole-Exome Sequencing Identifies Recessive WDR62 Mutations in Severe Brain Malformations," *Nature* 467 (2010): 207–10; X. Yi, "Sequencing of 50 Human Exomes Reveals Adaptation to High Altitude," *Science* 329 (2010): 75–78.

59. E. J. Topol, "The Resequencing Imperative," *Nature Genetics* 39 (2007): 439–40.

60. E. Hayden, "Genome Sequencing: The Third Generation," *Nature* 457 (2009): 768–69; C. Fuller, "The Challenges of Sequencing by Synthesis," *Nature Biotechnology* 27 (2009): 1013–23; T. Tucker, "Massively Parallel Sequencing: The Next Big Thing in Genetic Medicine," *American Journal of Human Genetics* 85, no. 2 (2009): 142–54; "Human Genome at 10: The Sequence Explosion," *Nature* 464 (2010): 671.

61. Hayden, "Genome Sequencing"; "Human Genome at 10."

62. "Biology 2.0."

63. Tucker, "Massively Parallel Sequencing"; "Human Genome at 10."

64. D. Pushkarev, "Single-Molecule Sequencing of an Individual Human Genome," *Nature Biotechnology* 27, no. 9 (2009): 777.

65. Andrew Pollack, "Dawn of Low-Price Mapping Could Broaden DNA Uses," *New York Times*, October 6, 2008, A1, www.nytimes.com/2008/10/06/business/06gene.html; R. Drmanac, "Human Genome Sequencing Using Unchained Base Reads on Self-Assembling DNA Nanoarrays," *Science* 327 (2010): 78–81.

66. Tucker, "Massively Parallel Sequencing"; genomics.xprize.org/.

67. Pollack, "Dawn of Low-Price Mapping"; Drmanac, "Human Genome Sequencing."

68. Davies, *The $1,000 Genome*, Chapter 1.

69. E. Ashley, "Clinical Assessment Incorporating a Personal Genome," *Lancet* 375 (2010):1525–35; L. Krieger, "Stanford's 'Molecular Autopsies' Hope to Help Grieving Families," *Mercury News* (San Jose, CA), February 7, 2011, www.mercurynews.com/science/ci_17314134.

70. Ley and Wilson, "DNA Sequencing"; Mardis, "Recurring Mutations Found."

71. W. Lee, "The Mutation Spectrum Revealed by Paired Genome Sequences from a Lung Cancer Patient," *Nature* 465 (2010): 473–77; H. Russnes, "Genomic Architecture Characterizes Tumor Progression Paths and Fate in Breast Cancer Patients," *Science* 2, no. 38 (2010): 38–747; P. Campbell, "The Patterns and Dynamics of Genomic Instability in Metastatic Pancreatic Cancer," *Nature* 467 (2010): 999–1005.

72. E. D. Pleasance, "A Small-Cell Lung Cancer Genome with Complex Signatures of Tobacco Exposure," *Nature* 463 (2010): 184–90.

73. Nicholas Wade, "Disease Cause Is Pinpointed with Genome," *New York Times*, March 11, 2010, www.nytimes.com/2010/03/11/health/research/11gene.html; J. Roach, "Analysis of Genetic Inheritance in a Family Quartet by Whole-Genome Sequencing," *Science* 328 (2010): 636–39.

74. J. R. Lupski, "Whole Genome Sequencing in a Patient with Charcot-Marie-Tooth Neuropathy," *New England Journal of Medicine* 362 (2010): 1181–91; S. Baranzani, "Genome, Epigenome and RNA Sequences of Monozygotic Twins Discordant for Multiple Sclerosis," *Nature* 464 (2010): 1351–56; M. N. Bainbridge, "Whole-Genome Sequencing for Optimized Patient Management," *Science Translational Medicine* 3, no. 87 (June 2011): 87re3; S. F. Kingsmore and C. J. Saunders, "Deep Sequencing of Patient Genomes for Disease Diagnosis: When Will It Become Routine?" *Science Translational Medicine* 3, no. 87 (June 2011): 87ps23; A. F. Rope, "Using VAAST to Identify an X-Linked Disorder Resulting in Lethality in Male Infants Due to N-Terminal Acetyltransferase Deficiency," *American Journal of Human Genetics* 89 (2011): 1–16; K. Davies, "VAAST Potential for New Genome Mutation Hunting Software," *Bio-IT World*, June 23, 2011.

75. M. N. Bainbridge, "Whole-Genome Sequencing."

76. A. F. Rope, "Using VAAST."

77. M. Lindhurst et al., "A Mosaic Activating Mutation in AKT1 Associated with the Proteus Syndrome," *New England Journal of Medicine* 365 (2011): 611–19.

78. "Genomes by the Thousand," *Nature* 467 (2010): 1026–27.

79. M. Herper, "Gene Machine," *Forbes*, December 13, 2010, www.forbes.com/forbes/2011/0117/features-jonathan-rothberg-medicine-tech-gene-machine.html; J. M. Rothberg, "An Integrated Semiconductor Device Enabling Non-Optical Genome Sequencing," *Nature* 475 (2011): 348–52.

80. "World Changing Ideas," *Scientific American* (December 2010): 42–53.

81. Ibid.; E. Zolfagharifard, "Dream Sequence: Real-Time DNA Testing," *Engineer*, November 15, 2010, www.theengineer.co.uk/dream-sequence-real-time-dna-testing/1006023.article.

82. R. Lifton, "Individual Genomes on the Horizon," *New England Journal of Medicine* 362 (2010): 1235–36; R. Resnick, "Implications of Exponential Growth of Global Whole Genome Sequencing Capacity," Genome Quest Industry, July 9, 2010, blog.genomequest.com/2010/07/implications-of-exponential-growth-of-global-whole-genome-sequencing-capacity.

83. M. Stratton, "The Cancer Genome," *Nature* 458, no. 9 (2009): 719–24.

84. Ibid.

85. Ibid.

86. G. Bollag, "Clinical Efficacy of RAF Inhibitor Needs Broad Target Blockade in *BRAF*-Mutant Melanoma," *Nature* 467 (2010): 596–99; K. Flaherty, "Inhibition of Mutated, Activated BRAF in Metastatic Melanoma," *New England Journal of Medicine* 363 (2010): 809–19.

87. R. Schilsky, "Personalized Medicine in Oncology: The Future Is Now," *Nature Reviews* 9 (2010): 363–65.

88. Davies, *The $1,000 Genome*, Chapter 12.

89. Allysia Finley, "A Geneticist's Cancer Crusade," *Wall Street Journal*, November 27, 2010, online.wsj.com/article/SB10001424052748703882404575519961343438740.html.

90. "International Team Halfway Through Effort to Map All Human Proteins," GenomeWeb, November 26, 2010, www.genomeweb.com/proteomics/international-team-halfway-through-effort-map-all-human-proteins.

91. S. Rosenberg, "Multicenter Validation of the Diagnostic Accuracy of a Blood-Based Gene Test for Assessing Obstructive Coronary Artery Disease in Nondiabetic Patients," *Annals of Internal Medicine* 153, no. 7 (2010): 425–34.

92. Alice Park, "Blood Test for Heart Attack," The Top 10 of Everything, *Time*, December 9, 2010, 197.

93. A. Katsnelson, "Epigenome Effort Makes Its Mark," *Nature* 467 (2010): 646; Andrew Pollack, "Beyond the Gene," *New York Times*, November 11, 2010, www.nytimes.com/indexes/2008/11/11/

science/index.html; B. T. Heijmans, "Persistent Epigenetic Differences Associated with Prenatal Exposure to Famine in Humans," *Procedures of the National Academies of Sciences* 105, no. 44 (2008): 17046–49; "Moving AHEAD with an International Human Epigenome Project," *Nature* 454 (2008): 711–15; R. Lister, "Humans DNA Methylomes at Base Resolution Show Widespread Epigenomic Differences," *Nature* 462 (2009): 315–22.

94. Heijmans, "Persistent Epigenetic Differences."

95. Lister, "Humans DNA Methylomes."

96. Eben Harrell, "The Human Epigenome, Decoded: Top 10 Science Discoveries," *Time*, December 8, 2009, 372.

97. Katsnelson, "Epigenome Effort"; A. Petronis, "Epigenetics as a Unifying Principle in the Aetology of Complex Traits and Diseases," *Nature* 465 (2010): 721.

98. L. Groop, "Open Chromatin and Diabetes Risk," *Nature Genetics* 42 (2010): 190–92; A. Kong, "Parental Origin of Sequence Variants Associated with Complex Diseases," *Nature* 462 (2009): 868–74; R. Barres, "Non-CpG Methylation of the PGC-1 Promoter Through DNMT3B Controls Mitochondrial Density," *Cell Metabolism* 10, no. 3 (2009): 189–98.

99. M. Skinner, "Metabolic Disorders: Fathers' Nutritional Legacy," *Nature* 467 (2010): 922–23.

100. "Don't Blame Your Genes," *Economist*, September 3, 2009.

101. M. Arumugam, "Enterotypes of the Human Gut Microbiome," *Nature* 473 (2011): 174–80.

102. J. Couzin-Frankel, "Bacteria and Asthma: Untangling the Links," *Science* 330 (2010): 1168–69.

103. C. Zaph, "Which Species Are in Your Feces?" *Journal of Clinical Investigation* 120, no. 12 (2010): 4182–85.

104. "The Ultimate Probiotic," The Daily Scan, GenomeWeb, December 14, 2010, www .genomeweb.com/blog/ultimate-probiotic; "Meta-Analysis Defines Three Human Gut Microbiome Subtypes," GenomeWeb, April 20, 2011, www.genomeweb.com/meta-analysis-defines-three -human-gut-microbiome-subtypes; Carl Zimmer, "Bacteria Divide People into 3 Types, Scientists Say," *New York Times*, April 20, 2011.

105. Ibid.

106. Rick Weiss, "What You Should Know Before You Spit into That Test Tube," *Washington Post*, July 20, 2008.

107. "Risky Business," editorial, *Nature Genetics* 39, no. 12 (2007): 1415.

108. D. Hunter, "Letting the Genome Out of the Bottle: Will We Get Our Wish?" *New England Journal of Medicine* 358 (2008): 105–7; J. Annes, "Risks of Presymptomatic Direct-to-Consumer Genetic Testing," *New England Journal of Medicine* 363 (2010): 100–101; J. Evans, "Preparing for a Consumer-Driven Genomic Age," *New England Journal of Medicine*, August 18, 2010, Policy and Reform, healthpolicyandreform.nejm.org/?p=11933.

109. Manolio, "Genome-wide Association Studies."

110. "Getting Personal," editorial, *Nature* 455 (2008): 1007.

111. B. Prainsack, "Misdirected Precaution," *Nature* 456 (2008): 34–35.

112. Amy Harmon, "Fear of Insurance Trouble Leads Many to Shun or Hide DNA Tests," *New York Times*, February 24, 2008, www.nytimes.com/2008/02/24/health/24iht-24dna.10330888.html.

113. Francis S. Collins, *The Language of Life: DNA and the Revolution in Personalized Medicine* (New York: Harper, 2010), 80–81.

114. Thomas Goetz, "Sergey's Search," *Wired* (July 2010): 104–42.

115. Davies, *The $1,000 Genome*, Chapter 13.

116. Thomas Goetz, "Your Life: Decoded," *Wired* (December 2007).

117. "The 50 Best Inventions of the Year," *Time*, November 10, 2008.

118. Goetz, "Sergey's Search."

119. C. Bloss, "Effect of Direct-to-Consumer Genomewide Profiling to Assess Disease Risk," *New England Journal of Medicine*, January 12, 2010, www.nejm.org/doi/pdf/10.1056/NEJM oa1011893.

120. Ibid.

121. R. C. Green, "Disclosure of APOE Genotype for Risk of Alzheimer's Disease," *New England Journal of Medicine* 361 (2009): 245–54; D. Grady, "Learning of Risk of Alzheimer's Seems to Do No Harm," *New York Times*, July 15, 2009, www.nytimes.com/2009/07/16/health/research/ 16dementia.html.

122. Prainsack, "Misdirected Precaution."

123. P. Ng, "An Agenda for Personalized Medicine," *Nature* 461, no. 8 (2009): 724.

124. Ibid.; European Society of Human Genetics, "Statement of the ESGH on Direct-to-Consumer Genetic Testing for Health-Related Purposes," Policy, *European Journal of Human Genetics* (2010): 1–3; T. Ray, "In Wake of 'Flawed' GAO Report, Consumer Genomics Firms Call for Regulatory Plan for DTC Industry," *GenomeWeb Pharmacogenomics Reporter*, July 28, 2010; "Standard Issues," *Nature* 466 (2010): 797; A. McGuire, "Regulating Direct-to-Consumer Personal Genome Testing," *Science* 330 (2010): 181–83.

125. J. Craig Venter, *A Life Decoded: My Genome, My Life* (New York: Viking, 2007), 132; A. El-Sohemy, "Coffee, CYP1A2 Genotype, and Risk of Myocardial Infarction" *Journal of the American Medical Association* 295 (2006): 1135–41.

126. Steven Pinker, "My Genome, My Self," *New York Times*, January 11, 2009, www.nytimes.com/2009/01/11/magazine/11Genome-t.html.

127. C. J. Bell, "Carrier Testing for Severe Childhood Recessive Disease by Next-Generation Sequencing," *Science Translational Medicine*, January 12, 2011; J. Couzin-Frankel, "New High-Tech Screen Takes Carrier Testing to the Next Level," *Science* 331 (2011):130–31.

128. Esther Dyson, "Full Disclosure," *Wall Street Journal*, July 25, 2007, A15.

129. Misha Angrist, *Here Is a Human Being: At the Dawn of Personal Genomics* (New York: Harper, 2010).

130. H. Thomson, "Glenn Close Reveals Her 'Fabulous Genes," *New Scientist*, November 13, 2010.

131. "What Lies Within," *Economist*, August 12, 2010; C. B. Do, "Web-Based Genome-Wide Association Study Identifies Two Novel Loci and a Substantial Genetic Component for Parkinson's Disease," *Public Library of Science Genetics*, June 23, 2011.

132. Davies, *The $1,000 Genome*, Chapter 7.

133. Ibid.

134. "What Lies Within"; Do, "Web-Based Genome-Wide Association Study."

135. Davies, *The $1,000 Genome*, Chapter 9.

136. Davies, *The $1,000 Genome*, Chapter 9; Prainsack, "Misdirected Precaution"; Angrist, *Here Is a Human Being*, 68.

137. Andrew Pollack, "Start-Up May Sell Genetic Tests in Stores," *New York Times*, May 10, 2010, www.nytimes.com/2010/05/11/health/11gene.html; Andrew Pollack, "Walgreens Delays Selling Personal Genetic Test Kit," *New York Times*, May 12, 2010, www.nytimes.com/2010/05/13/health/13gene.html.

138. Andrew Pollack, "F.D.A. Faults Companies on Unapproved Genetic Tests," *New York Times*, June 12, 2010, B2.

139. G. Kutz, *Direct-to-Consumer Genetic Tests*, United States Government Accountability Office, July 22, 2010, www.gao.gov/new.items/d10847t.pdf.

140. Topol, "Prescription for Progress?"; "What Lies Within"; Do, "Web-Based Genome-wide Association Study."

141. E. Singer, "Democratizing DNA Sequencing," *Technology Review*, December 8, 2010, www.technologyreview.com/biomedicine/26850/diygenomics.org/genomera.com.

142. Topol, "Prescription for Progress?"

143. M. Maves, "Molecular and Clinical Genetics Panel of the Medical Devices Advisory Committee," *American Medical Association*, February 23, 2011, www.ama-assn.org/ama1/pub/upload/mm/399/consumer-genetic-testing-letter.pdf; "Advisory Committee Tells FDA Clinical Genetic Tests Should Only Be Provided Through Docs, Not DTC," GenomeWeb, March 9, 2011, www.genomeweb.com/advisory-committee-tells-fda-clinical-genetic-tests-should-only-be-provided-thro.

144. Davies, *The $1,000 Genome*, Chapter 10.

145. Eric J. Topol, "What You Can Learn from a Gene Scan," *Wall Street Journal*, December 22, 2007, A10.

146. E. Marshall, Waiting for the Revolution," *Science* 331 (2011): 526–29.

147. El-Sohemy, "Coffee, CYP1A2 Genotype."

Chapter 6

1. B. J. Hillman and J. Goldsmith, "The Uncritical Use of High-Tech Medical Imaging," *New England Journal of Medicine* 363 (2010): 4–6.

2. S. J. Reiser, "Revealing the Body's Whispers: How the Stethoscope Transformed Medicine," *Technological Medicine: The Changing World of Doctors and Patients* 1 (2009): 1; "Reply to Dr. Graves' and Stokes' Remarks on Dr. Hope, in Reference to Auscultation," *London Medical Gazette* 1 (1838–39): 129–30.

3. J. Fauber, "St. Luke's Review Finds Almost 30% Echocardiograms Are Misread," *Journal Sentinel* (Milwaukee, WI), June 22, 2010, www.jsonline.com/features/health/96945709.html.

4. "Ultrasound," *Wikipedia*, n.d., en.wikipedia.org/wiki/Ultrasound.

5. J. M. Torpy, "JAMA Patient Page: Magnetic Resonance Imaging," *Journal of the American Medical Association* 302, no. 23 (2009): 2614.

6. "Functional Magnetic Resonance Imaging," *Wikipedia*, n.d., en.wikipedia.org/wiki/Functional _magnetic_resonance_imaging.

7. Ibid.; *White Paper: Initiative to Reduce Unnecessary Radiation Exposure from Medical Imaging*, Center for Devices and Radiological Health, U.S. Food and Drug Administration, February 1, 2010, www.fda.gov/Radiation-EmittingProducts/RadiationSafety/RadiationDoseReduction/ ucm199994.htm; R. Fazel, "Exposure to Low-Dose Ionizing Radiation from Medical Imaging Procedures," *New England Journal of Medicine* 361 (2009): 849–57.

8. J. Stokes, "Medicine: Smart Bot with X-ray Specs," *Wired* (January 2010).

9. *Initiative to Reduce Unnecessary Radiation*; Fazel, "Exposure to Low-Dose Ionizing"; Stokes, "Smart Bot."

10. Fazel, "Exposure to Low-Dose Ionizing"; D. J. Brenner, "Radiation Exposure from Medical Imaging: Time to Regulate?" *American Medical Association* 304, no. 2 (2010): 208–9; M. S. Lauer, "Elements of Danger—The Case of Medical Imaging," *New England Journal of Medicine* 361 (2009): 841–43; M. O. Baerlocher, "Discussion of Radiation Risks Associated with CT Scans with Patients," *American Medical Association* 304, no. 19 (2010): 2170–71; M. Marchione, "Biggest Radiation Threat Is Due to Medical Scans," MSNBC, June 2010, www.msnbc.msn.com/id/37623994 /ns/health-health_care; R. Smith-Bindman, "Is Computed Tomography Safe?" *New England Journal of Medicine* 363 (2010): 1–6; D. J. Brenner, "Medical Imaging in the 21st Century—Getting the Best Bang for the Rad," *New England Journal of Medicine* 362 (2010): 943–45.

11. J. K. Iglehart, "Health Insurers and Medical-Imaging Policy—A Work in Progress," *New England Journal of Medicine* 360 (2009): 1030–37.

12. Baerlocher, "Discussion of Radiation Risks."

13. E. Cardis, "The 15-Country Collaborative Study of Cancer Risk among Radiation Workers in the Nuclear Industry: Estimates of Radiation-Related Cancer Risks," *Radiation Research* 167 (2007): 396–416; L. Schenkman, "Second Thoughts About CT Imaging," *Science* 331 (2011): 1002–4; R. Miller, "Canadian Study Affirms Cancer Risk from Imaging," theheart.org, February 7, 2011.

14. Roni Caryn Rabin, "Hazards: For Children in E.R., a Big Increase in CT Scans," *New York Times*, April 7, 2011, www.nytimes.com/2011/04/12/health/research/12scan.html.

15. While on the subject of radiation risk and cancer, I will note, since this book heavily emphasizes the cell phone platform, that the relationship between use of cell phones and the incidence of brain cancer has been a major controversy. Cell phones emit radiofrequency electromagnetic fields (RF-EMF), a nonionizing form of radiation, which has not been shown to disrupt DNA. Nevertheless, a 2011 report from a subcommittee representing the World Health Organization concluded that RF-EMF is "possibly carcinogenic to humans," primarily based on review of a large case-control study. That study had many flaws, such as obtaining the history of cell phone use after a brain cancer was diagnosed, and was countered by another large 2011 study that showed no link. While the biologic underpinning for this association is certainly elusive, it remains possible that there is an exceptionally small risk in certain individuals who are both genetically predisposed to brain cancer and particularly sensitive to exposure of RF-EMF. Getting to the bottom of this question would likely require a prospective study with extensive follow-up duration of tens of

thousands of individuals who were also genomically sequenced. It would be certainly technically feasible someday, but until then the types of studies currently conducted would not likely resolve the controversy. See R. Baan et al., "Carcinogenicity of Radiofrequency Electromagnetic Fields," *Lancet*, June 22, 2011, doi:10.1016/S1470-2045(11)70147-4; Gautam Naik, "Study Sees No Cellphone-Cancer Ties," *Wall Street Journal*, July 28, 2011, online.wsj.com/article/SB1000142 4053111904800304576472232980823392.html.

16. Walt Bogdanich, "The Mark of an Overdose," *New York Times*, August 1, 2010; Walt Bogdanich, "Hospitals Performed Needless Double CT Scans, Records Show," *New York Times*, June 17, 2011.

17. Hillman and Goldsmith, "The Uncritical Use"; Fazel, "Exposure to Low-Dose Ionizing"; Brenner, "Radiation Exposure."

18. Bogdanich, "The Mark of an Overdose"; Bogdanich, "Hospitals Performed Needless."

19. Fazel, "Exposure to Low-Dose Ionizing."

20. Ibid.

21. N. M. Orme, "Incidental Findings in Imaging Research: Evaluating Incidence, Benefit and Burden," *Archives of Internal Medicine* 170, no. 17 (2010): 1525–32.

22. "CT Scans for Lung Cancer," editorial, *New York Times*, November 9, 2010.

23. Ibid.; G. Harris, "CT Scans Cut Lung Cancer Deaths, Study Finds," *New York Times*, November 4, 2010; "Lung Cancer Trial Results Show Mortality Benefit with Low-Dose CT: Twenty Percent Fewer Lung Cancer Deaths Seen Among Those Who Were Screened with Low-Dose Spiral CT Than with Chest X-ray," press release, National Cancer Institute, U.S. National Institutes of Health, November 4, 2010, www.cancer.gov/newscenter/pressreleases/NLSTresultsRelease; A. Gibbons, "The Promise and Pitfalls of a Cancer Breakthrough," *Science* 330 (2010): 900–901.

24. Ibid.

25. Lauer, "Elements of Danger"; A. J. Einstein, "Multiple Testing, Cumulative Radiation Dose, and Clinical Indications in Patients Undergoing Myocardial Perfusion Imaging," *American Medical Association* 304, no. 19 (2010): 2137–44; P. Kaul, "Ionizing Radiation Exposure to Patients Admitted with Acute Myocardial Infarction in the United States," *Circulation, American Heart Association* 122, no. 21 (2010): 2160–69.

26. Brenner, "Radiation Exposure"; Einstein, "Multiple Testing."

27. Einstein, "Multiple Testing."

28. Kaul, "Ionizing Radiation Exposure."

29. M. Patel, "Low Diagnostic Yield of Elective Coronary Angiography," *New England Journal of Medicine* 362 (2010): 886–95.

30. Gina Kolata, "Heart Scanner Stirs New Hope and a Debate," *New York Times*, November 17, 2004.

31. Ibid.; M. J. Garcia, "Noninvasive Coronary Angiography: Hype or New Paradigm?" *American Medical Association* 293, no. 20 (2005): 2531–33; M. J. Garcia, "Accuracy of 16-Row Multidetector Computed Tomography for the Assessment of Coronary Artery Stenosis," *American Medical Association* 296, no. 4 (2006): 403–11; J. M. Miller, "Diagnostic Performance of Coronary Angiography by 64-Row CT," *New England Journal of Medicine* 359 (2008): 2324–36; G. L. Raff, "Radiation Dose from Cardiac Computed Tomography Before and After Implementation of Radiation Dose-Reduction Techniques," *Journal of the American Medical Association* 301, no. 22 (2009): 2340–48.

32. Raff, "Radiation Dose from Cardiac."

33. H. W. Querfurth, "Alzheimer's Disease," *New England Journal of Medicine* 362 (2010): 329–44; G. Stix, "Alzheimer's: Forestalling the Darkness," *Scientific American* (June 2010); C. M. Clark, "Use of Florbetapir-PET for Imaging ß-amyloid Pathology," *Journal of the American Medical Association* 305, no. 3 (2011) : 275–83.

34. Querfurth, "Alzheimer's Disease"; Gina Kolata, "Promise Seen for Detection of Alzheimer's," *New York Times*, June 23, 2010.

35. Kolata, "Promise Seen for Detection"; Gina Kolata, "New Scan May Spot Alzheimer's," *New York Times*, July 12, 2010; Gina Kolata, "Rules Seek to Expand Diagnosis of Alzheimer's," *New York Times*, July 13, 2010; Gina Kolata, "Drug Trials Test Bold Plan to Slow Alzheimer's," *New York Times*, July 17, 2010; Gina Kolata, "Insights Give Hope for New Attack on Alzheimer's," *New York Times*, December 13, 2010; Gina Kolata, "Tests Detect Alzheimer's Risks, but Should Patients Be Told?" *New York Times*, December 17, 2010.

36. Kolata, "Promise Seen for Detection"; Kolata, "New Scan"; *Future Opportunities to Leverage the Alzheimer's Disease Neuroimaging Initiative: Workshop Summary* (Washington, DC: National Academies Press, 2010), www.nap.edu/catalog/13017.html.

37. Ibid.

38. Ibid.

39. R. Perrin, "Multimodal Techniques for Diagnosis and Prognosis of Alzheimer's Disease," *Nature* 461 (2009): 916–22.

40. Stix, "Alzheimer's"; Clark, "Use of Florbetapir-PET"; Kolata, "Promise Seen for Detection"; *Future Opportunities.*

41. Stix, "Alzheimer's"; Clark, "Use of Florbetapir-PET"; Kolata, "New Scan."

42. *Future Opportunities*; K. Blennow, "Biomarkers in Alzheimer's Disease Drug Development," *Nature Medicine* 16, no. 11 (2010): 1218–22; G. M. McKhann, "Changing Concepts of Alzheimer's Disease," *Journal of the American Medical Association* 305, no. 23 (2011): 2458–915; H. Hampel, "Biomarkers for Alzheimer's Disease: Academic, Industry and Regulatory Perspectives," *Nature Reviews Drug Discovery* 9 (2010): 560–74.

43. *Future Opportunities.*

44. Ibid.

45. Andrew Pollack, "Scientists Report First Blood Test to Diagnose Alzheimer's Disease," *New York Times*, October 15, 2007; www.satorisinc.com/news_details.html?id=2; Gina Kolata, "Two Tests Could Aid in Risk Assessment and Early Diagnosis of Alzheimer's," *New York Times*, January 18, 2011, www.nytimes.com/2011/01/19/health/research/19alzheimers.html.

46. "NIA Study Demonstrates ApoE's Potential as Blood-Based Protein Biomarker for Alzheimer's," GenomeWeb, December 24, 2010; K. Yaffe, "Association of Plasma ß-amyloid Level and Cognitive Reserve with Subsequent Cognitive Decline," *Journal of the American Medical Association* 305, no. 3 (2011): 261–66; M. M. Reddy, "Identification of Candidate IgG Biomarkers for Alzheimer's Disease via Combinatorial Library Screening," *Cell* 144 (2011) :132–42.

47. Ibid.

48. Ibid.; "Initial Analysis of ADNI Plasma Proteome Data Suggests Possible Protein Signatures for Alzheimer's," GenomeWeb, December 10, 2010.

49. Sanjay W. Pimplikar, "Alzheimer's Isn't Up to the Tests," *New York Times*, July 20, 2010.

50. F. Mangialasche, "Alzheimer's Disease: Clinical Trials and Drug Development," *Lancet* 9 (2010): 702–16; "Why Are Drug Trials in Alzheimer's Disease Failing?" editorial, *Lancet* 376 (2010): 658; Lauren Gravitz, "A Tangled Web of Targets," *Nature* 475 (2011): S9–S11.

51. Kolata, "Drug Trials Test"; Hampel, "Biomarkers"; M. Citron, "Alzheimer's Disease: Strategies for Disease Modification," *Nature Reviews Drug Discovery* 9 (2010): 387–98.

52. Blennow, "Biomarkers"; McKhann, "Changing Concepts"; Hampel, "Biomarkers"; Citron, "Alzheimer's Disease"; P. St George-Hyslop, "Alzheimer's Disease: Selectively Tuning Gamma-Secretase," *Nature* 467 (2010): 36–37.

53. M. Cerf, "On-line, Voluntary Control of Human Temporal Lobe Neurons," *Nature* 467 (2010): 1104–8.

54. G. Miller, "Science and the Law: fMRI Lie Detection Fails a Legal Test," *Science* 328 (2010): 1336–37; J. H. Fowler, "Biology, Politics, and the Emerging Science of Human Nature," *Science* 322 (2008): 912–14; H. Lau, "Neuroscience: Should Confidence Be Trusted?" *Science* 329 (2010): 1478–79; Tara Parker-Pope, "What Brain Scans Can Tell Us About Marriage," *New York Times*, June 4, 2010; U. Nili, "Fear Thou Not: Activity of Frontal and Temporal Circuits in Moments of Real-Life Courage," *Neuron* 66, no. 6 (2010): 949–62; S. Dehaene, "How Learning to Read Changes the Cortical Networks for Vision and Language," *Science* 330 (2010): 1359–64; K. Ressler, "Targeting Abnormal Neural Circuits in Mood and Anxiety Disorders: From the Laboratory to the Clinic," *Nature Neuroscience* 10, no. 9 (2007): 1116–24; R. Gollub, "For Placebo Effects in Medicine, Seeing Is Believing," *Science Translational Medicine* 3, no. 70 (2011): 70ps5.

55. D. Cyranoski, "Thought Experiment," *Nature* 469 (2011): 148–49; Benedict Carey, "Wariness on Surgery of the Mind," *New York Times*, February 14, 2011, www.nytimes.com/2011/02/15/health/15brain.html.

56. N. D. Schiff, "Behavioral Improvements with Thalamic Stimulation After Severe Traumatic Brain Injury," *Nature* 448 (2007): 600–603; M. Shadlen, "Neurology: An Awakening," *Nature* 448 (2007): 539–40.

57. A. Owen, "Putting Brain Training to the Test," *Nature* 465 (2010): 775–78; K. Bellstrom, "Bulking Up Your Brain," *Smart Money*, August 6, 2010, www.smartmoney.com/plan/careers/bulking-up-your-brain; Benedict Carey, "Brain Calisthenics for Abstract Ideas," *New York Times*, June 9, 2011.

58. Michael A. Blake and Mannudeep K. Kalra, eds., *Imaging in Oncology*, Cancer Treatment and Research 143 (New York: Springer, 2008).

59. G. Bollag, "Clinical Efficacy of a RAF Inhibitor Needs Broad Target Blockade in BRAF-Mutant Melanoma," *Nature* 467 (2010): 596–99; K. Flaherty, "Inhibition of Mutated, Activated BRAF in Metastatic Melanoma," *New England Journal of Medicine* 363 (2010): 809–19; P. B. Chapman et al., "Improved Survival with Vemurafenib in Melanoma with BRAF V600E Mutation," *New England Journal of Medicine* 364 (2011): 2507–16.

60. Ibid.

61. M. Dinan, "Changes in the Use and Costs of Diagnostic Imaging Among Medicare Beneficiaries with Cancer, 1999–2006," *Journal of the American Medical Association* 303, no. 16 (2010): 1625–631.

62. B. Fischer, "Preoperative Staging of Lung Cancer with Combined PET-CT," *New England Journal of Medicine* 361 (2009): 32–39; "A Quicker Way to Identify Skin Cancer," *Technology Review*, January 31, 2011, www.technologyreview.com/biomedicine/32236; N. Shute, "Beyond Mammograms," *Scientific American* (2011): 32–34.

63. Anthony Atala, *Printing a Human Kidney*, TED Talk, www.youtube.com/watch?v=6jyQXq0ZH4s.

64. "Surgeon Prints New Kidney on Stage," *Discovery News*, March 4, 2011, news.discovery.com/tech/surgeon-prints-new-kidney-on-stage-110304.html; "Printing a Kidney: A Glimpse at the Future," *Huffington Post*, April 17, 2011, www.huffingtonpost.com/2011/03/09/printing-a-kidney_n_832992.html; Anya Kamenetz, "Next Step in 3-D Printing: Your Kidney," *Fast Company*, March 3, 2011, www.fastcompany.com/1734436/next-step-in-3d-printing-your-kidneys.

65. A. Atala, "Tissue-Engineered Autologous Bladders for Patients Needing Cystoplasty," *Lancet* 367 (2006): 1241–46.

66. R. Ali, "DIY Eye," *Nature* 472 (2011): 42–43; Gautam Naik, "Lab-Made Trachea Saves Man," *Wall Street Journal*, July 8, 2011, online.wsj.com/article/SB10001424052702304793504576432093996469056.html.

67. "The Printed World," *Economist*, February 12, 2011, 77–79.

68. D. Bullock, "Sir, Your Liver Is Ready: Behind the Scenes of Bioprinting," *Wired*, July 11, 2010, www.wired.com/rawfile/2010/07/gallery-bio-printing; Johnny Ryan, "Manufacturing 2.0: Three-D Printers Are Coming to a Desktop Near You. Should Designers and Factories Be Worried?" *Fortune*, May 23, 2011.

Chapter 7

1. Vijay Vaitheeswaran, "A Very Big HIT," in "The World in 2011," special issue, *Economist*, November 22, 2010, 133–134.

2. D. Blumenthal, "Launching HITECH," *New England Journal of Medicine* 362 (2010): 382–85.

3. Institute of Medicine, *To Err Is Human: Building a Safer Health System* (Washington, DC: National Academies Press, 2000), Summary.

4. John Dorschner, "A Medical Enron," *Washington Post*, December 9, 2002, A22.

5. Robert Pear, "Group Asking U.S. for New Vigilance in Patient Safety," *New York Times*, November 30, 2009.

6. Institute of Medicine, *Crossing the Quality Chasm: A New Health System for the 21st Century* (Washington, DC: National Academies Press, 2001), Summary.

7. "Five Years After IOM Report on Medical Errors, Nearly Half of All Consumers Worry About the Safety of Their Health Care," press release, The Henry J. Kaiser Family Foundation, November 17, 2004, www.kff.org/kaiserpolls/pomr111704nr.cfm.

8. C. Landrigan, "Temporal Trends in Rates of Patient Harm Resulting from Medical Care," *New England Journal of Medicine* 363 (2010): 2124–34; J. Van Den Bos, "The $17.1 Billion Problem:

The Annual Cost of Measurable Medical Errors," *Health Affairs* 30, no. 4 (2011): 596–603; P. Pronovost, "A Road Map for Improving the Performance Measures," *Health Affairs* 30, no. 4 (2011): 569–73; Denise Grady, "Study Finds No Progress in Safety at Hospitals," *New York Times*, November 24, 2010, www.nytimes.com/2010/11/25/health/research/25patient.html.

9. Landrigan, "Temporal Trends."

10. B. Starfield, "Is US Health Really the Best in the World?" *Journal of the American Medical Association* 284, no. 4 (2000): 483–85.

11. T. Sheldon, "Dutch Study Shows that 40% of Adverse Incidents in Hospitals Are Avoidable," *British Medical Journal* 34 (2007): 925.

12. A. Jha, "Use of Electronic Health Records in U.S. Hospitals," *New England Journal of Medicine* 360 (2009): 1628–37.

13. C. DesRoches, "Electronic Health Records in Ambulatory Care: A National Survey of Physicians," *New England Journal of Medicine* 359 (2008): 50–60.

14. "HIT or Miss," *Economist*, April 16, 2009, www.economist.com/node/13438006.

15. Federal Communications Commission, *Connecting America: The National Broadband Plan*, www.broadband.gov.

16. *Report to the President Realizing the Full Potential of Health Information Technology to Improve Healthcare for Americans: The Path Forward*, Executive Office of the President, December 2010, www.whitehouse.gov/sites/default/files/microsites/ostp/pcast-health-it-report.pdf; Steve Lohr, "U.S. Tries Open-Source Model for Health Data Systems," *New York Times*, February 2, 2011, bits.blogs.nytimes.com/2011/02/02/u-s-tries-open-source-model-for-health-data-systems.

17. R. Steinbrook, "Personally Controlled Online Health Data: The Next Big Thing in Medical Care?" *New England Journal of Medicine* 358 (2008): 1653–56.

18. *Report to the President*; Lohr, "U.S. Tries Open-Source Model."

19. E. Poon, "Effect of Bar-Code Technology on the Safety of Medication Administration," *New England Journal of Medicine* 362 (2010): 1698–1707.

20. B. Chaudhry, "Systematic Review: Impact of Health Information Technology on Quality, Efficiency, and Costs of Medical Care," *Annals of Internal Medicine* 144, no. 10 (2006): 742–50.

21. S. T. Parente and J. S. McCullough, "Health Information Technology and Patient Safety: Evidence from Panel Data," *Health Affairs* 28 (2009): 357–60.

22. "Electronic Medical Records Not Always Linked to Better Care in Hospitals," press release, RAND Corporation, December 23, 2010, www.rand.org/news/press/2010/12/23.html; Katherine Hobson, "Electronic Medical Records Don't Improve Outpatient Care Quality: Study," *Wall Street Journal*, January 25, 2011, blogs.wsj.com/health/2011/01/25/electronic-medical-records-dont-improve-outpatient-care-quality-study.

23. Ibid.

24. Steve Lohr, "Seeing Promise and Peril in Digital Records," *New York Times*, July 16, 2011, www.nytimes.com/2011/07/17/technology/assessing-the-effect-of-standards-in-digital-health-records-on-innovation.html.

25. "Flying Blind," *Economist*, April 16, 2009, www.economist.com/node/13437966.

26. *Report to the President*; Lohr, "U.S. Tries Open-Source Model."

27. "Flying Blind."

28. "Health Information Technology Savings Dwarf Costs over the First 15 Years, Then Keep Growing," *RAND Review* (Spring 2009), www.rand.org/publications/randreview/issues/spring2009/cpiece.html.

29. "Medication Errors Injure 1.5 Million People and Cost Billions of Dollars Annually; Report Offers Comprehensive Strategies for Reducing Drug-Related Mistakes," press release, National Academies, July 20, 2006, www8.nationalacademies.org/onpinews/newsitem.aspx?recordid=11623.

30. "Society of Actuaries Study Finds Medical Errors Annually Cost at Least $19.5 Billion Nationwide," press release, Society of Actuaries, August 10, 2010, www.soa.org/news-and-publications/newsroom/press-releases/2010-08-09-med-errors.aspx.

31. "The Extormity Perpetual Investment Program," Extormity, n.d., extormity.com/index.php/perpetual-investment.

32. *Report to the President*; Lohr, "U.S. Tries Open-Source Model."

33. *Report to the President*; Tim Scott et al., *Implementing an Electronic Medical Record System: Successes, Failures, Lessons* (Oxford: Radcliffe Publishing, 2007).

34. G. Schiff and D. W. Bates, "Can Electronic Clinical Documentation Help Prevent Diagnostic Errors?" *New England Journal of Medicine* 362 (2010): 1066–69.

35. Danielle Ofri, "The Doctor vs. the Computer," *New York Times*, December 30, 2010, well.blogs.nytimes.com/2010/12/30/the-doctor-vs-the-computer.

36. Harris Meyer, "The Doctor (and His Scribe) Will See You Now," *Hospitals and Health Networks Magazine* (December 2010), www.hhnmag.com/hhnmag_app/jsp/articledisplay.jsp?dcrpath=HH NMAG/Article/data/12DEC2010/1210HHN_FEA_staffingissues&domain=HHNMAG.

37. Ibid.

38. Vaitheeswaran, "A Very Big HIT."

39. A. Jha, "Meaningful Use of Electronic Health Records," *Journal of the American Medical Association* 304, no. 15 (2010): 1709–10; D. Blumenthal, "The 'Meaningful Use' Regulation for Electronic Health Records," *New England Journal of Medicine* 363 (2010): 501–4.

40. Blumenthal, "Launching HITECH"; *Report to the President*; Lohr, "U.S. Tries Open-Source Model."

41. Blumenthal, "Launching HITECH."

42. Ibid.; *Report to the President*; Blumenthal, "Meaningful Use"; Institute of Medicine, *Digital Infrastructure for the Learning Health System* (Washington, DC: National Academies Press, 2010), 12–16.

43. Robert Pear, "Doctors and Hospitals Say Goals on Computerized Records are Unrealistic," *New York Times*, July 7, 2010, www.nytimes.com/2010/06/08/health/policy/08health.html.

44. Institute of Medicine, *Digital Infrastructure*; Milt Freudenheim, "Panel Set to Study Safety of Electronic Patient Data," *New York Times*, December 13, 2010, www.nytimes.com/2010/12/14/business/14records.html; Chad Terhune et al., "The Dubious Promise of Digital Medicine," *Business Week*, April 23, 2009, www.businessweek.com/magazine/content/09_18/b4129030606214.htm; "Electronic Medical Records Don't Boost Hospital Quality Measures," *Wall Street Journal*, December 28, 2010, blogs.wsj.com/health/2010/12/28/study-electronic-medical-records-dont-boost-hospital-quality-measures; Steve Lohr, "The Cloud Threat to the Software Business," *New York Times*, April 15, 2011, bits.blogs.nytimes.com/2011/04/15/the-cloud-threat-to-the-software-business; "Heads in the Cloud," *Economist*, April 2, 2011, 63; A. Fox, "Cloud Computing: What's in It for Me as a Scientist?" *Science* 331 (2011): 406–7.

45. Terhune, "The Dubious Promise."

46. Ibid.

47. Ibid.

48. "Electronic Medical Records."

49. Poon, "Effect of Bar-Code Technology"

50. Lohr, "The Cloud Threat"; "Heads in the Cloud"; Fox, "Cloud Computing."

51. Blumenthal, "Launching HITECH"; *Report to the President*.

52. Ibid.

53. Ibid.

54. Ibid.

55. S. Mangalmurti, "Medical Malpractice Liability in the Age of Electronic Health Records," *New England Journal of Medicine* 363 (2010): 2060–67.

56. Ibid.

57. T. Delbanco, "Open Notes: Doctors and Patients Signing On," *Annals of Internal Medicine* 153, no. 2 (2010): 121–25; Pauline W. Chen, "Should Patients Read the Doctor's Notes?" *New York Times*, July 27, 2010, D5; Laura Landro, "What the Doctor Is Really Thinking," *Wall Street Journal*, July 20, 2010, online.wsj.com/article/SB10001424052748704720004575377060985974450.html.

58. "HIT or Miss."

59. Delbanco, "Open Notes"; Chen, "Should Patients Read"; Landro, "What the Doctor."

60. B. Lagerqvist, "Long-Term Outcomes with Drug-Eluting Stents Versus Bare-Metal Stents in Sweden," *New England Journal of Medicine* 356 (2007): 1009–19; "Is That It, Then, for Blockbuster Drugs?" *Lancet* 365, no. 9440 (2004): 1100.

61. *Report to the President*; Lohr, "U.S. Tries Open-Source Model."

62. Michael R. Harrison, "The 28th Amendment: The Pursuit of Health," December 23, 2009, dev.thehastingscenter.org/Bioethicsforum/Post.aspx?id=4252.

63. See The Markle Foundation's website, www.markle.org; A. Krist, "A Vision for Patient-Centered Health Information Systems," *Journal of the American Medical Association* 305, no. 3 (2011): 300–301; "A 'Blue Button' to Help People Download Their Medical Records," *Wall Street Journal*, September 3, 2010, blogs.wsj.com/health/2010/09/03/a-blue-button-to-help-people-download-their-medical-records.

64. A. Krist, "A Vision for Patient-Centered Health Information Systems," *Journal of the American Medical Association* 305 (2011): 300–301.

65. "AMA & Markle Foundation Present PHR Survey Research at HIMSS," press release, Markle Foundation, March 3, 2010, www.markle.org/news-events/media-releases/ama-markle-foundation-present-phr-survey-research-himss.

66. Jeanette Borzo, "Tracking Your Health," *Wall Street Journal*, October 25, 2010, online.wsj.com/article/SB10001424052702304180804575188402688763416.html; "New Study Reveals Family Caregivers Want Web-Based and Mobile Technologies to Help Them Care for Their Loved Ones," press release, UnitedHealth Group, January 8, 2011, www.unitedhealthgroup.com/news room/news.aspx?id=15023bf6-4871-4a67-9d9d-d94f8f1722e9; "The Three Health Technologies Caregiver Want Most," *Wall Street Journal*, January 10, 2011.

67. Federal Communications Commission, "Enabling Health Information Technology: National Broadband Plan's Recommendations for Healthcare," presentation to the 2010 Indian Health Information Management Conference, May 13, 2010, www.ihs.gov/ihimc/documents/Graham-Jones%20IHS%20Presentation.pdf.

68. R. Kitzman, "Exclusion of Genetic Information from the Medical Record," *Journal of the American Medical Association* 304, no. 10 (2010): 1120–21.

69. Kevin Davies, *The $1,000 Genome: The Revolution in DNA Sequencing and the New Era of Personalized Medicine* (New York: Free Press, 2010), Kindle edition, Chapter 7.

70. *Report to the President*.

Chapter 8

1. M. Mutin et al., "Direct Evidence of Endothelial Injury in Acute Myocardial Infarction and Unstable Angina by Demonstration of Circulating Endothelial Cells," *Blood* 93 (1999): 2951–58.

2. E. Stern, "Label-Free Biomarker Detection from Whole Blood," *Nature Nanotechnology* 5 (2010): 138–42, www.nature.com/nnano/journal/v5/n2/abs/nnano.2009.353.html; L. Soleymani, "Programming the Detection Limits of Biosensors Through Controlled Nanostructuring," *Nature Nanotechnology* 4 (2009): 844–48, www.nature.com/nnano/journal/v4/n12/pdf/nnano.2009.276.pdf; B. Kim, "Nanomedicine," *New England Journal of Medicine* 363 (2010): 2434–43.

3. Y. Ling et al., "Implantable Magnetic Relaxation Sensors Measure Cumulative Expose to Cardiac Biomarkers," *Nature Biotechnology*, February 13, 2011.

4. Kit Eaton, "Heart Attack or Vicious Burrito? Embedded Sensor Knows," *Fast Company*, February 14, 2011.

5. Ling et al., "Implantable Magnetic Relaxation."

6. Kevin Davies, *The $1,000 Genome: The Revolution in DNA Sequencing and the New Era of Personalized Medicine* (New York: Free Press, 2010), Kindle edition, Chapter 12.

7. Marilynn Marchione, "Blood Test to Spot Cancer Gets Big Boost from J&J," Associated Press, January 2, 2011, www.signonsandiego.com/news/2011/jan/02/blood-test-to-spot-cancer-gets-big-boost-from-jj; Marilynn Marchione, "Blood Test to Spot Cancer Big Boost," MSNBC, January 3, 2011, www.msnbc.msn.com/id/40881967/ns/health-cancer.

8. Ibid.

9. R. Leary, "Development of Personalized Tumor Biomarkers Using Massively Parallel Sequencing," *Science Translational Medicine* 2, no. 20 (2010): 20ra14; L. Prokunina-Olsson, "Cancer Sequencing Gets a Little More Personal," *Science Translational Medicine* 2, no. 20 (2010): 20ps8.

10. Ibid.

11. Ibid.

12. S. Tomlins et al., "Urine TMPRSS2:ERG Fusion Transcript Stratifies Prostate Cancer Risk in Men with Elevated Serum PSA," *Science Translational Medicine* 3, no. 94 (August 2011): 94ra72.

13. N. Savage, "Spotting the First Signs," *Nature*, March 24, 2011.

14. Ibid.

15. Nicholas Wade, "New DNA Tests Aimed at Reducing Colon Cancer," *New York Times*, October 28, 2010.

16. T. M. Snyder et al., "Universal Noninvasive Detection of Solid Transplant Rejection," *Proceedings of the National Academy of Sciences*, March 28, 2011.

17. J. Bluestone, "Genetics, Pathogenesis and Clinical Interventions in Type 1 Diabetes," *Nature* 464 (2010): 1293–300; Shirley S. Wang, "Trying to Prevent Type 1 Diabetes," *Wall Street Journal*, June 7, 2011.

18. M. Heinig, "A Trans-Acting Locus Regulates an Anti-Viral Expression Network and Type 1 Diabetes Risk," *Nature* 467 (2010): 460–64.

19. Wang, "Trying to Prevent."

20. E. J. Topol, "Transforming Medicine via Digital Innovation," *Science Translational Medicine* 2, no. 16 (2010): 1–4; Anna Wilde Matthews, "So Young and So Many Pills," *Wall Street Journal*, December 28, 2010, D1.

21. S. Rappaport, "Environment and Disease Risk," *Science* 330 (2010): 460–61.

22. "The DNA Transistor," *Scientific American* 303 (December 2010).

23. G. Vogel, "Diseases in a Dish Take Off," *Science* 330 (2010): 1172–73; I. Itzhaki, "Modelling the Long QT Syndrome with Induced Pluripotent Stem Cells," *Nature* 471 (2011): 225–29; K. Brennand et al., "Modelling Schizophrenia Using Human Induced Pluripotent Stem Cells," *Nature* 473 (2011): 221–25.

24. Vogel, "Diseases in a Dish."

25. Brennand, "Modelling Schizophrenia."

26. F. Soldner et al., "Generation of Isogenic Pluripotent Stem Cells Differing Exclusively at Two Early Onset Parkinson Point Mutations," *Cell* 146 (2011): 1–14.

27. Itzhaki, "Modelling the Long QT Syndrome."

28. D. Huh, "Reconstituting Organ-level Lung Functions on a Chip," *Science* 328 (2010): 1662–67.

29. D.-H. Kim et al., "Epidermal Electronics," *Science* 333, no. 6044 (August 2011): 838–43.

30. Y. Lo, "Maternal Plasma DNA Sequencing Reveals the Genome-Wide Genetic and Mutational Profile of the Fetus," *Science* 2, no. 61 (2010): 61–74.

31. "Amniocentesis," *Wikipedia*, n.d., en.wikipedia.org/wiki/Amniocentesis; "Amniocentesis: Definition," Mayo Clinic, n.d., www.mayoclinic.com/health/amniocentesis/MY00155.

32. Lo, "Maternal Plasma DNA."

33. "Newborn 'Heel Stick' Screening (Newborn Genetic Screening)," Just Mommies, n.d., www.justmommies.com/articles/heel-stick-screening.shtml.

34. J. Gardy, "Whole Genome Sequencing and Social Network Analysis of a Tuberculosis Outbreak," *New England Journal of Medicine* 364 (2011): 730–39.

35. H. Waters, "New $10 Million X-Prize Launched for Tricorder-Style Medical Device," *Nature Medicine* 17, no. 7 (2011): 754.

Chapter 9

1. R. Horton, "Offline: If I Were a Rich Man," *Lancet* 376 (2010): 1972.

2. Chuck Salter, "The Doctor of the Future," *Fast Company* (May 2009): 66–70.

3. Trisha Torrey quoted in Nicholas Brody, "The Rise of the Empowered Patient," *Scientific American Pathways*, September 24, 2010, 4, www.sa-pathways.com/new-health-consumer/the-rise-of-the-empowered-patient/4.

4. Nick Bilton, *I Live in the Future & Here's How It Works: Why Your World, Work, and Brain Are Being Creatively Disrupted* (New York: Crown, 2010).

5. Angie C. Marek, "Medicine Without Doctors," *Smart Money*, September 7, 2010, www.smartmoney.com/personal-finance/health-care/medicine-without-doctors.

6. Avery Johnson, "Americans Cut Back on Visits to Doctor," *Wall Street Journal*, July 29, 2010, online.wsj.com/article/SB10001424052748703940904575395603432726626.html; Gardiner Harris, "More Physicians Say No to Endless Workdays," *New York Times*, April 1, 2011, www.nytimes.com/2011/04/02/health/02resident.html.

7. Ibid.

8. Atul Gawande, *The Checklist Manifesto: How to Get Things Right* (New York: Metropolitan Books, 2010), 32.

9. Abraham Flexner, *Medical Education in the United States and Canada*, Bulletin No. 4, New York City, 1910.

10. Molly Cooke, David. M. Irby, and Bridget C. O'Brien, *Educating Physicians: A Call for Reform of Medical School and Residency* (San Francisco: Jossey-Bass, 2010), Kindle edition, Introduction.

11. Jennifer Epstein, "Personalizing the M.D.," *Inside Higher Ed*, June 8, 2010, www.inside higher ed.com/news/2010/06/08/medical.

12. Horton, "Offline."

13. D. Barr, "The Art of Medicine: Science as Superstition: Selecting Medical Students," *Lancet* 376 (2010): 678–79.

14. V. Thurston, "The Current Status of Medical Genetics Instruction in U.S. and Canadian Medical Schools," *Academic Medicine* 82, no. 5 (2007): 441–45.

15. David A. Kaplan, "Bill Gates' Favorite Teacher," *CNN Money*, August 24, 2010, money.cnn .com/2010/08/23/technology/sal_khan_academy.fortune/index.htm; Bryant Urstadt, "Salman Khan: The Messiah of Math," *Bloomberg Businessweek*, May 19, 2011; Clive Thompson, "The New Way to Be a Fifth Grader," *Wired* (August 2011): 152.

16. Urstadt, "Salman Khan"; Don Tapscott and Anthony D. Williams, *Macrowikinomics: Rebooting Business and the World* (New York: Portfolio/Penguin, 2010), 147.

17. Tapscott and Williams, *Macrowikinomics*, 148.

18. Ibid.

19. Pauline W. Chen, "Rethinking the Way We Rank Medical Schools," *New York Times*, June 17, 2010, www.nytimes.com/2010/06/17/health/17chen.html; F. Mullan, "The Social Mission of Medical Education: Ranking the Schools," *Annals of Internal Medicine* 152, no. 12 (2010): 804–11.

20. "DTC for Docs-To-Be," The Daily Scan, GenomeWeb, June 8, 2010, www.genomeweb .com/blog/dtc-docs-be; Ruthann Richter, "Consumer Genomics Enters the Classroom," *Stanford Medical Magazine*, October 26, 2010, stanmed.stanford.edu/2010fall/article5.html; Lia Steakley, "Stanford Students Discuss Studying Their Own Genotypes," Scope, *Stanford News*, September 9, 2010, scopeblog.stanford.edu/archives/2010/09.

21. Ibid.; "A DNA Education," editorial, 465 (*Nature* 2010): 845–46.

22. Richter, "Consumer Genomics."

23. Melissa Healy, "As Genetic Testing Races Ahead, Doctors Are Left Behind," *Los Angeles Times*, October 24, 2009, articles.latimes.com/2009/oct/24/science/sci-genetic-tests24.

24. "U.S. Public Opinion on Uses of Genetic Information and Genetic Discrimination," *Genetics and Public Policy Center*, April 24, 2007, 2, www.dnapolicy.org/resources/GINAPublic_Opinion _Genetic_Information_Discrimination.pdf.

25. Rita Rubin, "Most Doctors Are Behind the Learning Curve on Genetic Tests," *USA Today*, October 24, 2010, www.usatoday.com/yourlife/health/medical/2010-10-25-Genetics24_CV _N.htm.

26. B. J. Hillman and J. C. Goldsmith, "The Uncritical Use of High Tech Medical Imaging," *New England Journal of Medicine* 363 (2010): 4–6.

27. Ralph Clayman, dean of UCI School of Medicine, personal communication, December 23, 2010.

28. Duff Wilson, "Using a Pfizer Grant, Courses Aim to Avoid Bias," *New York Times*, January 11, 2011, www.nytimes.com/2010/01/11/business/11drug.html.

29. Tapscott and Williams, *Macrowikinomics*, 146.

30. E. J. Topol, "Scorecard Cardiovascular Medicine: Its Impact and Future Directions," *Annals of Internal Medicine* 120 (1994): 65–70; Pennsylvania Health Care Cost Containment Council, *Pennsylvania's Guide to Coronary Artery Bypass Graft Surgery, 1994-1995*, May 1998, www.phc4 .org/reports/cabg/95/default.htm.

31. Denise Grady, "Consumer Reports Is Rating Surgical Groups," *New York Times*, September 7, 2010, www.nytimes.com/2010/09/08/health/08heart.html.

32. T. Ferris, "Public Release of Clinical Outcomes Data: Online CABG Report Cards," *New England Journal of Medicine* 363 (2010): 1593–95.

33. Anna Wilde Mathews, "Doctors Slam Insurers Over Their Ranking," *Wall Street Journal*, July 20, 2010, D2.

34. Claire Cain Miller, "Bringing Comparison Shopping to the Doctor's Office," *New York Times*, June 10, 2010, www.nytimes.com/2010/06/11/technology/11cost.html.

35. Mary Vanac, "Cleveland Clinic Backs Company That Helps Patients Shop for Care," *MedCity News*, June 11, 2010, www.medcitynews.com/2010/06/cleveland-clinic-backs-company-that-helps-patients-shop-for-care; A. Sinaiko, "Increased Price Transparency in Health Care: Challenges and Potential Effects," *New England Journal of Medicine* 364 (2011): 891–94; D. Cutler, "Designing Transparency Systems for Medical Care Prices," *New England Journal of Medicine* 364 (2011): 894–95.

36. Ferris, "Public Release of Clinical Outcomes Data"; M. Chassin, "Accountability Measures: Using Measurement to Promote Quality Improvement," *New England Journal of Medicine* 363 (2010): 683–88.

37. See Tapscott and Williams, *Macrowikinomics*, 185–87.

38. H. Luft, "Becoming Accountable: Opportunities and Obstacles for ACOs," *New England Journal of Medicine* 363 (2010): 1389–91; T. Lee, "Creating Accountable Care Organizations," *New England Journal of Medicine* 363 (2010): 1391; A. Sinaiko, "Patients' Role in Accountable Care Organizations," *New England Journal of Medicine* 363 (2010): 2583–85; T. L. Greaney, "Accountable Care Organizations: The Fork in the Road," *Health Policy and Reform: Remaking Health Care* (*New England Journal of Medicine*), December 22, 2010, healthpolicyandreform.nejm.org/?p=13451; R. Kocher, "Physicians Versus Hospitals as Leaders of Accountable Care Organizations," *New England Journal of Medicine* 363 (2010): 2579–81.

39. Lee, "Creating Accountable Care Organizations."

40. Sinaiko, "Patients' Role in Accountable Care Organizations."

41. Ibid.

42. C. DesRoches, "Physicians' Perceptions, Preparedness for Reporting, and Experiences Related to Impaired and Incompetent Colleagues," *Journal of the American Medical Association* 304, no. 2 (2010): 187–93.

43. "Physician Shortages to Worsen Without Increases in Residency Training," Association of American Medical Colleges, June 2010, https://www.aamc.org/download/153160/data/physician_shortages_to_worsen_without_increases_in_residency_tr.pdf; Tammy Worth, "Agencies Warn of Coming Doctor Shortage," *Los Angeles Times*, June 7, 2010, www.latimes.com/news/health/la-he-doctor-shortage-20100607,0,7762076.story; Suzanne Sataline, "Medical Schools Can't Keep Up," *Wall Street Journal*, April 12, 2010, online.wsj.com/article/SB10001424052702304506904575180331528424238.html.

44. "Many Children Lack Doctors, Study Finds," *New York Times*, December 19, 2010, www.nytimes.com/2010/12/20/us/20doctors.html.

45. Johnson, "Americans Cut Back on Visits to Doctor."

46. Ibid.

47. Pete Vanderveen, "How to Care for 30 Million More Patients," *Wall Street Journal*, July 19, 2010.

48. Anna Wilde Mathews, "When the Doctor Has a Boss," *Wall Street Journal*, November 8, 2010, online.wsj.com/article/SB10001424052748703856504575600412716683130.html.

49. Lee, "Creating Accountable Care Organizations."

50. Ibid.

51. Y. Zhou, "Improved Quality at Kaiser Permanente Through E-mail Between Physicians and Patients," *Health Affairs* 29 (July 2010): 1370–75.

52. "Public Still Pretty Clueless About Electronic Medical Records," Health Blog, *Wall Street Journal*, June 17, 2010, blogs.wsj.com/health/2010/06/17/public-still-pretty-clueless-about-electronic-medical-records; Kate Pickert, "The Doctor Is In—and Online," *Time*, August 9, 2010, 48; E. Boukus, "Physicians Slow to Email Routinely with Patients," Issue Brief, *Center for Studying Health System Change* 134 (October 2010): 3–4.

53. Boukus, "Physicians Slow to Email."

54. Katherine Chretien, "A Doctor's Request: Please Don't 'Friend' Me," *USA Today*, June 10, 2010, www.usatoday.com/news/opinion/forum/2010-06-10-column10_ST1_N.htm.

55. S. Mangalmurti, "Medical Malpractice Liability in the Age of Electronic Health Records," *New England Journal of Medicine* 363 (2010): 2060–67.

56. Ibid.

57. P. Hartzband, "Untangling the Web-Patients, Doctors, and the Internet," *New England Journal of Medicine* 362 (2010): 1063–66.

58. Brian Dolan, "ZocDoc Scoops Up $50 Million in Funding," *Mobi Health News*, August 2, 2011, mobihealthnews.com/12246/zocdoc-scoops-up-50-million-in-funding.

59. Pauline W. Chen, "Medicine in the Age of Twitter," *New York Times*, June 11, 2009, www.nytimes.com/2009/06/11/health/11chen.html; A. Mostaghimi and B. H. Crotty, "Professionalism in the Digital Age," *Annals of Internal Medicine* 154 (2011): 560–62; Chelsea Conaboy, "For Doctors, Social Media a Tricky Case," *Boston Globe*, April 20, 2011.

60. "Mayo Clinic Creates Center for Social Media," *Mayo Clinic*, July 27, 2010. www.mayoclinic.org/news2010-rst/5872.html; "Health Blog Q&A: Mayo Clinic's New Center for Social Media," *Wall Street Journal*, July 27, 2010, blogs.wsj.com/health/2010/07/27/health-blog-qa-mayo-clinics-new-center-for-social-media.

61. "Health Blog Q&A."

62. "AMA Policy: Professionalism in the Use of Social Media," American Medical Association, November 16, 2010, www.ama-assn.org/ama/pub/meeting/professionalism-social-media.shtml.

63. Chretien, "A Doctor's Request."

64. Daniel Lamas, "Friend Request," *New York Times*, March 11, 2010, www.nytimes.com/2010/03/14/magazine/14lives-t.html.

65. Ibid.

66. Chen, "Medicine in the Age of Twitter"; Mostaghimi and Crotty, "Professionalism in the Digital Age"; Conaboy, "For Doctors, Social Media."

67. See Tapscott and Williams, *Macrowikinomics*, 194.

68. C. Hawn, "Take Two Aspirin and Tweet Me in the Morning: How Twitter, Facebook, and Other Social Media Are Reshaping Health Care," *Health Affairs* 28, no. 2 (2009): 361–68.

69. Zhou, "Improved Quality at Kaiser Permanente."

70. Salter, "The Doctor of the Future."

71. Ibid.

72. Ibid.; see also Hello Health's website, hellohealth.com.

73. Jennifer Saranow Schultz, "Punishing Doctors Who Make You Wait," *New York Times*, June 29, 2010, bucks.blogs.nytimes.com/2010/06/29/punishing-doctors-who-make-you-wait; Katie Hafner, "Concierge Medical Care with a Smaller Price Tag," *New York Times*, January 31, 2011, www.nytimes.com/2011/02/01/health/01medical.html.

74. Ibid.; One Medical Group website, www.onemedicalgroup.com; Liz Kowalczyk, "More Doctors Gravitate Toward Boutique Practice," *Boston Globe*, April 17, 2011, articles.boston.com/2011-04-17/business/29428534_1_concierge-medicine-mdvip-boutique-practices.

75. Lesley Alderman, "The Doctor Will See You . . . Eventually," *New York Times*, August 2, 2011, www.nytimes.com/2011/08/02/health/policy/02consumer.html.

76. Kowalczyk, "More Doctors Gravitate Toward Boutique Practice."

77. Gardiner Harris, "News for Aspiring Doctors: The People Skills Test," *New York Times*, July 10, 2011, www.nytimes.com/2011/07/11/health/policy/11docs.html.

Chapter 10

1. A. Witty, "Research and Develop," *Economist*, November 22, 2011, www.economist.com/node/17493432.

2. S. Stovall, "Europe's Drug Regulator Says Innovation Must Pick Up," *Wall Street Journal*, December 15, 2010; Institute of Medicine, *Challenges for the FDA: The Future of Drug Safety, Workshop Summary* (Washington, DC: National Academies Press, 2007), www.nap.edu/catalog/11969.html; Jennifer Corbett Dooren, "Drug Approvals Slipped in 2010," *Wall Street Journal*, December 31, 2010, online.wsj.com/article/SB10001424052748704543004576052170335871018.html; "Health

Care," The World in Figures: Industries, *Economist*, November 22, 2010, www.economist.com/node/17509801; Duffy Wilson, "Patent Woes Threaten Drug Firms," *New York Times*, March 7, 2011.

3. "Health Care."

4. John C. Lechleiter, "America's Growing Innovation Gap," *Wall Street Journal*, July 9, 2010, online.wsj.com/article/SB10001424052748704111704575354863772223910.html.

5. "Health Care."

6. Ibid.

7. "Pharmaceuticals: Convergence or Conflict?" *Economist*, August 28, 2008, www.economist.com/node/12009882; H. Ledford, "Genzyme Deal Set to Alter Biotech Landscape," *Nature* 470 (2011): 449.

8. A. B. Engelberg, "Balancing Innovation, Access, and Profits: Market Exclusivity for Biologics," *New England Journal of Medicine* 361 (2009):1917–19; G. Walsh, "Biopharmaceutical Benchmarks 2010," *Nature* 28 (2010): 917–21.

9. V. Vaitheeswaran, "Generically Challenged," *Economist*, November 13, 2009, www.economist .com/node/14742621.

10. Malcolm Gladwell, "The Treatment," *New Yorker*, May 17, 2010, 69–79.

11. "Roche Digests Genentech: Back to the Lab," *Economist*, December 10, 2009.

12. E. J. Topol, "Comparison of Two Platelet Glycoprotein IIb/IIIa Inhibitors, Tirofiban and Abciximab, for the Prevention of Ischemic Events with Percutaneous Coronary Revascularization," *New England Journal of Medicine* 344 (2001): 1888–94.

13. D. Mukherjee, "Risks of Cardiovascular Events Associated with Selective COX-2 Inhibitors," *Journal of the American Medical Association* 286 (2001): 954–59.

14. T. Burton and G. Harris, "Note of Caution: Study Raises Specter of Cardiovascular Risk for Hot Arthritis Pills," *Wall Street Journal*, August 22, 2001.

15. Barbara Martinez et al., "Merck Pulls Vioxx from Market After Link to Heart Problems," *Wall Street Journal*, October 1, 2004, 1–7.

16. E. J. Topol, "Failing the Public Health: Rofecoxib, Merck, and the FDA," *New England Journal of Medicine* 351 (2004): 1707–8.

17. R. Gilmartin, "An Open Letter from Merck," Merck, September 30, 2004, www.merck .com/newsroom/vioxx/pdf/An_Open_Letter_From_Merck.pdf.

18. Eric J. Topol, "Good Riddance to a Bad Drug," *New York Times*, October 4, 2004, www .nytimes.com/2004/10/02/opinion/02topol.html.

19. Topol, "Failing the Public Health."

20. Alex Berenson, "For Merck, Vioxx Paper Trail Won't Go Away," *New York Times*, August 21, 2005, www.nytimes.com/2005/08/21/business/21vioxx.html; Snidgdha Prakash, *All the Justice Money Can Buy: Corporate Greed on Trial* (New York: Kaplan, 2011), 146–49.

21. Congressional Testimony of David J. Graham, MD, MPH, before Senate Finance Committee on Vioxx, November 18, 2004, www.disease-treatment.com/showthread.php?t=45989.

22. E. J. Topol, "Arthritis Medicines and Cardiovascular Events: 'House of Coxibs,'" *Journal of the American Medical Association* 293, no. 3 (2005): 366–68.

23. "Merck's $27 Billion Heart Attack," *Fortune*, November 1, 2004; "The Painkiller Panic," *Wall Street Journal*, December 23, 2004; R. Horton, "Vioxx, the Implosion of Merck, and Aftershocks at the FDA," *Lancet* 364 (2004):1995–96; Gardiner Harris, "Drug Safety System Is Broken, a Top F.D.A. Official Says," *New York Times*, June 9, 2005, www.nytimes.com/2005/06/09/politics/09fda.htm.

24. "Cleveland Clinic's Topol Loses Provost, Academic Chief Post," *Beasley Allen Legal News*, December 9, 2005, www.beasleyallen.com/news/Cleveland-Clinics-Topol-loses-Provost-Academic -Chief-Post; Michael Specter, *Denialism: How Irrational Thinking Hinders Scientific Progress, Harms the Planet, and Threatens Our Lives* (New York: Penguin Press, 2009), 44.

25. "Hiding the Data on Drug Trials," editorial, *New York Times*, June 1, 2005, www.nytimes .com/2005/06/01/opinion/01wed2.html; Alex Berenson, "Evidence in Vioxx Suits Shows Intervention by Merck Officials." *New York Times*, April 24, 2005, www.nytimes.com/2005/04/24/business/24drug.html; "Manipulating a Journal Article," editorial, *New York Times*, December 11,

2005, www.nytimes.com/2005/12/11/opinion/11sun2.html; K. Hill, "The ADVANTAGE Seeding Trial: A Review of Internal Documents," *Annals of Internal Medicine* 149 (2008): 251–58; Burton, "Note of Caution"; "Marketing Drugs to Unsuitable Patients," *New York Times*, January 28, 2005, www.nytimes.com/2005/01/28/opinion/28fri2.html.

26. Gardiner Harris, "Study Condemns F.D.A.'s Handling of Drug Safety," *New York Times*, September 23, 2006, A1, www.nytimes.com/2006/09/23/health/policy/23fda.html; Institute of Medicine, *The Future of Drug Safety: Promoting and Protecting the Health of the Public* (Washington, DC: National Academies Press, 2006).

27. A. Kesselheim, "Whistle-Blowers' Experience in Fraud Litigation against Pharmaceutical Companies," *New England Journal of Medicine* 362 (2010): 1832–40.

28. "Imatinib," *PubMed Health*, February 1, 2009, www.ncbi.nlm.nih.gov/pubmedhealth/PMH0000345.

29. G. Bollag, "Clinical Efficacy of a Raf Inhibitor Needs Broad Target Blockade in BRAF-Mutant Melanoma," *Nature* 467 (2010): 596–99; K. Flaherty, "Inhibition of Mutated, Activated BRAF in Metastatic Melanoma," *New England Journal of Medicine* 363 (2010): 809–19.

30. Andrew Pollack, "Scientists Cite Advances on Two Kinds of Cancer," *New York Times*, June 5, 2010, www.nytimes.com/2010/06/06/health/research/06cancer.html; Scott Gottlieb, "Two Steps Forward in the War Against Cancer," *Wall Street Journal*, June 9, 2010, A17, online.wsj.com/article/SB10001424052748703302604575294233359450658.html; F. S. Hodi, "Improved Survival with Ipilimumab in Patients with Metastatic Melanoma," *New England Journal of Medicine* 363 (2010): 711–23; P. D. Chapman, A. Hauschild, et al., "Improved Survival with Vemurafenib in Melanoma with BRAF V600E Mutation," *New England Journal of Medicine* 364 (2011): 2507–16; E. Kwak, "Anaplastic Lymphoma Kinase Inhibition in Non-Small-Cell Lung Cancer," *New England Journal of Medicine* 363 (2010): 1693–1703.

31. Pollack, "Scientists Cite Advances"; Kwak, "Anaplastic Lymphoma Kinase."

32. Hodi, "Improved Survival with Ipilimumab."

33. Pollack, "Scientists Cite Advances."

34. Vaitheeswaran, "Generically Challenged."

35. "High-Throughput Screening," *Wikipedia*, n.d., en.wikipedia.org/wiki/High-throughput_screening.

36. Gladwell, "The Treatment."

37. K. Rising, "Report Bias in Drug Trials Submitted to the Food and Drug Administration: Review of Publication and Presentation," *PLoS Medicine* 5, no. 11 (2008): e217.

38. Institute of Medicine, *Transforming Clinical Research in the United States: Challenges and Opportunities: Workshop Summary* (Washington, DC: National Academies Press, 2010), www.nap.edu/catalog/12900.html.

39. Ibid.

40. R. Steinbrook, "Controlling Conflict of Interest: Proposals from the Institute of Medicine," *New England Journal of Medicine* 360 (2009): 2160–163.

41. ARRA Bill, HR 1, 111th Congress, www.scribd.com/doc/14343855/ARRA-Bill; "Healthcare Reform Act Includes Gift Ban Mandates," MassDevice, March 29, 2010, www.massdevice.com/news/healthcare-reform-act-includes-gift-ban-mandates.

42. "The Path to Productive Partnerships," *Nature* 452 (2008): 665; J. Altshuler, "Opening Up to Precompetitive Collaboration," *Science Translational Medicine* 2 (2010): 10–13; M. Ratner, "Pfizer Reaches Out to Academia—Again," *Nature Biotechnology* 29 (2011): 3–4; D. Zinner, "Life-Science Research Within US Academic Medical Centers," *Journal of the American Medical Association* 302, no. 9 (2009): 969–76; M. Scudellari, "Clinical Drive Prompts Pharma and Academia to Partner Up," *Nature Medicine* 17, no. 3 (2011): 2.

43. Scudellari, "Clinical Drive Prompts Pharma"; Ron Winslow, "JP Morgan Healthcare: Sanofi-Aventis, UCSF in Research Pact," *Wall Street Journal*, January 13, 2011, blogs.wsj.com/health/2011/01/13/jp-morgan-healthcare-sanofi-aventis-ucsf-in-research-pact; "Yale-Gilead Partnership to Investigate Genetic, Molecular Basis of Cancer," GenomeWeb, March 30, 2011; S. C. Johnston, S. L. Hauser, et al., "Enhancing Ties Between Academia and Industry to Improve Health," *Nature Medicine* 17 (2011): 434–36; H. Ledford, "Drug Buddies," *Nature* 474 (2011): 433–34.

44. Johnston, Hauser, et al., "Enhancing Ties Between Academia and Industry"; Ledfored, "Drug Buddies."

45. Witty, "Research and Develop."

46. Tapscott and Williams, *Macrowikinomics*, 30.

47. Shirley S. Wang, "Drug Makers Will Share Data from Failed Alzheimer's Trials," *Wall Street Journal*, June 11, 2011, online.wsj.com/article/SB100014240527487036277045752987831538 84208.html; Jonathan Rockoff and Ron Winslow, "Drug Makers Refill Parched Pipelines," *Wall Street Journal*, July 11, 2011, online.wsj.com/article/SB1000142405270230349920457 6387 423702555648.html; "The Scientific Social Network," *Nature Medicine* 17 (2011): 137; Jonah Lehrer, "Sunset of the Solo Scientist," *Wall Street Journal*, February 5, 2011.

48. Ibid.

49. W. A. Anderson, "Changing the Culture of Science Education at Research Universities," *Science* 331 (2011): 152–53.

50. Ibby Caputo, "Probing Doctor's Ties to Industry," *Washington Post*, August 18, 2009, www.washingtonpost.com/wp-dyn/content/article/2009/08/17/AR2009081702090.html.

51. Zinner, "Life-Science Research Within US Academic Medical Centers."

52. L. DeFrancesco, "Drug User Fees Top $1 million," *Nature Biotechnology* 28, no. 10 (2010): 992.

53. Data from "FY 2011 Food and Drug Administration Congressional Justification," U.S. Food and Drug Administration, n.d., www.fda.gov/AboutFDA/ReportsManualsForms/Reports/ BudgetReports/ucm202301.htm.

54. Margaret Hamburg, "America's Innovation Agency: The FDA," *Wall Street Journal*, August 1, 2011, online.wsj.com/article/SB10001424053111904888304576474072017155038.html.

55. Jonathan D. Rockoff, "Prescription-Drug Sales Rise 5.1%," *Wall Street Journal*, April 2, 2010, online.wsj.com/article/SB10001424052702303395904575157752023093126.html.

56. "Almost Half of Americans Took a Prescription Drug in the Past Month," *Wall Street Journal*, September 2, 2010.

57. D. Ge, "Genetic Variation in IL28B Predicts Hepatitis C Treatment-Induced Viral Clearance," *Nature* 461 (2009): 399–401; V. Suppiah, "IL28B Is Associated with Response to Chronic Hepatitis C Interferon-a and Ribavirin Therapy," *Nature Genetics* 41 (2009): 1100–104.

58. Alex Berenson, "Cancer Drugs Offer Hope, but at a Huge Expense," *New York Times*, July 12, 2005, www.nytimes.com/2005/07/12/business/12cancer.html; L. Rapaport, "Lilly Erbitux Cancer Drug Not Worth Price, U.S. Scientists Say," *Bloomberg*, June 29, 2009, www.bloomberg.com/ apps/news?pid=newsarchive&sid=a477Nm93JyxM; Andrew Pollack, "The Work-Up-Costly Drugs Known as Biologics Prompt Exclusivity Debate," *New York Times*, July 22, 2009, www.nytimes.com/ 2009/07/22/business/22biogenerics.html; "Velcade's Wider OK Could Pose Problems for Revlimid," *BioWorld Today*, June 23, 2008, financial.tmcnet.com/news/2008/06/23/3513094.htm; Liz Szabo, "Cost of Cancer Drugs Crushes All but Hope," *USA Today*, July 11, 2006, www.usatoday.com/ news/health/2006-07-10-cancer-drugs_x.htm.

59. Berenson, "Cancer Drugs Offer Hope."

60. Pollack, "The Work-Up-Costly Drugs Known as Biologics."

61. Engelberg, "Balancing Innovation, Access, and Profits"; Walsh, "Biopharmaceutical Benchmarks 2010"; ibid.

62. Walsh, "Biopharmaceutical Benchmarks 2010."

63. A. Garber, "Satisfaction Guaranteed: 'Payment by Results' for Biological Agents," *New England Journal of Medicine* 356 (2007): 1575–77.

64. Institute of Medicine, *Transforming Clinical Research*.

65. Ibid.; D. Malakoff, "Spiraling Costs Threaten Gridlock," *Science* 322 (2008): 210–12.

66. Malakoff, "Spiraling Costs."

67. Ibid.

68. Hill, "The ADVANTAGE Seeding Trial"; D. Malakoff, "Allegations of Waste: The 'Seeding' Study," *Science* 322 (2008): 213.

69. Carl Elliott, "Useless Studies, Real Harm," *New York Times*, July 28, 2011, www.nytimes .com/2011/07/29/opinion/useless-pharmaceutical-studies-real-harm.html.

70. Institute of Medicine, *A Population-Based Policy and Systems Change Approach to Prevent and Control Hypertension* (Washington, DC: National Academies Press, 2010); C. Lanzani, "Adducin- and Ouabain-Related Gene Variants Predict the Antihypertensive Activity of Rostafuroxin; Part 2: Clinical Studies," *Science Translational Medicine* 2 (2010): 59–87.

71. M. McCarthy, "Genomics, Type 2 Diabetes, and Obesity," *New England Journal of Medicine* 363 (2010): 2339–50.

72. Walsh, "Biopharmaceutical Benchmarks 2010."

73. Bollag, "Clinical Efficacy of a Raf Inhibitor"; Flaherty, "Inhibition of Mutated, Activated BRAF."

74. D. Wilson, "Pfizer Hopes to Pair Cancer Drug, Gene Test," *New York Times*, October 28, 2010.

75. T. Ray, "Takeda to Use Zinfandel's TOMM40 Test in Alzheimer's Development Program for Actos," GenomeWeb, January 10, 2011.

76. McCarthy, "Genomics, Type 2 Diabetes, and Obesity."

77. Institute of Medicine, *Challenges for the FDA*.

78. J. Brownstein, "Digital Disease Detection: Harnessing the Web for Public Health Surveillance," *New England Journal of Medicine* 360 (2009): 2153–56.

79. Ibid.

80. J. Carey, "Good for What Ails Only You," *Businessweek*, February 1, 2010, 58–60.

81. L. Timmerman, "Dendreon Sets Provenge Price at $93,000, Says Only 2,000 People Will Get It in First Year," *Xconomy*, April 29, 2010, www.xconomy.com/seattle/2010/04/29/dendreon -sets-provenge-price-at-93000-says-only-2000-people-will-get-it-in-first-year.

82. Scott Gottlieb, "The FDA Is Evading the Law," *Wall Street Journal*, December 23, 2010, on-line.wsj.com/article/SB10001424052748704034804576025981869663212.html.

83. "An Overdose of Bad News," *Economist*, March 19, 2005; J. Greene, "Pharmaceutical Marketing and the New Social Media," *New England Journal of Medicine* 363 (2010): 2087–89.

84. Greene, "Pharmaceutical Marketing and the New Social Media."

85. "FDA Dings Novartis for Facebook Widget," *Wall Street Journal*, August 6, 2010, blogs .wsj.com/health/2010/08/06/fda-dings-novartis-for-facebook-widget.

86. "Social Networking Research Spawns a Healthcare Company Blog," *Wall Street Journal*, December 7, 2010, blogs.wsj.com/health/2010/12/07/social-networking-research-spawns-a-health care-company; "Case Study 1: Adoption of Januvia," MedNetworks Inc., n.d., www.mednetworks .com/case-studies.html.

87. "Case Study 1."

88. Duff Wilson, "Drug App Comes Free, Ads Included," *New York Times*, July 28, 2011, www.nytimes.com/2011/07/29/business/the-epocrates-app-provides-drug-information-and -drug-ads.html.

89. D. Sacks, "The Future of Advertising," *Fast Company*, November 17, 2010. www.fast company.com/magazine/151/mayhem-on-madison-avenue.html.

90. Andrew Pollack, "Trial Shows Cystic Fibrosis Drug Helped Ease Breathing," *New York Times*, February 23, 2011, prescriptions.blogs.nytimes.com/2011/02/23/vertex-says-cystic -fibrosis-drug-helped-patients-breatheasier; E. Dolgin, "Mutation-Specific Cystic Fibrosis Treat-ments on Verge of Approval," *Nature Medicine* 17 (2011): 396–97; Amy Harmon, "Drug to Fight Melanoma Prolonged Life in Trial," *New York Times*, January 19, 2011; Amy Harmon, "After Long Fight, Drug Gives Sudden Reprieve," *New York Times*, February 22, 2010, www.nytimes .com/2010/02/23/health/research/23trial.html; Amy Harmon, "New Drugs Stir Debate on Rules of Clinical Trials," *New York Times* September 10, 2010, www.nytimes.com/2010/09/19/health/ research/19trial.html.

91. Harmon, "Drug to Fight Melanoma"; Harmon, "After Long Fight"; Harmon, "New Drugs Stir Debate."

92. Ibid.

93. Harmon, "New Drugs Stir Debate."

94. B. A. Chabner, "Early Accelerated Approval for Highly Targeted Cancer Drugs," *New England Journal of Medicine* 364 (2011): 1087–89.

Chapter 11

1. Marshall McLuhan, *Understanding Media: The Extensions of Man* (New York: McGraw-Hill, 1964), 57.

2. Clay Shirky, *Cognitive Surplus: Creativity and Generosity in a Connected Age* (New York: Penguin, 2010), 191–92.

3. J. Nikles et al., "Prioritising Drugs for Single Patient (n-of-1) Trials in Palliative Care," *Palliative Medicine* 23 (2009): 623–24; S. Roberts, "Self-Experimentation as a Source of New Ideas: Ten Examples About Sleep, Mood, Health, and Weight," *Behavioral and Brain Sciences* 27 (2004): 227–88; O. Rascol, "A Proof-of-Concept, Randomized, Placebo-Controlled, Multiple Cross-overs (N-of-1) Study of Naftazone in Parkinson's Disease," *Fundamental and Clinical Pharmacology*, May 18, 2011 (online).

4. N. Marko, "Mathematical Modeling of Molecular Data in Translational Medicine: Theoretical Considerations," *Science Translational Medicine* 2, no. 56 (2010): 1–7.

5. K. M. Schneider, "Prevalence of Multiple Chronic Conditions in the United States Medicare Population," *Health and Life Quality Outcomes* 7 (2009): 82.

6. C. J. Bell et al., "Carrier Testing for Severe Childhood Recessive Diseases by Next-Generation Sequencing," *Science Translational Medicine* 3, no. 65 (2011): 65ra4; J. Couzin-Frankel, "New High-Tech Screen Takes Carrier Testing to the Next Level," *Science* 331 (2011): 130–31.

7. B. Starfield, "Is US Health Really the Best in the World?" *Journal of the American Medical Association* 284, no. 4 (2000): 483–85.

8. J. Van Den Bos et al., "The $17.1 Billion Problem: The Annual Cost of Measurable Medical Errors," *Health Affairs* 30, no. 4 (2011): 596–603; C. Landrigan, "Temporal Trends in Rates of Patient Harm Resulting from Medical Care," *New England Journal of Medicine* 363 (2010): 2124–34.

9. George Orwell, "How the Poor Die," November 1946, orwell.ru/library/articles/Poor_Die/english/e_pdie.

10. Elizabeth Svoboda, "Cisco's Virtual Doctor Will See You Now," *Fast Company*, April 20, 2011, www.fastcompany.com/magazine/155/the-virtual-doctor-will-see-you-now.html.

11. E. A. Balas, "Managing Clinical Knowledge for Health Care Improvement," in *Yearbook of Medical Informatics, 2000: Patient Centered Systems*, ed. J. Bemmel and A. Y. McCray (Stuttgart: Shattauer Verlagsgesellschaft mbH, 2000), 65–70; D. G. Contopoulos-Ioannidis et al., "Life Cycle of Translational Research for Medical Interventions," *Science* 321 (2008): 1298–99.

12. Mark C. Baker and Stewart Goetz, eds., *The Soul Hypothesis: Investigation into the Existence of the Soul* (New York: Continuum, 2011), Kindle edition, Chapter 6.

13. J. DeFelipe, "From the Connectome to the Synaptome: An Epic Love Story," *Science* 330 (2010): 1198–201; Ashlee Vance, "In Pursuit of a Mind Map, Slice by Slice," *New York Times*, December 27, 2010, www.nytimes.com/2010/12/28/science/28brain.html.

14. C. Zimmer, "100 Trillion Connections," *Scientific American* (January 2011).

15. Nick Bilton, *I Live in the Future & Here's How It Works: Why Your World, Work, and Brain Are Being Creatively Disrupted* (New York: Crown, 2010), Kindle edition, Chapter 3.

16. Ibid., Chapter 5.

17. Don Tapscott and Anthony D. Williams, *Macrowikinomics: Rebooting Business and the World* (New York: Portfolio/Penguin, 2010), 186; Daniel Reda, "Migraine Symptom Predicts Response to Imitrex," *CureTogether Blog*, January 11, 2011, curetogether.com/blog/2011/01/11/migraine-symptom-predicts-response-to-imitrex.

18. K. C. Bickart et al., "Amygdala Volume and Social Network Size in Humans," *Nature Neuroscience*, December 26, 2010.

19. Michael J. Fox, "Clinical Studies Can Lead to Cures: Volunteer to Fight Parkinson's," *Cleveland Plain Dealer*, July 31, 2011, www.cleveland.com/opinion/index.ssf/2011/07/clinical_studies_can_lead_to_c.html.

20. Malcolm Gladwell, "Small Change: Why the Revolution Will Not Be Tweeted," *New Yorker*, October 4, 2010.

21. Jonathan Franzen, "Liking Is for Cowards: Go for What Hurts," *New York Times*, May 28, 2011.

22. Gautam Naik, "Diabetes Cases Double to 347 Million," *Wall Street Journal*, June 27, 2011.

23. "It's a Smart World: A Special Report on Smart Systems," *Economist*, November 6, 2010.

24. David Gelernter, *Mirror Worlds, or, The Day Software Puts the Universe in a Shoebox: How It Will Happen and What It Will Mean* (New York: Oxford University Press, 1991), Prologue.

25. Steve Lohr, "Computers That See You and Keep Watch Over You," *New York Times*, January 1, 2011.

26. Raymond Tallis, "The Mind in the Mirror," *Wall Street Journal*, January 8, 2011.

27. Benedict Carey, "Genes as Mirrors of Life Experiences," *New York Times*, November 6, 2010.

28. Andy Clark, "Out of Our Brains," *New York Times*, December 12, 2010.

29. Nicholas Carr, *The Shallows: What the Internet Is Doing to Our Brains* (New York: W. W. Norton, 2010), 220.

30. Danielle Ofri, "Not on the Doctor's Checklist, but Touch Matters," *New York Times*, August 2, 2010, www.nytimes.com/2010/08/03/health/03case.html.

31. Ray Kurzweil, *The Singularity Is Near: When Humans Transcend Biology* (New York: Viking Press, 2005).

32. "It's a Smart World."

33. Ibid.

34. Scott James, Hispanics Rank High on Digital Divide," *New York Times*, June 17, 2011, www.nytimes.com/2011/06/17/us/17bcjames.html; C. Chin et al., "Microfluidics-Based Diagnostics of Infectious Diseases in the Developing World," *Nature Medicine*, July 31, 2011, doi: 10.1038/nm.2408; S. Reardon, "A World of Chronic Disease," *Science* 333 (2011): 558–59.

35. T. Ray, "Study Finds Docs Could Face Greater Malpractice Risk in Personalized Rx Era," GenomeWeb, June 29, 2011.

Afterword

1. L. Guterman, "Critical Partnerships: Wireless Medicine: A San Diego Success Story," *NCRR Reporter* 34, no. 1 (2010): 1–2; Jonathan Sidener, "Wireless Health Care, in S.D.: New Institute Will Be Nation's 1st in Its Field," *Union-Tribune* (San Diego), March 30, 2009, www.signonsandiego.com/news/2009/mar/30/1n30wireless23453-future-medicine-sd.

INDEX

Theranostics, 215
Thyroid cancer, 104
Tissue plasminogen activator (t-PA), 25–27, 211
Tissues, drug development and synthetic, 169–171
To Err Is Human, 142
TOMM40, 214
Topol, Susan, 41–42, 115
Torrent, 197
Torrey, Trisha, 178
Toumazou, Chris, 101
Toxic epidermal necrolysis, 92
Toyota, 76
t-PA. *See* Tissue plasminogen activator (t-PA)
Trachea, printing organs and synthetic, 140
Trait-o-matic, 120
Transcriptome, 228
Transcriptomics, 105–106
Transparency, in drug development process, 205
Transplant rejection, monitoring, 165–166
Trastuzumab, 103, 198, 203, 212, 218
Trick or Treatment? (Singh), 41
Tricorder X-Prize for mobile diagnostics, 173
"Truth Wears Off, The" (Lehrer), 31
Tumor necrosis factor, 203
Tumor necrosis factor blockers, 219
Twain, Mark, 35
20/20 (television program), 44
23andMe, 108, 109, 113, 238
Twins, epigenomics and, 106
Twitter, 9, 10, 11, 48, 190–191, 220, 238
Type 1 diabetes, 85, 93, 166–167
Type 2 diabetes, 85, 86, 106–107, 213–214

Ultrasound, 122–125, 138
Ultraviolent A (UVA), 39
UnitedHealth, 158, 192
United States, adoption of electronic medical record keeping in, 143–144
Unstable angina, 162
Urogenital Microbiome Consortium, 108
U.S. News & World Report (magazine) hospital rankings, 50–51
Utility, of personal genomics, 114, 116
Uveal melanomas, 94

Vaccines, 219, 231–232
Vagelos, Roy, 201
Valentine, Shelley, 247
Vanderveen, Pete, 188
Variable number tandem repeats (VNTR), 173
Vectibix, 210
Vein opening procedure, 43
Velcade model, 211, 218
V600E mutation, 103, 137–138, 139, 212, 214, 225
Venter, Craig, 78, 95, 116
Ventricular arrhythmias, 67
Venuti, Michael, 169
Vertex Pharmaceuticals, 223
Vest, Charles, 184
Veterans Health Administration (VHA), 148–149
Viagra, 36
Videoconferencing, 192
ViiV Healthcare, 206
Vioxx, 92, 149, 199–202, 210, 215, 216, 217, 218
Virtualization of real world, 239–240
Virtual medical practice, 193–194
Virtual office visits (v-visits), 75–76, 192
Vital signs, wireless monitoring of, 68–70
Vitamin B$_{12}$, 38
Vitamin D, 38
Vitamin E, 30–31, 38
Vitamins, 37–38
VNTR. *See* Variable number tandem repeats (VNTR)
Volker, Nicholas, xii, 77–78, 95, 99
Von Hoff, Daniel D., 51–52
Vscan, 122–124, 184, 192
Vytorin, 32

Wade, Nicholas, 10
Wald, Nicholas, 29
Walgreens, 118–119
Warfarin, 40, 46, 90–91, 120, 233
Warner Lambert, 197
Waskal, Sam, 103
Watson, James, 104, 116
Watson genome, 95
Watson IBM computer, xiii
WeAre.Us, 11
WebMD, 47, 241
WebMD Health Manager, 157–158